TIME SERIES AND ECONOMETRIC MODELLING

THE UNIVERSITY OF WESTERN ONTARIO SERIES IN PHILOSOPHY OF SCIENCE

A SERIES OF BOOKS
IN PHILOSOPHY OF SCIENCE, METHODOLOGY,
EPISTEMOLOGY, LOGIC, HISTORY OF SCIENCE
AND RELATED FIELDS

VOLUME 36

ADVANCES IN THE STATISTICAL SCIENCES

Festschrift in Honor of Professor V. M. Joshi's 70th Birthday

VOLUME III

TIME SERIES AND ECONOMETRIC MODELLING

Edited by

IAN B. MacNEILL and GARY J. UMPHREY

Department of Statistical and Actuarial Sciences,
The University of Western Ontario

Associate editors:

RICHARD A. L. CARTER

Department of Economics, The University of Western Ontario

A. IAN McLEOD

Department of Statistical and Actuarial Sciences,
The University of Western Ontario

AMAN ULLAH

Department of Economics, The University of Western Ontario

D. REIDEL PUBLISHING COMPANY

A MEMBER OF THE KLUWER ACADEMIC PUBLISHERS GROUP

DORDRECHT / BOSTON / LANCASTER / TOKYO

Library of Congress Cataloging in Publication Data

Time series and econometric modelling.

(Advances in the statistical sciences; v. 3)
(The University of Western Ontario series in philosophy of science; v. 36)
1. Time-series analysis—Congresses. 2. Econometric models—Congresses. I. MacNeill, Ian B., 1931– . II. Umphrey, Gary J., 1953– . III. Series. IV. Series: University of Western Ontario series in philosophy of science; v. 36.
QA276.A1A39 vol. 3 519.5 s 86–29672
[QA280] [519.5′5]
ISBN-13: 978-94-010-8624-0 e-ISBN-13: 978-94-009-4790-0
DOI: 10.1007/978-94-009-4790-0

Published by D. Reidel Publishing Company,
P.O. Box 17, 3300 AA Dordrecht, Holland.

Sold and distributed in the U.S.A. and Canada
by Kluwer Academic Publishers,
101 Philip Drive, Assinippi Park, Norwell, MA 02061, U.S.A.

In all other countries, sold and distributed
by Kluwer Academic Publishers Group,
P.O. Box 322, 3300 AH Dordrecht, Holland.

TABLE OF CONTENTS

CONTENTS OF THE OTHER VOLUMES OF THE JOSHI FESTSCHRIFT

VOLUME I

Applied Probability, Stochastic Processes, and Sampling Theory

VOLUME II

Foundations of Statistical Inference

VOLUME IV

Stochastic Hydrology

VOLUME V

Biostatistics

VOLUME VI

Foundations of Actuarial Science

PREFACE

On May 27–31, 1985, a series of symposia was held at The University of Western Ontario, London, Canada, to celebrate the 70th birthday of Professor V. M. Joshi. These symposia were chosen to reflect Professor Joshi's research interests as well as areas of expertise in statistical science among faculty in the Departments of Statistical and Actuarial Sciences, Economics, Epidemiology and Biostatistics, and Philosophy.

From these symposia, the six volumes which comprise the "Joshi Festschrift" have arisen. The 117 articles in this work reflect the broad interests and high quality of research of those who attended our conference. We would like to thank all of the contributors for their superb cooperation in helping us to complete this project.

Our deepest gratitude must go to the three people who have spent so much of their time in the past year typing these volumes: Jackie Bell, Lise Constant, and Sandy Tamowski. This work has been printed from "camera ready" copy produced by our Vax 785 computer and QMS Lasergraphix printers, using the text processing software TEX. At the initiation of this project, we were neophytes in the use of this system. Thank you, Jackie, Lise, and Sandy, for having the persistence and dedication needed to complete this undertaking.

We would also like to thank Maria Hlawka-Lavdas, our systems analyst, for her aid in the layout design of the papers and for resolving the many difficult technical problems which were encountered. Nancy Nuzum and Elly Pakalnis have also provided much needed aid in the conference arrangements and in handling the correspondence for the Festschrift.

Professor Robert Butts, the Managing Editor of *The University of Western Ontario Series in Philosophy of Science* has provided us with his advice and encouragement. We are confident that the high calibre of the papers in these volumes justifies his faith in our project.

In a Festschrift of this size, a large number of referees were needed. Rather than trying to list all of the individuals involved, we will simply say "thank you" to the many people who undertook this very necessary task for us. Your contributions are greatly appreciated.

Financial support for the symposia and Festschrift was provided by The University of Western Ontario Foundation, Inc., The University of Western Ontario and its Faculties of Arts, Science, and Social Science, The UWO Statistical Laboratory, and a conference grant from the Natural Sciences and Engineering Research Council of Canada. Their support is gratefully acknowledged.

Finally, we would like to thank Professor Joshi for allowing us to hold the conference and produce this Festschrift in his honor. Professor Joshi is a very modest man who has never sought the limelight. However, his substantial contributions to statistics merit notice (see Volume I for a bibliography of his papers and a very spiffy photo). We hope he will accept this as a tribute to a man of the highest integrity.

INTRODUCTION TO VOLUME III

As recently as two decades ago, the methodology of time series consisted largely of spectral analysis and standard regression. These methods have continued to be important, but many new models, together with their concomitant analyses, have since emerged. This has resulted both in a large and continuing expansion of the time series "community", and in a burgeoning of the scope of application; a major use of time series methods is the analysis of economic data. The twenty-five articles which comprise this volume and which discuss both old and new models are organized so that papers on time series appear at the beginning and are followed by those on econometric models.

The large expansion of the collection of available models has introduced a new and vexing problem for time series and econometric modellers, namely, that of model selection. Although this problem has been approached from a number of directions, much attention has been given recently to the study of criteria, such as the AIC, which can be used to compare models of widely differing character. The first four papers in this volume, by Akaike, Duong, Hannan and Teräsvirta, deal with this problem of model selection.

The level of complexity of a model useful for describing a random mechanism is related to the extent of the generated data set: a small amount of data demands the property of simplicity in a model, whereas a large amount permits consideration of more complexity. How the statistician is to make decisions regarding the appropriate level of complexity is the subject of the paper by Hannan; the model selection procedure he discusses for linear systems involves the use of criteria such as the AIC. Akaike also considers the problem of model selection, particularly as it applies to the selection of models which demonstrate practical usefulness as opposed to similarity to the mechanism which generated the data. Duong discusses the application of subset selection procedures to model selection criteria with the aim of obtaining a small subset of models containing the true model with specified high probability; this brings to model selection the notion of the confidence interval. Teräsvirta discusses smoothing restrictions in regression and provides generalizations of model selection criteria for use in choosing smoothing parameters.

Related to the concept of model selection is that of model adequacy, different aspects of which are discussed by Abraham and Kedem. Kedem discusses a goodness-of-fit test based on a small number of higher order crossings, whereas Abraham considers the problem of detection and rejection of outliers.

Inferential problems are considered in another set of papers. Hui and Li discuss the use of both shrinkage and empirical Bayes estimators for parameters of moving average models of multi-item inventory systems. Kheoh and McLeod establish the low efficiency of certain consistent estimators in ARMA model estimation. Sutradhar, MacNeill and Sahrmann discuss the concept of time-series valued experimental designs, and propose tests that fit into the standard ANOVA paradigm. Space-time models are formulated and discussed by Aroian.

Many statistical agencies, including Statistics Canada, use concurrent seasonal adjustment procedures to reduce the size of revisions as new data are accumulated. By applying various measures of filter revision to the frequency response function of seasonal adjustment filters, Dagum explores the problem of how often the concurrent seasonal adjustment filter of X-11-ARIMA should be revised.

Fourier methods form the focus of another set of papers. Stoffer and Panchalingam survey Walsh-Fourier spectral methods and provide sequency domain analyses of binary data. Jensen and Mansinha discuss the properties of self-similar fractal stochastic processes and explore their use in modelling flicker noise processes that arise in geophysics. Feuerverger discusses properties of the empirical characteristic function and applies them to nonparametric testing for independence in multivariate data.

The second half of this volume deals with topics of importance to econometrics. Granger, in a time series paper, points out that many macroeconomic variables have a typical spectral shape, a shape consistent with the property that one difference will produce a stationary series. This paper considers a number of other models which generate series having typical spectral shape.

Significant impetus for development of certain areas of statistical inference comes from econometrics. This is illustrated by the papers of Phillips; Singh, Ullah and Carter; Vinod; and Zinde-Walsh and Ullah. Although this work has particular relevance for econometrics, its usefulness extends into surrounding areas of statistics.

Phillips has introduced fractional matrix calculus as a new tool for the study of multivariate distributions useful in econometrics. The methodology provided in this paper unifies the theories for finite sample and asymptotic distributions. Robustness of tests in regression models is the subject of the paper by Zinde-Walsh and Ullah. They consider problems of numerical and inferential robustness and provide bounds on critical values of various statistics that guarantee robustness of test conclusions. Distributional assumptions in econometrics often may be questionable. It is possible to avoid problems caused by such misspecification by using nonparametric techniques. Singh, Ullah and Carter present nonparametric estimation procedures for multivariate densities and apply them to several econometrics problems. Vinod, as well as presenting a comprehensive review of available alternatives for constructing confidence intervals for ridge regression paramaters, also discusses a new alternative.

Models of particular interest to econometrics are central to the papers of Dufour, Maasoumi, Peters and Power. Rational expectation models, introduced a decade ago, are now widely studied by econometricians. For such models, Power discusses asymptotic properties of single equation errors-in-variables estimators. Inference about the vector of covariances between the stochastic exploratory variables and the error term of a structural equation is the subject discussed by Dufour. Peters investigates the finite sample moments of ordinary least square estimators for a simple dynamic model when the error term is small. Maasoumi develops small sample approximations to the moments of the three-stage least square reduced form estimator in a general linear simultaneous equation model.

A principal use of regression is forecasting, with forecast uncertainty generally assessed in terms of a normality assumption which may not hold true. To circumvent this assumption, Veall applies the bootstrap to the problem of estimation of the probability distribution of the forecast error.

The use of the mean squared-error of forecast in testing for structural shift in parameters is investigated by Tsurumi; sampling and Bayesian distributions for the statistic are derived.

The contents of the twenty-five papers in this volume, encapsulated above, reveal something of the panoramic breadth of applications and methodology of time series analysis and econometrics. The importance of the content, and the care taken in presentation of this content, make it the editors pleasure to thank the authors of the articles in this volume. We expect all practitioners and researchers in time series and econometric modelling will find valuable the new results presented in these papers, and will appreciate the efforts of the authors in carrying out the always difficult task of reviewing and summarizing the body of knowledge in the various fields covered in this volume.

E. J. Hannan [1]

APPROXIMATION OF LINEAR SYSTEMS

1. INTRODUCTION

The classical paradigm of statistics assumes that data is generated by a stochastic process whose structure is entirely known save for a fixed number of parameters. Of course there are many departures from this assumption and consequent statistical methods are useful over a much wider range than the paradigm suggests. In time series analysis the paradigm is rarely relevant, which partly explains the wide use of Fourier methods, which are non-parametric. The other major part of time series analysis is that based on autoregressive-moving average (ARMA) models. Much of the literature associated with these acts as if the data are actually generated by such a model though this attitude has been modified in some of the systems and control literature (Rissanen, 1983) and in the work of Akaike and Shibata (see Shibata, 1980). Here the point of view will, also, be taken that the ARMA models are fitted as approximations. A complete treatment will be impossible because of the space available and also the state of the development of the ideas.

In order to set the scene consider a vector stationary process

$$y(t) = \sum_0^\infty K(j)\varepsilon(t-j), \quad K(0) = I_s, \qquad \sum \| K(j) \|^2 < \infty,$$

$$E\{\varepsilon(s)\varepsilon(t)'\} = \delta_{st}\Sigma, \qquad \Sigma > 0. \tag{1.1}$$

It is assumed that $y(t)$, $\varepsilon(t)$, have s components and that the $\varepsilon(t)$ are the linear innovations, i.e. $\varepsilon(t) = y(t) - y(t \mid t-1)$ where $y(t \mid t-1)$ is the best predictor (in the least squares sense) of $y(t)$ from $y(s)$, $s \leq t$. Then

$$k(z) = \sum_0^\infty K(j)z^{-j}.$$

---

[1] Department of Statistics, I.A.S., Australian National University, GPO Box 4, Canberra, ACT 2601, Australia

I. B. MacNeill and G. J. Umphrey (eds.), Time Series and Econometric Modelling, 1–12.

is not only analytic for $|z| \geq 1$ but also $\det(k) \neq 0$, $|z| > 1$. Put

$$\mathcal{H} = [K(i+j-1)]_{i,j=1,2,\ldots}$$

for the infinite Hankel matrix and let H_o be composed of a minimal number, n, of rows of $\mathcal{H}$ (all of which rows are in ℓ_2). Of course $n = \infty$ is the standard case. Let H_1 be composed of the first s rows of $\mathcal{H}$ and $H_o = [K, H_2]$ where K consists of the first s columns of H_o. Put $\varepsilon_t' = [\varepsilon(t)', \varepsilon(t+1)', \ldots]$ and $x(t \mid t-1) = H_o\varepsilon_{t-1}$. Since H_1, H_2 are composed of rows of $\mathcal{H}$ then $H_1 = HH_o$, $H_2 = FH_o$ for suitable H, F and then

$$x(t+1 \mid t) = Fx(t \mid t-1) + K\varepsilon(t), \qquad y(t) = Hx(t \mid t-1) + \varepsilon(t).$$

If n is finite then $k(z) = H(zI_n - F)^{-1}K + I_s$, which is rational and, conversely, if k is rational then n is finite. Here we propose to study approximations to (1.1) in general (i.e. for $n = \infty$) by systems for which $\hat{n}$, chosen from the data $y(t)$, $t = 1, \ldots, T$, is finite. The integer n is called the order or McMillan degree so that we are concerned with approximating a general system by one of finite order. We shall use n_o for the true order when emphasis is needed.

This point of view has consequences, of course. It makes the asymptotic theory more difficult since, as is almost obvious, $\hat{n}$ will increase with T. Also the method of maximum likelihood (ML) fails since, clearly, this method will choose n indefinitely large. (However the method reappears as will be seen.) Thus some new criterion is needed. This problem has been considered by Akaike (1969) and Rissanen (1983). In the case of the latter the criterion introduced is that of a minimal description length of the data on the basis of a model. Via some approximations and on the basis of an assumption that the $\varepsilon(t)$ are Gaussian he arrives at the criterion

$$\log\det \hat{\Sigma}_n + d(n)\log T/T, \tag{1.2}$$

where $d(n)$ is the number of system parameters fitted, i.e., parameters other than those specifying Σ, and where $\hat{\Sigma}_n$ is the ML estimate of Σ when the order is n. Akaike, on the basis of prediction theory, suggested

$$\log\det \hat{\Sigma}_n + 2d(n)/T. \tag{1.3}$$

More generally we could take

$$\chi(n) = \log\det \hat{\Sigma}_n + d(n)C_T/T, \qquad C_T/T \to 0, \qquad C_T > 1, \tag{1.4}$$

for some prescribed sequence C_T. The conditions on C_T are necessary for a reasonable procedure. Of course $\hat{n}$ is determined by minimising one of these formulae, subject to $\hat{n} \leq N_T$, for N_T yet to be prescribed.

In the next section some results will be presented relating to these methods when the models considered are autoregressions. In Section 3 general ARMA approximations will be discussed, albeit far from completely. In Section 4 a brief discussion of problems will be given.

2. AUTOREGRESSIVE APPROXIMATION

The model fitted is now of the form

$$\sum_{0}^{h} \hat{\Phi}_h(j) y(t-j) = \hat{\varepsilon}_h(t), \qquad \hat{\Phi}_h(0) = I_s. \tag{2.1}$$

The $\hat{\Phi}_h(j)$ might be estimated from the Yule-Walker relations,

$$\sum_{0}^{h} \hat{\Phi}_h(j) C(j-k) = \delta_{ok} \hat{\Sigma}_h, \qquad k = 0, 1, \ldots, h,$$

$$C(j) = \frac{1}{T} \sum_{1}^{T-j} y(t) y(t+j)', \qquad j \geq 0; \qquad C(-j) = C(j)'.$$

Mean corrections would be made in practice. These Yule-Walker equations can be solved by a recursion on h (Whittle, 1963) that will be described in Section 3. They can give badly biased estimates, for even quite large T (Tjostheim and Paulsen, 1983). There are many modifications of them (see Friedlander, 1982) and we briefly discuss this, again in Section 3. Putting $\Gamma(j) = E\{y(t)y(t+j)'\}$ the $\hat{\Phi}_h(j)$ can be regarded as estimates of $\Phi_h(j)$ where

$$\sum_{0}^{h} \Phi_h(j) \Gamma(j-k) = \delta_{ok} \Sigma_h, \qquad k = 0, 1, \ldots, h; \qquad \Phi_h(0) = I_s.$$

The quantity $d(n)$ is now hs^2, as is obvious. We shall have

$$\sum_{0}^{h} \Phi_h(j) y(t-j) = \varepsilon_h(t), \qquad E\{\varepsilon_h(t)\varepsilon_h(t)'\} = \Sigma_h,$$

where $\varepsilon_h(t)$ is the error in predicting $y(t)$ optimally from $y(t-1), \ldots, y(t-h)$. If $\varepsilon_h(t) = \varepsilon(t)$ then $n = hs$, but that is not assumed.

Ergodicity ensures that $C(j) - \Gamma(j) \to 0$, a.s. (almost surely). Hence $\hat{\Phi}_h(j) \to \Phi_h(j)$, a.s., $j = 1, \ldots, n$, for each fixed h. Thus this must hold also when h increases sufficiently slowly. We wish to improve this weak statement. Such an improvement must rest on an improvement of the result concerning $C(j) - \Gamma(j)$ and for this purpose additional conditions must be introduced. The best predictor of $y(t)$ From $y(s)$, $s \leq t$ is $E\{y(t) \mid \mathcal{F}_{t-1}\}$ where $\mathcal{F}_t$ is the σ-algebra determined by $y(s)$, $s \leq t$. Since $\varepsilon(t) = y(t) - y(t \mid t-1)$,

$$E\{\varepsilon(t) \mid \mathcal{F}_{t-1}\} = E\{y(t) \mid \mathcal{F}_{t-1}\} - y(t \mid t-1).$$

Thus we impose

$$\begin{aligned} &E\{\varepsilon(t) \mid \mathcal{F}_{t-1}\} = 0, \\ &E\{\varepsilon(t)\varepsilon(t)' \mid \mathcal{F}_{-\infty}\} = \Sigma, \\ &E\{(\varepsilon_j(t))^4\} < \infty, \qquad j = 1, \ldots, s. \end{aligned} \tag{2.2}$$

The first is reasonable for a theory of approximation of linear systems since if $y(t \mid t-1)$ and $E\{y(t) \mid \mathcal{F}_{t-1}\}$ are very different the method is inappropriate because the best predictor is far from the best linear predictor and a linear model should not be used. The later part of (2.2) is minor. However for some purposes the second part has to be strengthened to

$$E\{\varepsilon(t)\varepsilon(t)' \mid \mathcal{F}_{t-1}\} = \Sigma, \tag{2.3}$$

which is not so minor. For some purposes also we need the condition

$$E\{\mid \varepsilon_j(t) \mid^4 \log_+ \mid \varepsilon_j(t) \mid\} < \infty, \qquad j = 1, \ldots, s, \tag{2.4}$$

where by $\log_+ x$ we mean $\log x$ for $x > 1$ and zero for $x \leq 1$.

So far as $k(z)$ is concerned we require

$$\det\{k(z)\} \neq 0, \quad \mid z \mid \geq 1, \quad \sum_0^\infty j^{1/2} \parallel K(j) \parallel < \infty, \tag{2.5}$$

when it follows that we may put

$$k(z)^{-1} = \sum_0^\infty \Phi(j) z^{-1}, \qquad \sum_0^\infty j^{1/2} \parallel \Phi(j) \parallel < \infty.$$

The following theorem then results. We use $O(\cdot)$ to indicate an order relation that holds a.s. If the relation holds only in probability we write $O_p(\cdot)$.

Theorem 1. *If $y(t)$ is generated by* (1.1) *and* (2.2), (2.4) *hold then, for* $H_T = o\{(T/\log T)^{1/2}\}$

$$\max_{1\leq j\leq h} \| \hat{\Phi}_h(j) - \Phi_j(j) \| = O\left\{(\log T/T)^{1/2}\right\} + o(1)\sum_{h+1}^{\infty} \| \Phi(j) \|, \text{a.s.}$$

where the order relations hold a.s. and uniformly in $h \leq H_T$. If (2.3) *holds then the second term on the right may be deleted and without* (2.3) *this is true for $s = 1$ provided*

$$\lim_{j\to\infty} \sup j \| k(j) \| < \infty.$$

This result is proved by Hannan and Kavalieris (1984, 1986).

For $s = 1$ the result is very satisfactory, since the last condition is minor, and the result seems the best possible. In interpreting the result we may use the extension of a result due to Baxter (1967), namely, for $c < \infty$,

$$\sum_{0}^{h} \| \Phi_h(j) - \Phi(j) \| \leq c\sum_{h+1}^{\infty} \| \Phi(j) \| . \tag{2.6}$$

Using (2.6), under the conditions of the theorem,

$$\| \sum_{0}^{h} \hat{\Phi}_h(j)e^{ij\omega} - \sum_{0}^{\infty} \Phi(j)e^{ij\omega} \| = O\left\{h(\log T/T)^{1/2}\right\} + O\left\{\sum_{h+1}^{\infty} \| \Phi(j) \|\right\},$$

which shows how the estimated frequency response function converges to the true one.

The other things that an asymptotic theory should do is to determine how $\hat{h}$, determined from (1.4), behaves and how accurate are the $\hat{\varepsilon}_h(t)$, from (2.1), as estimates of the $\varepsilon(t)$. The latter are covered by the following.

Theorem 2. *If $y(t)$ is generated by* (2.1) *with* (2.2), (2.3), (2.4), (2.5) *holding and*

$$\bar{c}_T(\log\log T)^2 \leq h \leq H_T = o\{(T/\log T)^{1/2}\}, \qquad \bar{c}_T \uparrow \infty,$$

then

$$\frac{1}{T}\sum_{1}^{T} \{\hat{\varepsilon}_h(t) - \varepsilon(t)\}\{\hat{\varepsilon}_h(t) - \varepsilon(t)'\}$$
$$= \{(sh/T)\Sigma + \Sigma_h - \Sigma\}\{1 + o(1)\}\, o_p(h/t)$$

and if

$$c(\log T)^2 \leq h \leq H_T = o\left\{(T/\log T)^{1/2}\right\}, \qquad c > 0,$$

then the term $o_p(h/T)$ may be replaced by $o(h/T)$.

This theorem is proved by Hannan and Kavalieris (1986).

The evaluation shows how accurate the $\hat{\varepsilon}_h(t)$ are. The condition $h/(\log\log T)^2 \to \infty$ is not onerous since, as the next theorem shows, even for BIC and $y(t)$ generated by an ARMA process (i.e. n is finite) then $\hat{h}$ increases as $\log T$. The next theorem deals with $\hat{h}$.

Theorem 3. *Under the same conditions as for Theorem 2,*

$$\log\det\hat{\Sigma}_h + hs^2C_T/T = \log\det\left\{\frac{1}{T}\sum_1^T \varepsilon(t)\varepsilon(t)'\right\} + hs^2(C_T-1)/T$$
$$+\operatorname{tr}\left\{\gamma^{-1}\left(\Sigma_h - \Sigma\right)\right\}\{1+o_p(1)\}.$$

This result is essentially due to Shibata (1980) for $s=1$ and $\varepsilon(t)$ Gaussian. It shows that $\hat{h}/h^* \to 1$, in probability where h^* minimises $hs^2(C_T - 1)/T + \operatorname{tr}\{\Sigma^{-1}(\Sigma_h - \Sigma)\}$. In case $C_T = 2$, $s=1$ this is $h/T + \left(\frac{\sigma_h^2}{\sigma^2} - 1\right)$ and when the process is actually ARMA then $h^* = \log T/(-2\log\rho_0)$ where ρ_0 is the modulus of a zero of $k(z)$ nearest to $|z| = 1$. This result is also essentially due to Shibata (1980).

This concludes our discussion of the asymptotic theory of autoregressive approximation. No form of central limit theorem has been given. Such a theorem, say about the $\hat{\Phi}_h(j)$, to be useful would have to be uniform in h, asserting, for example, that $T^{1/2}$ by any linear combination of the elements of $\hat{\Phi}_h(j) - \Phi_h(j)$ satisfied the central limit theorem, for the length of the vector of coefficients uniformly bounded, and h bounded by some function of t increasing sufficiently slowly. Such a theorem can be established, but we shall not discuss it here.

3. APPROXIMATION BY RATIONAL FUNCTIONS

This section will be concerned with algorithms and how they might be constructed and not with theory because there is little theory available at the moment, except for the case where n_o, the true order, is finite. (See, however, Taniguchi, 1980.) In order to confine the account within reasonable bounds let us return to $\mathcal{H}$ and make some observations. Let $M(n)$ be the set of all $\mathcal{H}$, i.e. of all $k(z)$, for which $\mathcal{H}$ is of rank n (and for which $k(z)$ is analytic and has $\det(k) \neq 0$, for $|z| \geq 1$). Then, as is well known, $M(n)$ is an analytic manifold. An open dense set in $M(n)$ is $U(n)$ where $U(n)$ consists of those $\mathcal{H}$ for which H_o can be taken to be comprised by the first n rows of $\mathcal{H}$. The set of all $k(z)$ in $U(n)$ can be represented in ARMA form as $k(z^{-1}) = a(z)^{-1}b(z)$, $a(z) = \sum A(j)z^j$, $b(z) = \sum B(j)z^j$ where, if $n = ps + q$, $0 \leq q < s$,

$$A(0) = \begin{bmatrix} I_q & 0 \\ * & I_{s-q} \end{bmatrix} = B(0), \qquad A(1) = \begin{bmatrix} * & 0 \\ * & * \end{bmatrix}, \tag{3.1}$$

$$A(p+1),\ B(p+1) = \begin{bmatrix} * \\ 0 \end{bmatrix}, \qquad A(j) = B(j) = 0, \qquad j > p+1.$$

Here all partitions are after the qth row or column and $*$ entries indicate freely varying submatrices. The set of all fully varying coefficients in $A(j)$, $B(j)$ will be called τ. All $A(j)$, $B(j)$ not listed in (3.1) are freely varying. Since $U(n)$ is open and dense in $M(n)$ it is not unreasonable to confine one's attention to this set. This deserves discussion of course and will not be universally agreed with, but to give a fuller account would require us to introduce such concepts as Kronecker indices (see Kailath, 1980) as well as other coordinate neighbourhood in $M(n)$ and this could not possibly be done within the scope of this paper. It must be remembered that here we are approximating to $\mathcal{H}_0$, the true Hankel matrix, and are *not* asserting that $n_0 < \infty$. In such an approximation procedure the choice of the systems, for $n < \infty$, as approximants is already arbitrary. They are chosen, partly, for mathematical convenience. In such a context it is not unreasonable to further confine ourselves. The algorithms here outlined could be used in the more general context where, for example, Kronecker indices are to be determined but we cannot discuss that though it could be a preferable procedure.

In this context, where n as well as τ has to be determined, the computational task becomes great. One procedure is to use the Gaussian likelihood and to optimise that for each n to be examined, choosing n by (1.4). The likelihood can be constructed using the Kalman filter apparatus for both the likelihood and its derivatives. Such a procedure must be iterative. An alternative is to choose n at each iteration, i.e. to make each iteration a step in a Gauss-Newton procedure that views n as a parameter to be determined along with τ. This can reduce the calculations if the variation in n at each iteration can be handled by a recursion on n. Consider the approximation to $-2T^{-1}$ (log likelihood) afforded by

$$\frac{1}{2\pi}\int_{-\pi}^{\pi} \operatorname{tr}\left\{\Sigma^{-1}k(e^{i\omega})^{-1}I(\omega)k^*(e^{i\omega})^{-1}\right\}d\omega + \log\det\Sigma,$$

$$I(\omega) = w(\omega)w(\omega)^*, \qquad w(\omega) = T^{-1/2}\sum_1^T y(t)e^{it\omega}. \tag{3.2}$$

Reducing the likelihood with respect to Σ we obtain

$$\log\det\left\{\frac{1}{2\pi}\int_{-\pi}^{\pi} k(e^{i\omega})^{-1}I(\omega)k^*(e^{i\omega})^{-1}d\omega\right\} + s. \tag{3.3}$$

Since $\hat{k}$ is an already available estimate of k, let

$$k^{-1} = \hat{k}^{-1} + \left(\partial_\tau \hat{k}^{-1}\right)(\tau - \hat{\tau}), \tag{3.4}$$

where $\partial_\tau \hat{k}^{-1}(\tau - \hat{\tau}) = \sum \partial_i \hat{k}^{-1}(\tau_i - \hat{\tau}_i)$, τ_i is the ith component of τ and $\partial_i \hat{k}^{-1}$ indicates differentiation followed by evaluation at $\hat{k}$. Let us use $\hat{k}^{-1}y(t)$, for example, to mean that $\hat{k}^{-1}$ is interpreted as a lag operator with $z^{-1}y(t) = y(t-1)$. Then

$$\left(\hat{k} - \partial_\tau \hat{k}^{-1}\hat{\tau}\right) y(t) = \hat{\eta}(t) - \hat{\xi}(t) + \hat{\varepsilon}(t)$$

and, recalling that $\hat{k} = \hat{a}^{-1}\hat{b}$,

$$\hat{\varepsilon}(t) = \hat{k}^{-} y(t), \qquad \hat{\eta}(t) = \hat{b}^{-1}y(t), \qquad \hat{\xi}(t) = \hat{b}^{-1}\hat{\varepsilon}(t). \tag{3.5}$$

These may be calculated recursively using the Toeplitz assumption inherent in (3.2), namely $y(t) = 0$, $t \leq 0$. Also

$$\partial_\tau \hat{k}^{-1}\tau y(t) = \hat{b}^{-1}(a_\tau - 1)y(t) - \hat{b}^{-1}(b_\tau - 1)\hat{\varepsilon}(t).$$

Thus (3.3) becomes after the use of (3.4):

$$\log\det\left\{\frac{1}{2\pi}\int_{-\pi}^{\pi} w_e(\omega)w_e(\omega)^* d\omega\right\} + s,$$

where

$$w_e(\omega) = T^{-1/2}\sum_1^T e(t)e^{it\omega}$$

and

$$\begin{aligned} e(t) = \hat{\eta}(t) - \hat{\xi}(t) + \hat{\varepsilon}(t) + \hat{b}^{-1}\{a_\tau - A_\tau(0)\}\, y(t) \\ - \hat{b}^{-1}\{b_\tau - A_\tau(0)\}\,\hat{\varepsilon}(t) + \hat{b}^{-1}(A_\tau(0) - I_s)\{y(t) - \hat{\varepsilon}(t)\}. \end{aligned}$$

(See (3.1) for $A_\tau(0)$.) Thus we are reduced to a regression procedure. The calculation may be effected by a sequence of steps that we briefly describe before going on to detail the calculations for $s = 1$.

Let us indicate iterative stages by a superscript.

(0) Take $\hat{k}^{(0)}$ to be the estimate of k obtained from the autoregressive procedure described in Section 2. Thus

$$\hat{k}^{(0)^{-1}} = \sum_0^{\hat{h}} \hat{\Phi}_{\hat{h}}(j) z^{-j}.$$

(1) Next consider $n = sp$ and investigate $p = 0, 1, 2, \ldots$. For this sequence of cases the calculation may be done by a recursion of p. The calculation becomes onerous as s increases since it involves a vector $v(t)$ (see below for $s = 1$) that has $2s^2$ components. (The coefficients of the τ_i in $\hat{b}^{-1}(a_\tau - I_s)y(t)$, $\hat{b}^{-1}(b_\tau - I_s)\hat{\varepsilon}(t)$. Now $A_\tau(0) = I_s$, for $n = ps$.) This "curse of dimensionality" plagues all estimation once s is large. However the calculation at the first iteration, i.e. to determine $\hat{k}^{(1)}$, is much simpler and does not suffer from these difficulties, because in $\hat{k}^{(0)}$, $\hat{b} = I_s$ so that the recursive procedure at the first iteration involves only $2s$ components in the vector $v(t)$. Thus one procedure would be to determine $\hat{p}^{(1)}$ and examine, at later iterations, only a few values of p near $\hat{p}^{(1)}$. For details see Hannan and Kavalieris (1984).

(2) Once $\hat{p}$ is chosen it may be sufficient to examine $n = ps + q$, $0 \leq q < s$, at each iteration, only for a few values of p near $\hat{p}$.

(3) The value of C_T in (1.4) must be chosen. The common choices are $C_T = 2$, $C_T = \log T$ and there is evidence, that $C_T = 2$ has virtues (Shibata, 1980; see however Lütkepohl, 1985).

(4) One must not examine values of p that are too large in relation to T and certainly they must be $o(T^{1/2})$.

To conclude this section we give more detail for $s = 1$, which is an important case. Now $q \equiv 0$ and since there is no dimensionality problem the full recursion may as well be done at each iteration. We recall the algorithm of Whittle (1963) which we use with a vector of observations that we call $v(t)$, of ν components. Put

$$C_v(t) = \frac{1}{T}\sum_{s=1}^{T_t} v(s)v(s+t)' = C_v(-t)', \qquad t \geq 0.$$

Then the algorithm computes as follows:

$$\begin{aligned} F_h(j) &= F_{h-1}(j) + F_h(h)\tilde{F}_{h-1}(h-j), \\ \tilde{F}_h(j) &= \tilde{F}_{h-1}(j) + \tilde{F}_h(h)F_{h-1}(h-j), \\ \tilde{F}_h(0) &= F_h(0) = I_\nu, \qquad j = 1, 2, \ldots, h-1, \end{aligned}$$

$$\begin{aligned} F_h(h) &= -\Delta_{h-1}\tilde{S}_{h-1}^{-1}, \\ \tilde{F}_h(h) &= -\Delta'_{h-1}S_{h-1}^{-1}, \\ \Delta_h &= \sum_0^h F_h(j)C_v(j-h-1), \end{aligned}$$

$$S_h = \left\{ I_\nu - F_h(h)\tilde{F}_h(h) \right\} S_{h-1},$$
$$\tilde{S}_h = \left\{ I_\nu - \tilde{F}_h(h)F_h(h) \right\} \tilde{S}_{h-1},$$
$$S_0 = \tilde{S}_0 = C_v(0).$$

At the autoregressive stage, (0), $v(t) = y(t)$ and $\nu = s$. At later stages $\nu = 2s^2$ and for $s = 1$, in particular $V(t) = \left(\hat{\eta}^{(i)}(t), \hat{\xi}^{(i)}(t)\right)$, where $i = 0, 1, 2, \ldots$ indexes the iterations. For $i = 0$ then $\hat{\eta}^{(0)}(t) = y(t)$, $\hat{\xi}^{(0)}(t) = \hat{\varepsilon}_{\hat{h}}(t)$, and so on. We now identify h with n, or equivalently with p, since $n = ps$ and $s = 1$. Now calling $\alpha(j)$, $\beta(j)$, $j = 1, \ldots, p$, the coefficients in $a(z)$, $b(z)$ we have the recursion, using a superscript to indicate the iterative stage and a subscript for the recursive stage:

$$\begin{bmatrix} \hat{\alpha}_p^{(i)}(j) \\ \hat{\beta}_p^{(i)}(j) \end{bmatrix} = \begin{bmatrix} \hat{\alpha}_{p-1}^{(i)}(j) \\ \hat{\beta}_{p-1}^{(i)}(j) \end{bmatrix} + \tilde{F}_{p-1}(p-j)' \begin{bmatrix} \hat{\alpha}_p^{(i)}(p) \\ \hat{\beta}_p^{(i)}(p) \end{bmatrix}, \quad j = 1, \ldots, p-1,$$

$$\begin{bmatrix} \hat{\alpha}_p^{(i)}(p) \\ \hat{\beta}_p^{(i)}(p) \end{bmatrix} = \tilde{S}_{p-1}^{-1} \sum_0^{p-1} \tilde{F}_{p-1}(j) \frac{1}{T} \sum_t v(t-p+j)$$
$$\left\{ \hat{\eta}^{(i-1)}(t) + \hat{\varepsilon}^{(i-1)}(t) - \hat{\varepsilon}^{(i-1)}(t) \right\},$$

$$\left(\hat{\sigma}_p^{(i)}\right)^2 = \left(\hat{\sigma}_{p-1}^{(i)}\right)^2 - \left(\hat{\alpha}_p^{(i)}(p), \hat{\beta}_p^{(i)}(p)\right) \tilde{S}_{p-1} \left(\hat{\alpha}_p^{(i)}(p), \hat{\beta}^{(i)}(p)\right)',$$

$$\left(\hat{\sigma}_0^{(i)}\right)^2 = \frac{1}{T} \sum_t \left\{ \hat{\eta}^{(i-1)}(t) + \hat{\varepsilon}^{(i-1)}(t) - \hat{\xi}^{(i-1)}(t) \right\}^2.$$

Of course $\hat{p}^{(i)}$ is chosen to minimise

$$\log \left(\hat{\sigma}_p^{(i)}\right)^2 + 2pC_t/T, \qquad p = 0, 1, \ldots.$$

The summations, $\sum_t$, are over $0 \le t$ for which $\hat{\eta}(t)$, $\hat{\xi}(t)$ etc. are non-null and these ranges are determined by the rule $y(t) = 0$, $t \le 0$, $t \ge T$. When this is done we are acting consistently with the minimisation of (3.2). However it is known that this "Toeplitz" convention leads to biases. There is a large literature that tries to mitigate its effects. See Friedlander (1982) for references.

4. SOME FURTHER COMMENTS

Except when $s = 1$ these procedures, or indeed any procedures for fitting ARMA models, do not seem to have been widely used except in cases where there are fairly considerable fixed constraints based on physical understanding.

In the ad hoc fitting procedures here considered the procedure may constitute only a stage towards a final result, such as spectral estimation. Though the consideration of all rational transfer function systems has appeal it is also computationally costly and the cost may become considerable if the program is to run when the best fitting structure is near to unstable, e.g. if $\det b_0(z)$ is near to zero at $|z| = 1$ or $a_0(z)$, $b_0(z)$ are near to non-coprime. (For example, the possibility that $b_0(z)$ is zero for $|z| < 1$ will have to be taken care of in (3.5). For $s = 1$ this is easily done.)

It is conceivable that consideration of something less than the full set of rational transfer function models will be better in some circumstances. In any case one is concerned with an approximation to the truth and the choice of the model set from which to obtain the approximation is in the hands of the investigator and is a reflection of his skill. The use of autoregressive approximation is a manifestation of these considerations.

Much remains to be done.

REFERENCES

Akaike, H. (1969), "Fitting autoregressive models for prediction". *Annals of the Institute of Statistical Mathematics* **21**, 243–247.

Baxter, G. (1962), "An asymptotic result for the finite predictor". *Mathematica Scandinavica* **10**, 137–144.

Friedlander, B. (1982), "Lattice methods for spectral estimation". *Proceedings of the IEEE* **70**, 990–1017.

Hannan, E. J., and L. Kavalieris (1984), "Multivariate linear time series models". *Advances in Applied Probability* **16**, 492–561.

Hannan, E. J., and L. Kavalieris (1986), "Regression, autoregression models". *Journal of Time Series Analysis* **7**, 27–49.

Kailath, T. (1980), *Linear Systems*. Englewood Cliffs, New Jersey: Prentice Hall.

Lütkepohl, H. (1985), "Comparison of criteria for estimating the order of a vector autoregressive process". *Journal of Time Series Analysis* **6**, 35–52.

Rissanen, J. (1983), "A universal prior for integers and estimation by minimum description length". *Annals of Statistics* **11**, 416–431.

Shibata, R. (1980), "Asymptotically efficient estimation of the order of the models for estimating parameters of a linear process". *Annals of Statistics* **8**, 147–164.

Taniguchi, M. (1980), "On selection of the order of the spectral density model for a stationary process". *Annals of the Institute of Statistical Mathematics* **32**,

401–419.

Tjostheim, D., and J. Paulsen (1983), "Bias of some commonly used time series estimates". *Biometrika* **70**, 389–400.

Whittle, P. (1963), "On the fitting of multivariate autoregressions and the approximate canonical factorisation of a spectral density matrix". *Biometrika* **50**, 129–134.

Hirotugu Akaike [1]

SOME REFLECTIONS ON THE MODELLING OF TIME SERIES

ABSTRACT

To respond properly to the increasing demand for efficient data processing procedures in diverse areas of application, emphasis must be placed on the advancement of time series modelling. The progress of the art of time series modelling, or of statistical modelling in general, may be accelerated by explicit recognition of the fact that the subject is essentially concerned with the proper use of false models. In this paper the implication of this point of view is illustrated by examples, including Bayesian models for seasonal adjustment and for estimation of changing spectrum, and non-Gaussian autoregressive models for robust analysis of a system with sporadic impulsive disturbances.

1. INTRODUCTION

With the progress of data acquisition techniques and computing facilities the demand for efficient data processing procedures is increasing rapidly. Proper response to this societal need is vital for the future development of time series analysis.

In time series literature, the main emphasis has been placed on the development of exact mathematical analysis of statistical procedures under simplifying assumptions such as Gaussianity and stationarity. These assumptions are usually accepted as being satisfied by a time series. This common attitude on the part of time series analysts is not precisely in agreement with the usual understanding that a statistical model is an approximation to the structure of a stochastic system that generated the observed data.

[1] The Institute of Statistical Mathematics, 4-6-7 Minami-Azabu Minato-Ku, Tokyo 106, Japan

I. B. MacNeill and G. J. Umphrey (eds.), Time Series and Econometric Modelling, 13–28.

In this paper we stress a particular point of view which considers that the role of a statistical model is to provide proper incentive for the generation of necessary algorithms for efficient data processing. This means that it is not necessarily the similarity of the model to the unknown generating mechanism of a time series but rather the practical utility of the resulting data processing procedure that justifies the use of a time series model. We will call this the instrumental point of view of statistical modelling.

Once this point of view is accepted, statisticians can freely develop new time series models based on their perceptions of required data processing procedures. The conventional structural point of view that considers a statistical model to be the representation of a real stochastic structure has severely limited the range of the development of time series models.

The situation is similar to the case of the development of general Bayesian modelling. The idea that a model should be the representation of a unique structure for each particular application has severely limited the possibility of developing practical, useful Bayesian models. From our present point of view a statistical model is only a representation of the framework through which we look at a given set of data. Thus the introduction of a model is justified only if it leads to the development of a useful data processing procedure. The practical utility can be judged through the accumulation of experiences of the application to real problems.

In the present paper the potential of the instrumental point of view is demonstrated by some examples of time series modelling. The first one is concerned with Bayesian modelling for seasonal adjustment. This example clarifies the unjustified nature of the usual expectation that the selection of a proper model from a set of models will lead to a satisfactory result. Such an expectation can be justified only when the set contains models that show sufficiently good fit to the data. This is illustrated by the comparison of predictive performances and likelihoods of some models.

In the second example an extended form of Bayesian modelling for the analysis of a quickly changing spectrum is discussed. In this model observed values are allowed to come into the definition of the "prior" distribution. The performance of the procedure based on this model shows that the model provides a reasonable basis for the analysis of a system with quickly changing characteristics. This shows that even a formal extension of the Bayesian model to allow observed variables to enter into the "prior" distribution may produce useful algorithms for data analysis.

The third example is concerned with the modelling of the autoregressive (AR) process to control the effect of sporadic large impulsive inputs on the estimation of the structure of the autoregression. After discussion of the use of a mixed Gaussian distribution for the residual series, the use of a Cauchy distribution is considered. The analysis shows that, in maximum likelihood computations, the Cauchy model discounts the effect of residuals with large absolute values on the estimation of autoregressive coefficients, at least as compared with the Gaussian model. A numerical example shows the bias and robustifying effect of the model when the true structure is Gaussian.

2. PREDICTIVE EVALUATION OF TIME SERIES MODELS

The seasonal adjustment procedure BAYSEA (Akaike, 1980a) is based on the ordinary representation

$$y(n) = T(n) + S(n) + I(n),$$

where $y(n)$, $T(n)$, and $S(n)$ denote the original observation, trend, and seasonal component, respectively. The data distribution is defined by assuming the i.i.d. type Gaussian structure for the distribution of the irregular component $I(n)$. The prior distribution of the trend and seasonal components is defined by assuming that a Gaussian distribution of these components controls the smoothness of their behavior. The posterior mode, which defines the final estimates of the trend and seasonal components, is obtained by the solution of

$$\min \| y - T - S \|^2 + r \| FT + GS \|^2,$$

where y, T, and S denote the vectors of $y(n)$, $T(n)$, and $S(n)$ $(n = 1, 2, \ldots, N)$, $\| \cdot \|$ denotes the Euclidean norm; and F and G denote properly defined constant matrices. The positive parameter r is determined by maximizing the likelihood of the Bayesian model.

The use of this type of constrained least squares procedure for the solution of an ill-posed problem that defies direct application of ordinary least squares method has been well-known. However, the performance of the procedure is controlled by the choice of the parameter r; the application of Bayesian modelling for the choice of this parameter was first discussed by Akaike (1980b) to illustrate the instrumental use of a Bayesian model. This example shows typically that a prior distribution could be considered as an artificial construction to produce a reasonable data processing procedure rather than the representation of the "state of the mind of the analyst", which is rather difficult to specify.

In the same case of the original BAYSEA procedure, a homogeneous spherical Gaussian distribution was assumed for the differences, of some order, of the trend. This model is obviously unsatisfactory for the purpose of prediction of the trend, as it produces only a very limited form of prediction: for example, flat and linear trends for models based on first and second order

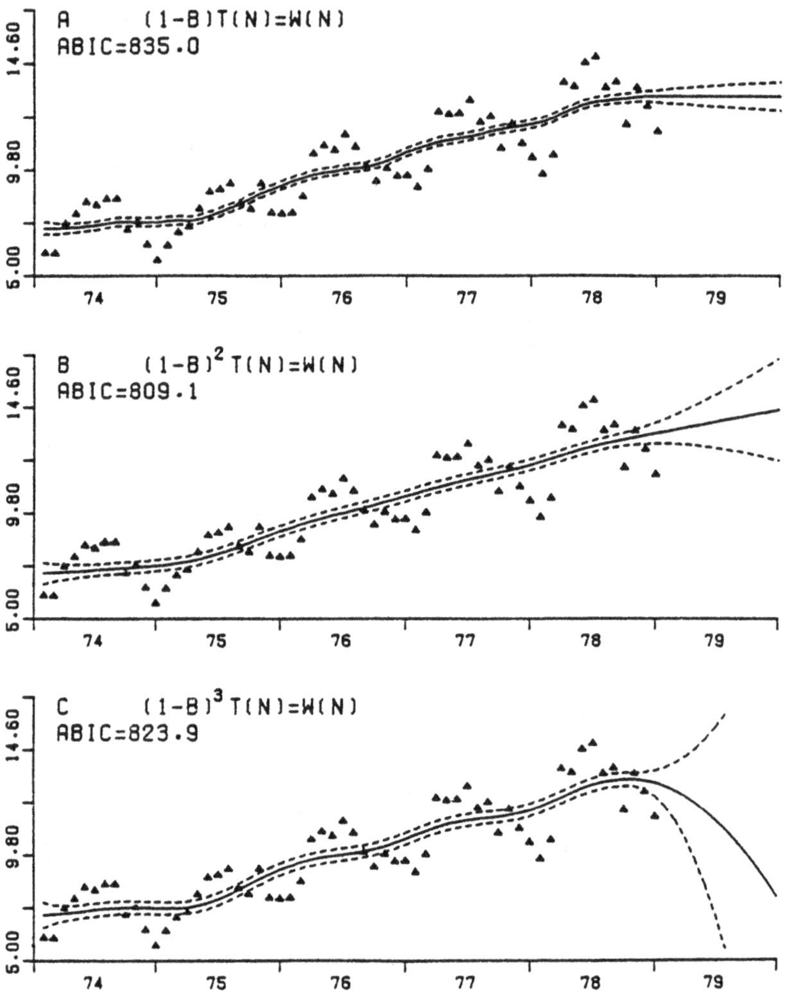

Figure 1. *Examples of the trend prediction by the basic models of BAYSEA.*

differences, respectively. Figure 1 shows examples of trend prediction with corresponding bands of twice the posterior standard deviations.

The number denoted by ABIC means minus twice the logarithm of the likelihood of the Bayesian model defined by the integral of the likelihood of each Gaussian model with respect to the prior distribution, and represents the badness-of-fit of each model to the observed data. The first order difference based model shown in Figure 1A produces the narrowest band of posterior standard deviations but shows poor fit to data with a linear trend.

One natural idea is to modify the prior distribution of the trend component by considering an incomplete differencing, i.e., by assuming a first order AR model for the trend after proper differencing. First order AR models for first and second order differences of the trend provide interpolating models of the three models of Figure 1. Some examples of the results obtained by these models are illustrated in Figure 2, with observations for the last year added to check for goodness-of-prediction.

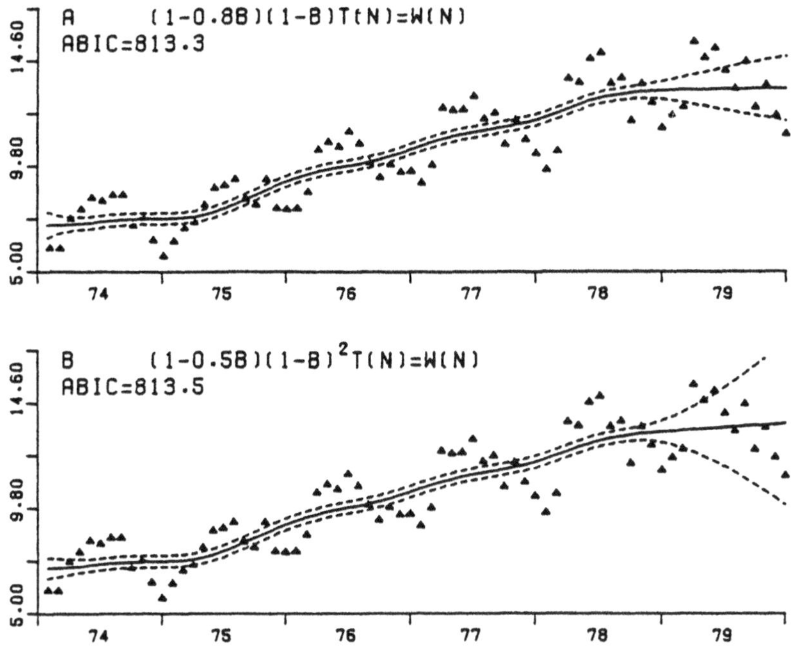

Figure 2. *Examples of the trend prediction by the AR modelling.*

The result shows that, although there are slight increases in the ABIC, both models produce twelve months ahead predictions that are in better agreement with the actual trend of observations than that obtained by the model of Figure 1B. This suggests that the lower value of the ABIC of the latter is due mainly to the clearly linear trend in the preceding years and thus that the model may not be useful when change in the trend is possible.

3. POSTERIOR DISTRIBUTION AS THE CONDITIONAL LIKELIHOOD

The example discussed in the preceding section shows the necessity of care in developing a Bayesian model for forecasting. In particular, it has shown that the posterior distribution must be carefully checked by comparing it with those of other models representing different possibilities, with proper attention being paid to the variation of likelihoods within the models. This means that there are situations where one should consider the posterior distribution of a parameter simply as the representation of the relative likelihood of each possible parameter value under the assumption of the Bayesian model.

In Bayesian modelling the posterior distribution satisfies the relation

$$p(X,A) = p(A \mid X)p(X),$$

where $p(X,A)$ denotes the simultaneous probability (density) of (X,A) before observing X, $p(A \mid X)$ denotes the posterior probability (density) of A, and $p(X)$ represents the likelihood of the Bayesian model defined by the integral of $p(X,A)$ with respect to the measure of the parameter A. The distribution $p(X,A)$ provides the basic framework through which one looks at the data, and the present observation shows that, during the process of model building, one should keep in mind the relative nature of the posterior distribution. If the situation demands, one should check the necessity of modifying $p(X,A)$ by considering the likelihoods of possible alternative models.

4. ESTIMATION OF QUICKLY CHANGING AUTOREGRESSION

At the 1979 meeting of the International Statistical Institute the author presented a paper on the construction of composite time series models (Akaike, 1979). In that paper a procedure for the estimation of a quickly changing spectrum was introduced which was realized by minimizing $L = SSR + Q$ where SSR and Q were defined by

$$SSR = \sum_{p=1}^{P} \sum_{i=(p-1)K+1}^{pK} [y(i) - \sum_{m=1}^{M} a(m:p)y(i-m) - a(0:p)]^2$$

and

$$
\begin{aligned}
Q = & c\sum_{i=1}^{K}\left[\sum_{m=1}^{M} a(m:1)y(i-m)\right]^2 \\
& + a\sum_{p=2}^{P}\sum_{i=(p-1)K+1}^{pK}\left[\sum_{m=1}^{M} a(m:p)(y(i-m)-y(i-m-K))\right]^2 \\
& + b\sum_{p=1}^{P}\sum_{i=(p-1)K+1}^{pK}\left[\sum_{m=1}^{M} ma(m:p)y(i-m)\right]^2 \\
& + dPKa(0:1)^2 + ePK\sum_{p=2}^{P}[a(0:p)-a(0:p-1)]^2 ,
\end{aligned}
$$

where $a(m:p)$ denotes the mth coefficient of autoregression for the pth span of data with length K $(p = 1, 2, \ldots, M)$.

The procedure requires the choice of positive constants a, b, c, d, and e. The constants a and b control the smoothness of predicted values in time and frequency domains, respectively, c controls the sizes of the predicted values for the initial span, and d and e control the size of the initial value and the smoothness of the constant term of the regression, respectively.

The structure of the procedure is quite similar to that of BAYSEA but the term Q that defines the prior distribution depends on the observations. This dependence on the observed values gives the impression that the modelling is quite un-Bayesian.

However, a Bayesian model may be viewed simply as a specification of a simultaneous distribution of the observation and the parameters. Thus, if by a proper normalization, $\exp(-gL)$ $(g > 0)$ defines a probability distribution over the set of possible observations and coefficients of autoregression, its integral with respect to the latter will define the likelihood of the model and the corresponding conditional distribution of the autoregressive coefficients will give the posterior distribution.

Based on this observation an experimental computer program called LOCCAR (locally constant AR-model fitting) was developed by slightly modifying the definition of Q in the equation $L = SSR + Q$ to make the procedure operative even for the case with $K = 1$ (Akaike et al., 1985). The term Q is then given by:

$$Q = \sum_{p=1}^{H+1} \Big\{ c[a(0:p) + \sum_{m=1}^{M} a(m:p)y(M+(p-1)K-m)]^2$$
$$+ a \sum_{i=1}^{M} [a(0:p+1) - a(0:p) + \sum_{m=1}^{M} (a(m:p+1)$$
$$- a(m:p))y(M+(p-1)K-m+i)]^2 \Big\}$$
$$+ a \sum_{p=2}^{P} \sum_{i=pK-J}^{pK} I(i>K)[a(0:p) - a(0:p-1)$$
$$+ \sum_{m=1}^{M} (a(m:p) - a(m:p-1))y(i-m)]^2$$
$$+ b \sum_{p=1}^{P} [\sum_{m=1}^{M} ma(m:p)]^2 + dPKa(0:1)^2$$
$$+ ePK \sum_{p=1}^{P} [a(0:p) - a(0:p-1)]^2,$$

where H denotes the integer part of $(M-1)/K$, $I(i>K)=1$, for $i>K$, 0, otherwise, and J denotes $\max(K, M)$.

Figure 3 shows the EEG record of some brain waves and also the spectra obtained by applying LOCCAR with $M=5$ and $K=4$. It can be seen that the procedure responds to the change of the spectrum at an early stage of the recording. The unspecified parameters were adjusted by maximizing the likelihood of the model within a finite set of alternatives. Results with some simulated data demonstrated satisfactory performance of the procedure.

The distribution assumed in this model may be viewed as a representation of a Bayesian structure with a prior distribution that adapts to past experience. Thus it is not free from the serious problem of the choice of the speed of adaptation. Numerical results show that the choice of this speed is realized by the maximization of the likelihood of the model, at least for the purpose of the analysis of a past record.

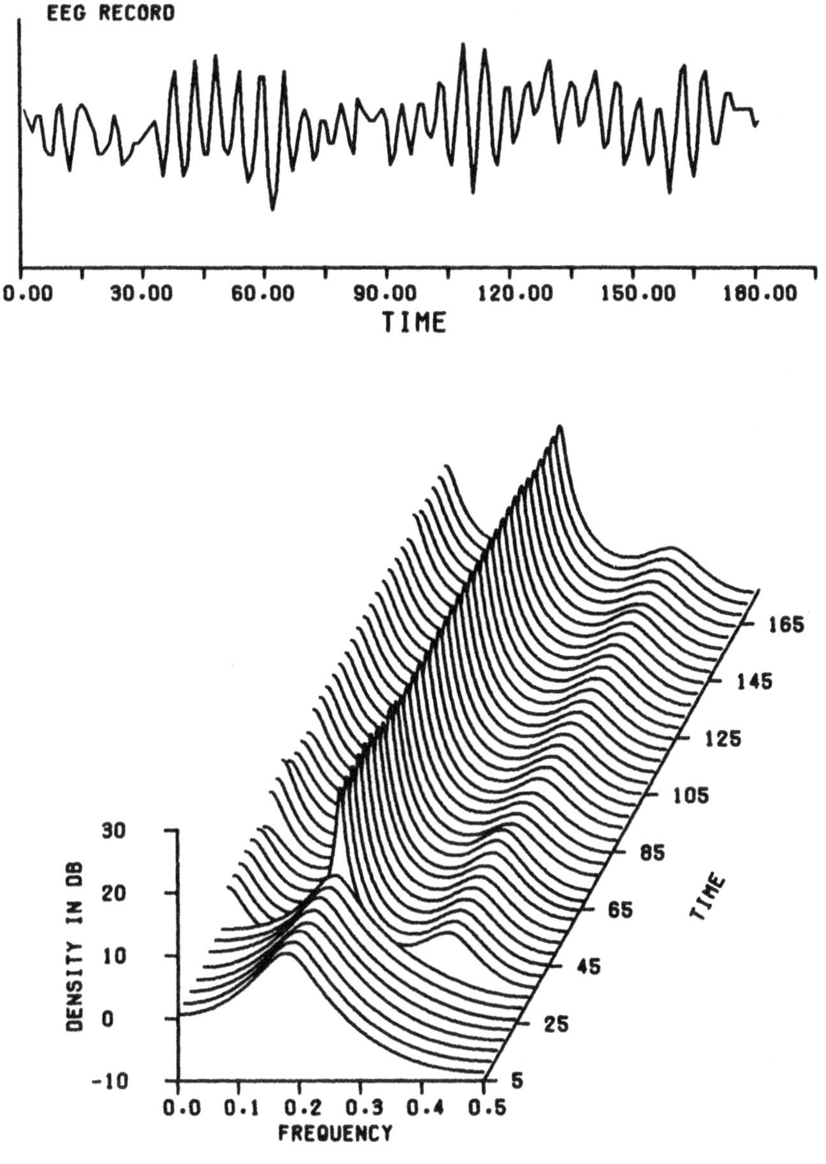

Figure 3. *Estimation of a changing spectrum by the locally constant AR modelling.*

5. AUTOREGRESSION WITH CAUCHY DISTRIBUTED ERROR TERMS

One of the most popular time series models is the Gaussian autoregressive model. This model is defined by

$$y(i) = \sum_{m=1}^{M} a(m)y(i-m) + z(i),$$

where the error term $z(i)$ is assumed to be independent of the past values $y(i-m)$ and $z(i-m)$, $m = 1, 2, \ldots$, and identically distributed as Gaussian with mean 0 and variance v. The popularity of this model is, to a large extent, due to the simplicity of the concomitant estimation procedure, which is usually realized by the method of least squares.

However, there are certain situations where the assumption of Gaussianity is not appropriate. One such example is the case where the basic process is disturbed by some sporadic input. To model this situation precisely it is necessary to specify the distributional property of the input process. Here we consider the use of a heavy-tailed distribution for $z(i)$, to compensate for the effect of the sporadic input.

First we consider the use of a mixture of the original Gaussian distribution and another Gaussian distribution with mean 0 and variance w much larger than v. In particular, we consider the combination $C(z \mid r; v, w) = rG(z \mid v) + (1-r)G(z \mid w)$ defined with a weight r $(0 < r < 1)$, where $G(z \mid v)$ denotes the density at z of a Gaussian variable with mean 0 and variance v. The likelihood of the model is given by

$$L(a, v, w) = \prod_{i=1}^{N} [rG(z(i) \mid v) + (1-r)G(z(i) \mid w)].$$

Expansion of the product into the sum of the products of proper combinations of the two terms inside the square brackets yields:

$$L(a, v, w) = \sum_{k=0}^{N} r^k (1-r)^{(N-k)} \{S(I, J : k)[\prod_{i \in I} G(z(i) : v) \prod_{j \in J} G(z(j) : w)]\},$$

where $S(I, J : k)$ denotes the summation over the possible mutually exclusive sets I and J of i and j such that the number of elements within I is equal to k and that of J is $N - k$.

This representation shows that the likelihood is composed of the weighted sum of the likelihoods of models, each one of which assumes a particular sequence of locations of wild inputs. By considering these models separately, one can develop detailed analyses of the nature of the input process. Thus there is a possibility of performing the Bayesian analysis by assuming a prior distribution for the distribution of wild inputs over time. This type of modelling has been adopted for the analysis of outlying observations by Kitagawa and Akaike (1982).

In the above representation of the likelihood the enumeration of all the possible locations of wild inputs quickly causes insurmountable complexity of the likelihood computation when N is large. If one is interested only in obtaining a reasonable estimate of the autoregression, then it is desirable to have a structure that is insensitive to the location of wild inputs. Such a model can be obtained by assuming a single heavy-tailed distribution for the error term.

The likelihood of a model defined by assuming the Cauchy distribution

$$C(z \mid c) = 1/\{\pi c[1 + (z/c)^2]\}$$

for the residuals is given by

$$L(a, c) = \prod_{i=1}^{N} C(z(i) : c),$$

where

$$z(i) = y(i) - \sum_{m=1}^{M} a(m)y(i - m).$$

One obtains estimates of these parameters by maximizing the likelihood with respect to the (vector) parameter $a(\cdot)(= (a(1), a(2), \ldots, a(M)))$ and c. The maximum likelihood computation can be performed by a numerical optimization procedure.

To derive some insight into the nature of the estimator of the autoregressive parameter $a(\cdot)$ we consider the structure of a single step of the Newton-Raphson procedure for the maximization of $L(a, c)$. We define NLL, twice the negative log likelihood, by

$$NLL = 2N \log(\pi) + 2 \sum_{i=1}^{N} \log(z(i)^2 + d) - N \log(d),$$

where $d = c^2$. The quantities required for the computation of the Hessian and gradient are given by

$$\begin{aligned}
D(NLL/d) &= 2S[1/Q(i)] - N/d \\
D(NLL/a(m)) &= -4S[z(i)y(i-m)/Q(i)] \\
DD(NLL/d,d) &= -2S[1/Q(i)^2] + N/d^2 \\
DD(NLL/a(m),d) &= 4S[z(i)y(i-m)/Q(i)^2] \\
DD(NLL/a(m),a(l)) &= 4S[\{y(i-m)y(i-l)/Q(i)\}\{1-2z(i)^2/Q(i)\}],
\end{aligned}$$

where S denotes the summation over $i = 1, 2, \ldots, N$, $Q(i) = z(i)^2 + d$, and $D(NLL/a(m))$ and $DD(NLL/a(m), a(l))$ denote partial derivatives of the first and second orders of NLL with respect to $a(m)$, and $a(m)$ and $a(l)$, respectively.

Consider the situation where d is given. Then the elements of the gradient and Hessian are given by

$$g(m) = D(NLL/a(m)) = -4S[z(i)y(i-m)/Q(i)],$$

and

$$\begin{aligned}
H(m,l) &= DD(NLL/a(m), a(l)) \\
&= 4S[\{y(i-m)y(i-l)/Q(i)\}\{(-z(i)^2 + d)/Q(i)\}].
\end{aligned}$$

From this one may note that when $z(i)^2$ is very small compared with d, then the contribution from $z(i)$ to $g(m)$ is given approximately by $-4z(i)y(i-m)/d$, and it is given by $-4\{z(i)y(i-m)/d\}\{d/z(i)^2\}$ when $z(i)^2$ is very large compared with d.

Consider the contribution to the gradient from the observation with $z(i)^2$ much larger than d, which controls the scale of the Cauchy distribution. The last result shows that this is significantly discounted compared with $4z(i)$ $y(i-m)/d$ which approximates the contribution from the observation with $z(i)^2$ much smaller than d.

Similarly it can be seen that the contribution from the observation with small $z(i)^2$ to the element $H(m,l)$ is approximately equal to $\frac{4y(i-m)y(i-l)}{d}$. This result and the former result for the gradient, show that the contribution from observations with small $z(i)^2$ to the one-step correction by the Newton-Raphson procedure is similar to that under the assumption of the Gaussian distribution, while the contribution corresponding to an observation with large $z(i)^2$ is discounted.

The present observation suggests that the maximum likelihood estimate of $a(\cdot)$ will be insensitive to an input $z(i)$ with large amplitude; i.e., the estimates of the autoregressive coefficients will show some kind of robustness against a wild input and that they will approximate the Gaussian estimates when wild inputs are absent.

The result of a numerical experiment with the present procedure is illustrated in Figure 4. The theoretical transfer function is defined by

$$A(f) = \left(1 - \sum_{m=1}^{M} a(m) \exp(-i2\pi m f)\right)^{-1}$$

and the figures are on the scale of 20 log $| A(f) |$. The estimates by the fourth order AR models were obtained by applying the minimum AIC procedure to the models up to order $M = 10$, fitted to the first 300 observations. The observations were generated from the stationary Gaussian process defined by the frequency response function and the innovation process $z(i)$ which is Gaussian with mean 0 and variance appropriate to simulate a real record of an earthquake. The simulated record is shown in Figure 5, below the actual seismic record.

The lower two estimates were obtained using the 340 observations with the additional last 40 input values $z(i)$ generated from a zero mean Gaussian distribution with standard deviation five times that of the original. To see the effect of these abnormal inputs the coefficients of autoregression $a(m)$ were replaced by $(0.9)^m a(m)$ to generate the observations $y(i)$ for the span of i that corresponded to the span of wild inputs.

It can be seen from the estimates that the effect of the abnormal input is quite significant for the Gaussian estimate, while the Cauchy estimate is rather insensitive. However, the Cauchy estimates show a systematic deviation or bias from the theoretical transfer function with lower values in the very low frequency range.

The likelihood of the Gaussian model is either higher or lower than that of the Cauchy model for either the first 300 observations or for the 340 observations, respectively. Taking into account the fact that the Cauchy model is never a real representative of the generating process of the data, this result demonstrates the use of the expected log likelihood as the basic criterion of fit of instrumental models with different distributional characteristics. The fact that the log likelihood can provide a reasonable criterion of fit of a false "model" is based on the fact that its expectation forms an "objective" criterion of fit of a model to the true distribution of the data, whatever this latter might be; see, for example, Akaike (1985).

The present model has been applied to an actual earthquake record which is shown in Figure 5. The maximum likelihood of the Gaussian model is higher for the span of the data where only the effect of the stationary microtremor is recorded, while that of the Cauchy model is higher for the total span of data where the arrival of the P-wave of the earthquake is

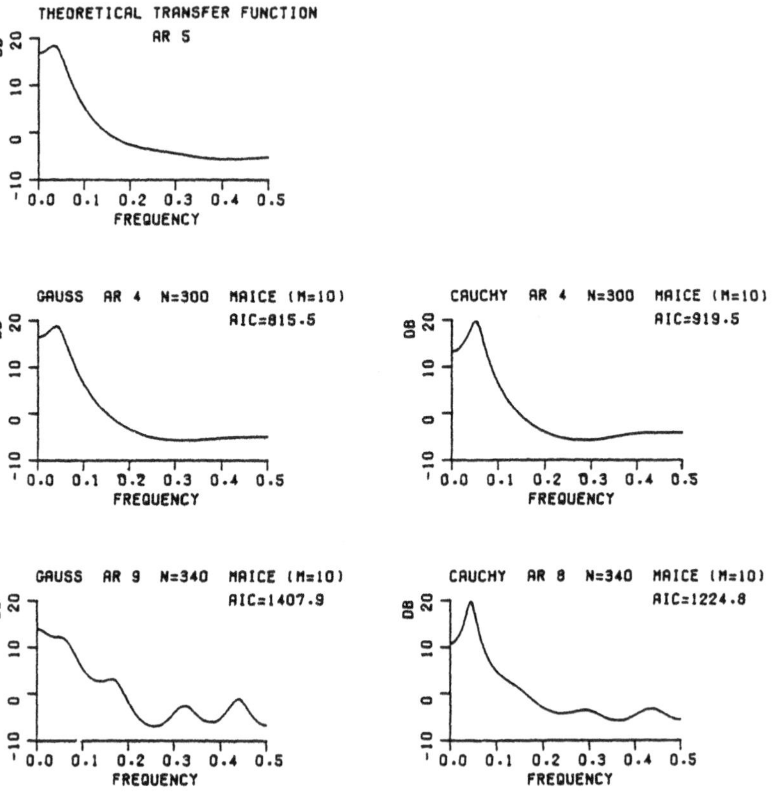

Figure 4. *Comparison of the robustness of the AR estimates based on the assumption of the Gaussian and Cauchy distributions of error terms.*

observed. This demonstrates the possibility of the combined use of the two models for the detection of abnormal inputs.

6. CONCLUDING REMARKS

In this paper, emphasis has been placed on the instrumental point of view, which considers a statistical model as an artificial structure for the development of useful data processing procedures. The numerical results demonstrate the potential of this point of view for the advancement of modelling of time series.

The most serious problem with the application of this point of view is

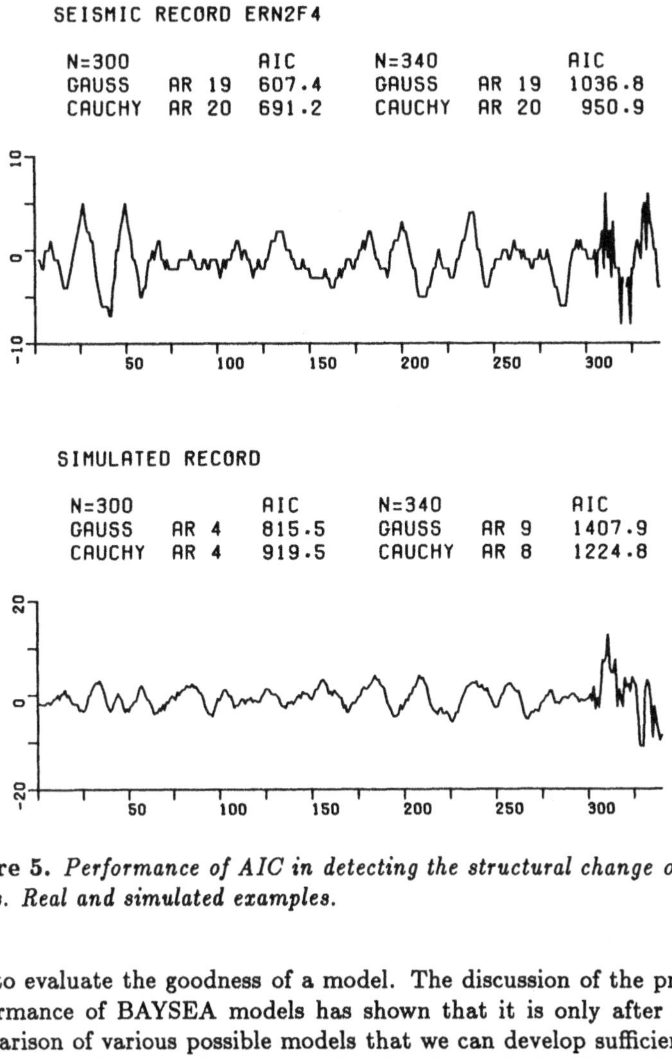

Figure 5. *Performance of AIC in detecting the structural change of a time series. Real and simulated examples.*

how to evaluate the goodness of a model. The discussion of the predictive performance of BAYSEA models has shown that it is only after detailed comparison of various possible models that we can develop sufficient confidence in applying a model or procedure for the analysis of a particular set of data. In such a situation, as the number of possible models increases, it becomes impossible to reach a reasonable decision within a limited amount of time without a proper criterion of fit of a model. The use of a large number of models is becoming the practice rather than the exception with the development of efficient computing procedures.

The instrumental point of view introduces much freedom into Bayesian modelling. The example of the LOCCAR procedure suggests the practical utility of a procedure which can formally be viewed as the Bayes procedure defined with a prior distribution that depends on data.

The concept of the expected log likelihood, of which estimates can be obtained from the NLL statistic, does play a vital role in developing criteria such as AIC and ABIC. The use of such a criterion becomes mandatory when the comparison is concerned with models with different distributional properties, as in the case of the comparison of the Gaussian and Cauchy AR models.

The importance of developing new structural models based on knowledge of a particular area of application should never be overlooked. Nevertheless, it is hoped that the discussion in this paper has demonstrated the potential of the instrumental point of view, when it is equipped with a criterion of fit. This approach opens up the possibility of developing a proper use of "false" models representing various ways of looking at statistical data, including time series.

ACKNOWLEDGMENTS

The author is grateful to Emiko Arahata for preparing the graphical outputs presented in this paper. This work was partly supported by a Grant-in-Aid No. 58450058 by the Ministry of Education, Science and Culture.

REFERENCES

Akaike, H. (1970), "On the construction of composite time series models." *Proceedings of the 42nd Session of the International Statistical Institute* **1**, 411–422.

Akaike, H. (1980a), "Seasonal adjustment by a Bayesian modelling." *Journal of Time Series Analysis* **1**, 1–13.

Akaike, H. (1980b), "Likelihood and the Bayes procedure." In *Bayesian Statistics*, ed. J.M. Bernardo, H. M. DeGroot, D. V. Lindley and A. F. M. Smith, pp. 143–166. Valencia, Spain: University Press.

Akaike, H. (1985), "Prediction and entropy." In *A Celebration of Statistics*, ed. A. C. Atkinson and S. E. Fienberg, pp. 1–14. New York: Springer.

Akaike, H., T. Ozaki, M. Ishiguro, Y. Ogata, G. Kitagawa, Y. Tamura, E. Arahata, K. Katsura, and Y. Tamura (1985), "TIMSAC-84, Part 1 and 2." Computer Science Monographs, No.'s 22 and 23, The Institute of Statistical Mathematics, Tokyo.

Kitagawa, G., and H. Akaike (1982), "A quasi Bayesian approach to outlier detection." *Annals of the Institute of Statistical Mathematics* **34**, 389–398.

Quang Phuc Duong [1]

MODEL SELECTION AND FORECASTING: A SEMI-AUTOMATIC APPROACH

ABSTRACT

Econometric or time-series forecasts in the telecommunications industry (e.g., service demand, revenue) are an important element in a number of decision making processes; i.e., staffing, budgeting, tariff setting and other marketing strategies. Since these forecasts are only one of a number of inputs to decision making, no optimality criterion can be defined. The absence of an optimality criterion and the large number of series involved makes the selection of models an even more difficult exercise. Usually, the selection process is subject to two validation procedures: first, statistical tests on historical data to ensure inclusion of meaningful explanatory variables and proper fit, and second, tests of the model's ability to allow the evaluation of the impact of future (hypothetical) market conditions and/or internal or external (e.g., Government) policies.

In this paper, a two-stage 'semi-automatic' selection criterion, which produces a subset of feasible and 'ranked' models using an internal validation procedure is suggested. The criterion used in the first stage is based on the performance of competing models in predicting observed data points (forward validation); the selected subset is then validated at the second stage through subjective assessments (scenario validation).

1. INTRODUCTION

In this paper some problems are addressed which are faced by many time series modellers and forecasters in selecting the 'best' model among a set of competitors using model selection criteria such as the AIC-type criteria. It is argued here that in most practical situations, a single model is not sufficient for the purpose of analysis, and hence a set containing several useful models

[1] Corporate Management Sciences, Bell Canada, Montreal, Quebec H2Z 1S4

I. B. MacNeill and G. J. Umphrey (eds.), Time Series and Econometric Modelling, 29–46.

should be considered. The necessity of considering a collection of models arises in the case of a non-stationary time series; for example, one with gradually changing characteristics (Akaike, 1979a). Also, results of some simulations (MacNeill and Duong, 1982) have shown that wrong models could be chosen by some of these model selection criteria when the time series undergoes even modest changes in its structure (parameter or order changes). Similar difficulties arise when one attempts to model short series, in which case the selection statistic may not be reliable due to small sample variability. An attractive solution for this problem has been discussed by Kitagawa and Gersch (1985a,b), who suggested using smoothness priors in the form of constraints on the AR model parameters. A similar approach has been suggested by Duong (1981).

In the context of forecasting, the problem of choosing among several available forecasting models has also received much attention lately. Makridakis *et al.* (1982) evaluated the accuracy of 24 major forecasting methods on a collection of 1,001 time series. A comprehensive classification of the more commonly used methods and their accuracy is described by Mahmoud (1984). Given the current proliferation of forecasting methods, a decision-maker will often be presented with a set of k forecasts, or more generally, a set of k mathematical models which could generate forecasts. The question facing him is the following: should he attempt to select the 'best' forecast, or should he try to combine the k forecasts in some way? Both alternatives (selection versus synthesis) require optimality criteria, which in practice might be difficult to define and evaluate.

For the two situations in hand, model selection via a minimum-selection criterion or choice of a forecasting method (which one may or may not be able to characterize by a model selection criterion), we suggest a two-stage, 'semi-automatic' approach, which produces a subset of feasible and 'ranked' models. In the first stage, the performance of the competing models in predicting observed data points is used to rank the models; a subjective selection procedure is then used in the second stage to validate and, if necessary, combine the 'good' models. While the first stage is essentially automatic in nature, except for the specification of the degree of uncertainty we are willing to tolerate, the second stage requires subjective assessments of the various scenarios represented by the selected model(s).

In Section 2, a method for ranking models based on model selection criteria is discussed. Section 3 discusses the use of forecasting performances as ranking measures in the more general situation when nonparametric and/or judgmental forecasting techniques are considered. The proposed approach is then applied in Section 4 to an econometric model from the telecommunications industry.

2. MODEL SELECTION CRITERIA

The selection of models in regression analysis and time series analysis has been discussed extensively in the statistical literature during the last decade or so. For the case of regression analysis, measures such as Mallows' C_P (see Mallows, 1973) provide a method for assessing the adequacy of a p-predictor model. For time series data, Akaike's FPE (Final Prediction Error) was suggested and then generalized to the celebrated AIC (Akaike Information Criterion), defined as follows:

$$\text{AIC} = -2\log\ (\text{maximum likelihood}) + 2\ (\text{number of parameters}).$$

In a nutshell, AIC is a measure of the goodness-of-fit of the model, either in terms of the Kullback-Leibler information measure, or in terms of the expected log-likelihood ratio. Similar criteria (BIC, Schwarz's criterion, Hannan and Quinn's ϕ, Parzen's CAT, etc.) were later developed, using different approaches. These model selection criteria all represent a marked departure from the more traditional hypothesis-testing framework which has been found to be quite satisfactory in large sample situations, and when the tested models are nested. In the latter situation, it could also be shown that there is a connection between AIC and an F-test (see Soderstrom, 1977). Although in principle, non-nested hypotheses could be entertained, several theoretical difficulties remain to be solved, not the least of which is the choice of a significance level. This choice becomes irrelevant with AIC-type model selection criteria. Furthermore some of these criteria can be given a Bayesian interpretation, with an arbitrary prior probability over the set of competing models, thus adding to their popularity (see Akaike, 1979b).

Not withstanding their many merits, which include ease of computation and usefulness in comparing among non-nested and non-linear models, AIC-type model selection criteria cannot avoid the small-sample problem inherent in all estimation procedures based on the maximum-likelihood principle. More fundamentally, one might rightly question to what extent the term 'best' is well defined in certain situations. A possible answer is provided through consideration of the concept of "subset of models", very much like that of "confidence interval" for point estimation. Duong (1984) suggested the ranking and selection approach to this problem. The main thrust of this approach is the use of specific procedures for determining which models should be retained for further consideration, in order to have some control over the degree of confidence that the correct model will be included. In essence, the models will be ranked as usual with respect to some computed model selection criterion, and all models 'close' enough to the best Minimum-AIC model will be considered as valid competitors and will be included for further consideration in the subset of selected models. To carry out this procedure, one needs to specify beforehand the desired probability of correct selection, that is, the probability that the true model is included. For AR models, it could be shown that the size of the selected subset depends strongly on the difference between the maximum and minimum orders among the competing models. This point is important in many practical situations, for example, when costs are directly linked to model complexity.

As already pointed out in the introduction, there could be a host of situations where we do not have enough confidence in either the data (short data span, inaccuracies, etc.) or even the class of models being considered (e.g., it is only an approximation to some underlying law) to feel comfortable with one single 'best' model; that is, we feel that the available evidence is not sufficient to discriminate among the competing models. It is then necessary to weigh this evidence against our degree of confidence. Although the ranking and subset selection approach to be presented below is not without some disadvantages, especially in terms of computing the threshold value, we believe that it is the most natural one for this situation; the proposed procedure explicitly views the modelling exercise as an attempt to rank the competing models by how close they seem to fit the data, and hence, how close these are to the true model. In summary, the procedure explicitly contains the two main ingredients of any real decision-making situation:

(i) a selection strategy with the aim of choosing the 'best' alternative, and

(ii) a method for incorporating our subjective assessment of the data and/or requirements of the study through specification of the probability of correct selection.

The suggested procedure is now briefly described. For more details, see Duong (1984).

Assume that one has a series of length T, and consider the problem of selecting the "best" AR model among a set of competing models of order $0, 1, 2, \ldots, K$, using AIC as the selection criterion. The models are chosen as follows:

(R): Retain model k (autoregression of order k) in the selected subset if and only if

$$\mathrm{AIC}(k) - c < \min_{0 \le j \le k} \mathrm{AIC}(j) \qquad (k = 0, 1, \ldots, K)$$

where $\mathrm{AIC}(k) = T \log_e \hat{\sigma}^2(k) + 2k$, and $c > 0$ is a constant to be determined. $\hat{\sigma}^2(k)$ is the M.L.E. of the residual variance for the kth order AR model.

The selection is said to be a correct selection (CS) if the true model of order k_0 is retained in the selected subset. Let $pr(CS; R)$ be the probability of correct selection under rule R.

Then

$$\begin{aligned} pr(CS;R) &= pr\{\text{AIC}(k_0) - c < \min_{0\leq j\leq K} \text{AIC}(j)\} \\ &= pr\{\text{AIC}(k_0) - c < \text{AIC}(k) \ (k=0,1,\ldots,K,\ k\neq k_0)\} \\ &= pr\{(\text{AIC}(k_0) - c < \text{AIC}(k),\ K \geq k \geq k_0) \text{ and} \\ &\quad (\text{AIC}(k_0) - c < \text{AIC}(k),\ 0 \leq k \leq k_0)\}. \end{aligned}$$

Let $A = \{\text{AIC}(k_0) - c \leq \text{AIC}(k),\ K \geq k \geq k_0\}$. Then an argument similar to Shibata's (1976) can be used to show that for $c \geq 0$ fixed as $T \to \infty$, we have

$$pr\{\text{AIC}(k_0) - c < \text{AIC}(k),\ 0 \leq k \leq k_0 | A\} = 1.$$

Hence, for T sufficiently large,

$$pr(CS;R) = pr\{\text{AIC}(k_0) - c < \text{AIC}(k),\ k_0 < k \leq K\}.$$

Note that by replacing 2 in the definition of AIC by $\log T$, one can get $pr(CS;R) = 1$.

This procedure could, of course, be generalized to ARIMA models, and for other model selection criteria. Note that in terms of FPE, this procedure could be modified as follows: Retain model k in the selected subset if and only if

$$FPE(k) < \min_{0\leq j\leq K} FPE(j)e^{c/T}.$$

As the value c would be difficult to compute in most cases, more experience with this procedure is required. Some simulations have been carried out and the results discussed by Duong (1984).

Once the subset of 'good' models has been formed, some refinements could be carried out to now rank the models in the subset on a goodness-of-prediction test (if forecasting is the primary goal) and to combine these models in some optimal way. Of course, a model could be chosen for reasons other than forecasting, such as spectral density estimation. The goodness-of-fit criterion must hence be tailored to the purpose of the analysis. We see this as a necessary validation step, which could be carried out simultaneously with the selection step through the minimization of some loss structure which trades off the predictive ability of the model against its goodness-of-fit. A simple loss function could be derived by weighting the model selection criterion and the predictive performance of the model:

$$L(\text{model}_k) = \beta\text{AIC}(k) + (1-\beta)\sum_{i\epsilon J}(\hat{y}_i(k) - y_i)^2, \quad 0 \leq \beta \leq 1,$$

where J denotes the "training" data set used to fit the models, and $\hat{y}_i(k)$ is the predicted value from model k. In Section 4, an example is used to illustrate how the "training" data could be chosen in some reasonable way. In general, predictions are much more model dependent than data fitting, and hence it is possible to choose a suitably small "training" data set to which the competing models are fit; this will facilitate the choice of a preferable model, since, hopefully, we would then have a large predictive data set to validate the selected models. A similar approach was taken by Hjorth (1982) using a forward validation procedure in connection with model selection criteria. Ribaric (1984) applied the same approach to the more general problem of parsimonious data fitting.

More often than not, forecasts are obtained from nonparametric techniques or are judgmental in nature, in which cases the procedure (R) above must be modified. In the next section, a slightly modified procedure is given and illustrated. The second stage of the procedure, that is, how one goes about assessing the worth of the selected models and forecasts, is also discussed.

3. RANGE FORECAST

In many situations, several forecasts have been made of the same event. They result either from the use of different forecasting techniques or from opinions of different individuals. As pointed out by Bates and Granger (1969), the forecasts which are usually discarded by the decision-maker following the selection of the "best" one, may contain useful independent information. It seems then reasonable to combine the available forecasts and assign the weights in such a way that the composite forecast would have a lower mean square error of prediction than each of its components. Bates and Granger (1969) and Dickinson (1973, 1975), among others, have suggested and studied alternative ways of computing these weights. This problem is very similar to that of selecting the shrinkage factor in James-Stein-type estimators, or in the time series context, the signal-to-noise variance ratios. Duong (1981) has given a detailed discussion on these more general problems.

The optimal weights will depend on unknown quantities (in this case, the forecast error variances and their correlations), which have to be estimated from the observed data. Combining forecasts and, in general, estimating by the James-Stein procedure could be considered as methods for estimating parameters and pooling information from different sources. Bayesian interpretations of these methods have been extensively discussed in the last decade. See Bunn (1975) for a discussion of the forecasting problem. A natural step to consider for certain types of problems is the use of ranking and subset selection procedures, especially when the classical hypothesis testing approach is not appropriate. Generally known as multiple decision procedures (e.g., Gibbons *et al.*, 1977), this new research area has been developing rapidly, with applications in almost every field of statistics. The procedure (R) described in the previous section is, to our knowledge, the first use of this approach in model selection problems.

We now adapt the procedure (R) for the selection of a subset of forecasts from a potentially large set of competing forecasts. An example is then given to illustrate the procedure.

Assume that there are k forecasting techniques in competition. Also assume that they are unbiased, in the sense that they do not consistently overestimate (or underestimate) the true value. We also impose the following simplifying restrictions:

(i) There is no correlation between forecast errors.

(ii) The number of forecast values is the same (m) for all k techniques.

As one may suspect, these two restrictions are due to computational difficulties, more than for any theoretical reason. Let e_{ir} $(i = 1, 2, \ldots, k,\ r = 1, 2, \ldots, m)$ denote the error in the rth forecast value using the ith forecasting technique, and let $(i = 1, 2, \ldots, k)$ denote the (unknown) error variance of the ith forecasting technique. Let

$$\sigma^2_{\{1\}} \leq \sigma^2_{\{2\}} \leq \cdots \leq \sigma^2_{\{k\}}$$

denote the (unknown) ordered values. For simplicity, assume that $\sigma^2_{\{1\}} < \sigma^2_{\{2\}}$, so that the following problem is well defined:

Based on sample estimates, $\hat{\sigma}^2_i = \sum_{r=1}^{m} e^2_{ir}/m$, $i = 1, 2, \ldots, k$, *identify the best forecasting technique, that is the one with the smallest error variance,* $\sigma^2_{\{1\}}$.

The goal here is to divide the set of k forecasting techniques into two distinct subsets in such a way that one, the selected subset, contains the best technique with a high probability, at least P^*, a pre-specified value, and the other one does not. The selection rule is simply:

(R') *For each* $i = 1, 2, \ldots, k$, *retain the* i*th forecasting technique in the selected subset if and only if* $\hat{\sigma}^2_i \in I$, *where*

$$I = \left[\hat{\sigma}_{\{1\}}, \hat{\sigma}^2_{\{1\}}/c\right]$$

and where $0 \leq c \leq 1$ *is a constant to be determined.*

The selected subset is never empty, but one usually prefers not to include too many competitors, subject to the condition that the probability of correct selection, that is including the forecast with the smallest error variance, is at least P^*. The values of c will depend on P^*, m and k.

By following the same technique as used by Gupta and Sobel (1962a,b), and denoting by $pr\{CS;R\}$ the probability of correct selection under rule R, it can be shown that

$$\inf_{\Omega} pr\{CS;R\} = \left\{ \frac{\min_{j=1,2,\ldots,p} \chi_j^2(m)}{\chi_0^2(m)} \geq c \right\},$$

where Ω is the parameter space $\{\sigma_{\{i\}}^2 > 0 \ (i = 1,2,\ldots k)\}$ and where $\chi_j^2(m)$ $(j = 0,1,2,\ldots,p;\ p = k-1)$ denote k independent-chi-square random variables with a common number of degrees of freedom m. Hence, the determination of c such that $pr\{CS;R\} \geq P^*$, is equivalent to the determination of a lower percentage point of the Studentized smallest chi-square statistic with m degrees of freedom for all k chi-squares. Gupta and Sobel (1962b) have constructed tables for the largest c-values satisfying the requirement $pr\{CS;R\} \geq P^*$. Note that in the case when m is an even integer, the result can be obtained in the form of a finite series.

An example

The following table (Table 1) is taken from Bates and Granger (1969) and gives the actual values and forecasts, using two different methods, of the 1948–1965 U.K. output indices for the gas, electricity and water sector, as published in *National Income and Expenditure,* 1966. It is obvious here that the exponential forecast procedure yields a better result than does the linear forecast procedure. It is certainly not clear whether the latter should be considered. Using the ranking approach described above, one would use the linear forecast if

$$\sum_r e_{1r}^2 < \sum_r e_{2r}^2 / c.$$

Considering only the latest forecasts (1958–1965), the c-values are determined from Table 3 of Gupta and Sobel (1962b) for $k = 2$, $m = 8$, at different P^* values as follow:

$$P^* = 0.90, \qquad c = 0.3862$$
$$P^* = 0.95, \qquad c = 0.2909$$
$$P^* = 0.99, \qquad c = 0.1659$$

Since $\sum_{r=1}^{8} e_{2r}^2 = 51.02$ and $\sum_{r=1}^{8} e_{1r}^2 = 242.60$, we could reach the following conclusions:

Table 1. *Individual Forecasts of Output Indices for the Gas, Electricity and Water Sector in U.K.*

Year	Actual index (1958 = 100)	Linear forecast	Exponential forecast
1948	58		
1949	62		
1950	67	66.0	66.3
1951	72	71.3	71.9
1952	74	76.5	77.4
1953	77	79.2	80.3
1954	84	81.9	83.2
1955	88	89.0	88.6
1956	92	91.6	93.7
1957	96	96.0	98.5
1958	100	100.2	103.2
1959	103	104.3	107.8
1960	110	108.1	112.1
1961	116	112.9	117.4
1962	125	118.0	123.3
1963	133	124.2	130.2
1964	137	130.9	137.8
1965	145	137.0	145.0
Sum of squared errors		263.2	84.2

(i) Since $\sum_{r=1}^{8} e_{2r}^2/0.3862 = 132.107$, the linear forecast is excluded;

(ii) Since $\sum_{r=1}^{8} e_{2r}^2/0.2909 = 175.38$, the linear forecast is excluded;

(iii) Since $\sum_{r=1}^{8} e_{2r}^2/0.1659 = 307.53$, the linear forecast is included.

Hence the linear forecast should only be considered as a valid competitor if one imposes a high probability (0.99) of correct selection. Dickinson (1973) also questioned the use of the linear forecast in this case. As already mentioned, the idea behind the multiple decision procedures approach is fundamentally different from that for hypothesis testing or estimation. The goal of the experimenter here is to rank his forecasts, assuming from the outset that they are unequal. In situations where one is interested less in the forecasted values than in the problem of assessing the "goodness" of different competitive methods, which may result from differences in assumptions, theories, expertises, types of information used, etc., this formulation may be more realistic.

Recent works by Makridakis *et al.* (1982) have focused on various ways of evaluating the accuracy of forecasting methods, including those resulting from combining known methods. Although accuracy is certainly an important feature for short-range forecasts, it is almost a futile task to try to improve the accuracy of long-range forecasts. By long-range forecasting, one means forecasting over a time horizon where "large changes in the environment may be expected to occur" (Armstrong, 1978). This is certainly the case for many macro-economic forecasts, whereby economic scenarios are used as input to produce a set of long-range forecasts, each reflecting a different view about future economic situations.

Chen and Kung (1984) used the term "range forecasting" to refer to the range of multiple hypothesized scenarios, as compared to the width of some error band (e.g., variance) for a single-valued forecast. For an econometric model, "range forecasting" could be carried out by either one of the two methods:

(a) A range of forecasts is obtained by having different input scenarios for the independent variables; the model coefficient is, however, estimated only once from the available data.

(b) The historical data is divided into k (possibly overlapping) periods which are thought to represent the k scenarios being entertained. The model is re-evaluated in each case to produce corresponding forecasts.

Figure 1 depicts a typical situation with various economic scenarios, such as expansion, downturn, significant technological improvements, and price changes.

We have extended the use of range-forecasting to include, in addition to econometric models, time series models (ARIMA models, non-linear models, non-stationary autoregressions to model trends for "Long-Memory" series, etc.) and smoothing techniques. Among the latter methods, the recently developed class of generalized Holt exponential smoothing methods with autoregressive-damping (AD) parameters to model linear trends could also be considered as an attempt to incorporate scenarios into the forecasting process. However, this class was intended to be part of an automatic forecasting system, with the AD parameter chosen on the basis of *a priori* information about the behaviour of the future trend. For example, the choice of an AD parameter between 0 and 1 would correspond to the trend being damped

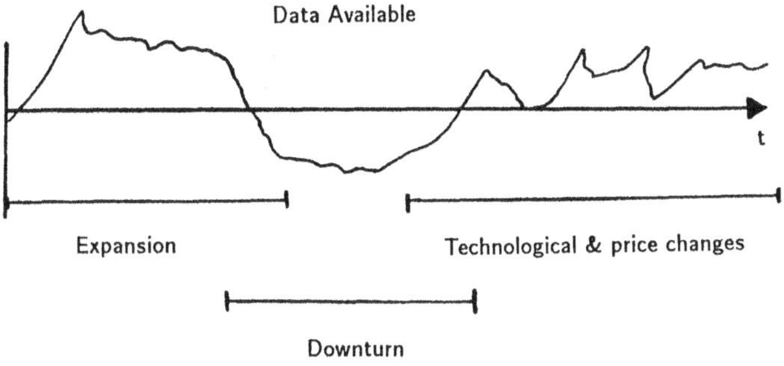

Figure 1. *Economic scenarios from historical data.*

out at rates according to the choice of parameter (Gardner and McKenzie, 1985). Given the sensitivity of forecasts to the choice of the AD parameter, it is natural to consider a subset of forecasts to reflect this sensitivity.

Posing similar challenges are forecasts of a volatile series such as Special Services Circuits (for example, Voice and Data services offered by a telephone company other than ordinary message telephone service, such as foreign exchanges, tie lines, and off-premise extensions). A five-year forecast is generally required for various planning operations. Due to the large number of series to be forecast and to the high degree of volatility of special services (see Figure 2 for an example), a time-dependent method using Kalman-filter techniques was applied. However, forecasts were "smoothed" with smoothing parameter α. The computational details are given by Duong (1983). For various values of the parameter, which reflect subjective expectations about service growth, comparisons with the manual (or "current view") forecasts were made on a randomly selected set of 100 series each from both Voice and Data services.

It should be noted that the five-year forecasts are reviewed approximately every year, and revisions result from the evaluation input information given by groups such as Marketing, Sales, and Rates. For illustrative purposes, we have treated the current forecasts as monthly forecasts (for the next 5 months). Since underprovisioning in this case has different implications than overforecasting the service demands, the following loss measure was used as an aid to the choice of "good" forecasts, that is, those close to the current view:

$$\text{DAVEG} = \sum_{\substack{\text{over}\\ \text{sampled}\\ \text{series}}} \Big(\sum_{\substack{\text{over 5}\\ \text{years}}} (\text{KFF} - \text{CV})/5 \Big)/100,$$

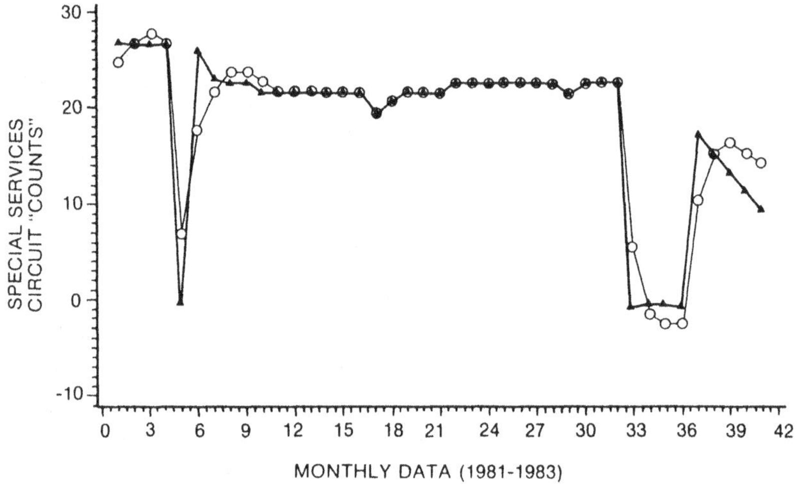

Figure 2. *Special Services Forecasts. Triangles represent data points (last 5 points are forecasts); circles are Kalman-filter estimates.*

where DAVEG is the average difference, KFF is the Kalman-filter forecast, and CV is the current view. Table 2 summarizes the results. With the interpretation of this loss measure as described above, the following general conclusions could be drawn:

- the Data series are in general less volatile than the Voice series, as expected.
- it is quite plausible that the range $\alpha \in [05\ ,1]$ for the Data series will do a good job in duplicating the manual forecasts; the Voice series certainly requires careful analysis. For example, in Figure 2 we have chosen the Kalman-filter parameters to reflect our assessment: in this case, an expected decrease.

It should be obvious by now that we strongly advocate the use of range-forecasting, at least in situations where it is called for, namely when the forecast horizon is long enough to include potential changes which might have an impact on the series, and for noisy series with volatile trends. The "range" could be obtained by considering the most extreme forecasts from the selected subset. The important feature to remember here is that although forecasts were generated and selected in an (almost) automatic fashion, the final forecasts should always be subject to manual adjustments before they

Table 2. *Smoothed Kalman-Filter Forecasts*

	DAVEG	
α	Data Series	Voice Series
0.0	–2.40	–2.40
0.1	–1.40	1.60
0.2	–1.20	2.46
0.3	–0.83	3.30
0.5	–0.60	5.00
1	0.44	4.70*

Notes: The data-files from which the samples were drawn contain approximately 22,000 Voice series and 5,000 Data series.

*For this case, four (4) selected series were subjectively excluded from the computations (forecasts differ from the most current 12-month average by more than 100).

are used. Examples of causes for these adjustments in the telephone industry include marketing strategies, sales projections, and anticipated rate changes. Furthermore, there might be different groups of users for the same forecast(s); this would result in different adjustment methods reflecting different needs and interests. Figure 3 illustrates the kind of forecasting strategy we have in mind.

In summary, we have suggested a semi-automatic approach to forecasting either when the series are volatile or when the forecast horizon is expected to include significant changes. A range of forecasts is produced through ranking and subset selection methods with input/output scenarios to assess and validate the selected forecasts.

4. AN APPLICATION TO TELEPHONE SERVICES DEMAND

Like many businesses and public organizations, telephone companies rely heavily on econometric models to provide forecasts for many policy analyses and corporate decisions. The models are used in a variety of applications, ranging from general planning activities, such as budgeting, network provi-

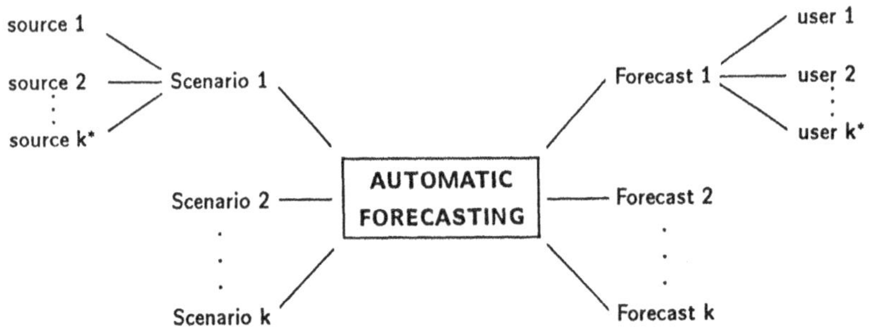

Figure 3. *Schema of semi-automatic forecasting system.*

sioning, and marketing strategies, to providing support for tariff revisions associated with rate applications. In most cases, state-of-the-art time series and econometric modeling techniques are used. However, there are certain unique features of the telecommunications environment which make these methods difficult to apply, such as technological advances, decisions by the regulatory commission, market changes to new services being offered and competition. All these imponderables need to be taken into account when constructing forecast scenarios; this is in addition to considering the usual economic/social factors such as strikes, recessions, and seasonality patterns. Each factor could have major impact on service demand and revenue. The construction of these scenarios is illustrated next.

Figure 4 is a quarterly series representing network access service (on a log-scale) for a certain region during 1971–1984. To obtain forecasts for this series, an econometric model was developed using as the main explanatory variable the ratio of Gross Domestic Product over Population Size (15 years or older). Figure 5 is a graph of the logarithm of this series. Besides seasonality, two special events were also accounted for using dummy variables: the strike in 1979, and the economic slowdown in the first quarter of 1982. There are three main periods, identified by Kiss (1983) from the historical data, which correspond to three possible forecast scenarios:

(i) 1971–1975: This was a period of fast growth due to technological improvement and high demand for telephone services.

(ii) 1976–1980: There was a slowdown in demand which coincided with, and was in part caused by, worldwide economic problems and some political uncertainties in Canada (1976–1978); there seems to have been some modest gains in the 1979–1980 period, due mainly to the introduction of digital electronic switches (DMS).

(iii) 1981–1984: Except for the first quarter of 1982 where the recession effect was unusually severe, there are some indications of a steady growth in demand for that period due mainly to expansion of the network and addition of new services.

It is interesting to note that the explanatory variable (Figure 5) also exhibits the same growth pattern as the telephone demand series, indicating a very strong link between general economic performance and telephone demand. While this link is hardly surprising, it helps develop and validate certain scenarios based on a modest to strong growth for this economic indicator (dotted lines on Figure 5); this in turn will be used to develop the forecast range for the telephone demand series. However, final forecasts are not presented here for confidentiality reasons.

Finally, it must be emphasized that the above forecasts would almost certainly be revised to account for important regulatory decisions which were expected in 1985; these include rate increases and competition for long distance services.

5. CONCLUSIONS AND EXTENSIONS

In this paper, the problem of selecting a subset of "good" models using model selection criteria has been extended to include models developed for forecasting purposes. The concept of "range" for economic forecasting was also discussed and illustrated.

As useful extensions to this approach, we are now investigating the following two problems:

(i) The use of AIC-type criteria to detect structural changes in time series models (e.g., changes in the order of an AR model).

(ii) The extension of the concept of range forecasting to independent observations; for example, estimation of the mean under different distributional assumptions.

ACKNOWLEDGMENTS

The author wishes to thank Dr. R. W. Shorrock and Mr. I. Bougas for many helpful comments and suggestions, and Ms. C. Mac Donald for computational assistance. However, opinions expressed in this paper are those of the author only; they do not necessarily reflect the position or policy of Bell Canada.

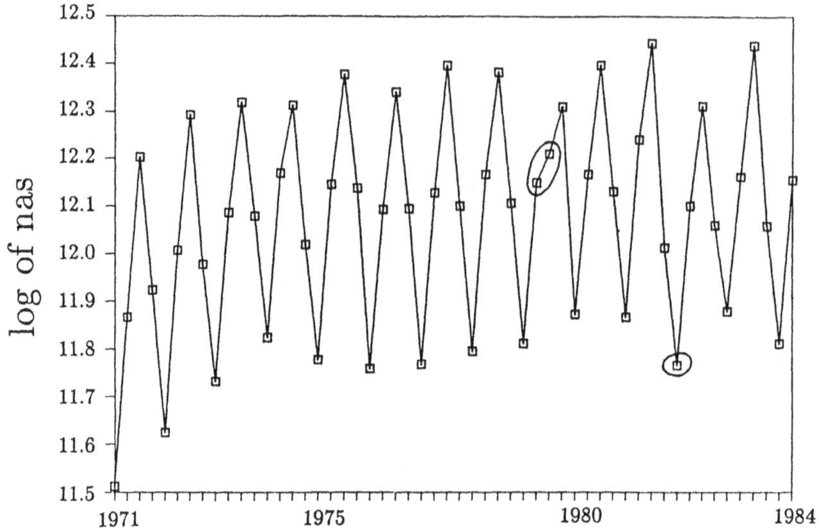

Figure 4. *Log of Network Access Service (Quarterly series 1974–1984).*

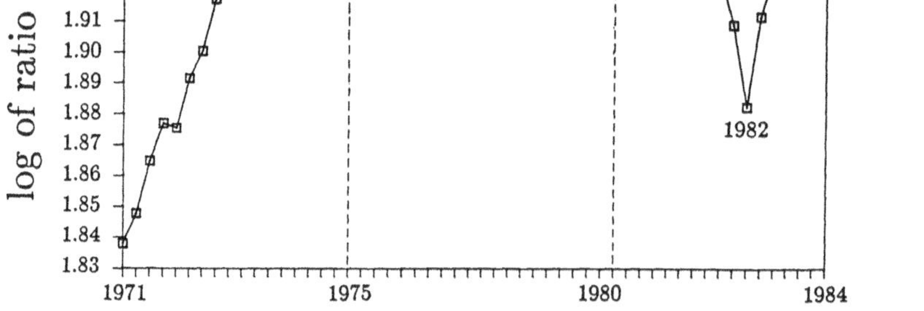

Figure 5. *Log of ratio of GDP over Population 15+ (Quarterly series 1971–1984).*

REFERENCES

Akaike, H. (1979a), "On the construction of composite time series models". *Proceedings of the 42nd Session, International Statistical Institute.*

Akaike, H. (1979b), "Bayesian extension of the minimum AIC procedure of autoregressive model fitting". *Biometrika* **66**, 237–242.

Armstrong, J. S. (1978), *Long-Range Forecasting: From Crystal Ball to Computer.* New York: Wiley and Sons.

Bates, J. M., and C. J. Granger (1969), "The combination of forecasts". *Operational Research Quarterly* **20**, 451–468.

Bunn, D. W. (1975), "A Bayesian approach to the linear combination of forecasts". *Operational Research Quarterly* **26**, 325–329.

Chen, K., and S. H. Kung (1984), "Synthesis of qualitative and quantitative approaches to long-range forecasting". *Technological Forecasting and Social Change* **26**, 255–266.

Dickinson, J. P. (1973), "Some statistical results in the combination of forecasts". *Operational Research Quarterly* **24**, 253–260.

Dickinson, J. P. (1975), "Some comments on the combination of forecasts". *Operational Research Quarterly* **26**, 205–210.

Duong, Q. P. (1981), "Shrinkage, order selection, and threshold". Ph.D. thesis, The University of Western Ontario.

Duong, Q. P. (1983), "On the use of the Kalman Filter Algorithm in special services forecast". Technical Memorandum, Bell Canada.

Duong, Q. P. (1984), "On the choice of the order of autoregressive models: a ranking and selection approach". *Journal of Time Series Analysis* **5**, 145–157.

Gardner, E. S., Jr., and E. McKenzie (1985), "Forecasting trends in time series". *Management Science* **31**, 1237–1246.

Gibbons, J. D., I. Olkin, and M. Sobel (1977), *Selecting and Ordering Populations.* New York: Wiley and Sons.

Gupta, S. S., and M. Sobel (1962a), "On selecting a subset containing the population with the smallest variance". *Biometrika* **49**, 495–507.

Gupta, S. S., and M. Sobel (1962b), "On the smallest of several correlated F statistics". *Biometrika* **49**, 509–523.

Hjorth, U. (1982), "Model selection and forward validation". *Scandinavian Journal of Statistics* **9**, 95–105.

Kiss, F. (1983), "Productivity gains in Bell Canada". In *Economic Analysis of Communications. Theory and Practice,* ed. L. Courville, A. de Fontenay and R. Dobell, pp. 85–113. Amsterdam: North-Holland.

Kitagawa, G., and W. Gersch (1985a), "A smoothness priors time-varying AR coefficient modeling of nonstationary covariance time series". *IEEE Transactions on Automatic Control* **AC-30**, 48–56.

Kitagawa, G., and W. Gersch (1985b), "A smoothness priors long AR model method for spectral estimation". *IEEE Transactions on Automatic Control* **AC-30**, 57–65.

MacNeill, I. B., and Q. P. Duong (1982), "Effect on AIC-type criteria of change of parameters at unknown time". In *Time Series Analysis: Theory and Practice 1*, ed. O. D. Anderson, pp. 65–80. Amsterdam: North-Holland.

Mahmoud, E. (1984), "Accuracy in forecasting". *Journal of Forecasting* **3**, 139–159.

Makridakis, S., *et al.* (1982), "The accuracy of extrapolation (time series) methods: results of a forecasting competition." *Journal of Forecasting* **1**, 111–153.

Mallows, C. L. (1973), "Some comments on C_P". *Technometrics* **15**, 661–675.

Ribaric, M. (1984), *Computational Methods for Parsimonious Data Fitting.* Vienna: Physica–Verlag.

Shibata, R. (1976), "Selection of the order of an autoregressive model by Akaike's Information Criterion". *Biometrika* **63**, 117–126.

Soderstrom, T. (1977), "On model structure testing in system identification". *International Journal of Control* **26**, 1–18.

Timo Teräsvirta [1]

SMOOTHNESS IN REGRESSION: ASYMPTOTIC CONSIDERATIONS

ABSTRACT

Smoothness restrictions may be used in linear models, for instance, when the data set does not contain enough information to allow sufficiently accurate estimation of parameters. In this connection the problem of weighting the sample and non-sample information by a smoothing parameter arises. Suppose we want to choose the smoothing parameter in such a way that the mean squared prediction error (MSPE) of the model will be minimized. Starting from this assumption, an equation for the minimizer is derived, and its asymptotic solution is shown to be finite and unique. Next, it is proposed that common model selection criteria such as the AIC be generalized to be used for the estimation of the smoothing parameter. Asymptotically, several generalized criteria, called smoothing criteria in the paper, yield exactly the unique value which minimizes the MSPE. In fact, these are generalizations of the model selection criteria which are optimal according to the definition of Shibata (1981). Application of the above theoretical results to ridge regression is discussed. The effect of autocorrelated errors on the estimation of the smoothing parameter is also considered, and cases where autocorrelation may have little influence on the coefficient estimates of the model are indicated.

1. INTRODUCTION

Smoothness restrictions may be used in linear models when the data set itself does not contain enough information to allow sufficiently accurate estimation of parameters. In that context, the problem of weighting the sample and non-sample information, i.e., choosing an optimal degree of smoothing,

[1] University of California, San Diego and Research Institute of the Finnish Economy, Helsinki

I. B. MacNeill and G. J. Umphrey (eds.), Time Series and Econometric Modelling, 47–64.

arises. A similar problem appears in non-parametric regression although its original motivation is usually different; smoothness restrictions are not necessarily a consequence of the paucity of data. Rather, they stem from the belief that the underlying "true" relationship between the dependent and the independent variable can be represented by a smooth curve; see, for example, Shiller (1984). The degree of smoothing is a compromise between an expression of this prior belief and the original variability of the data.

A popular example of smoothness restrictions is polynomial distributed lag estimation. Shiller (1973) discussed the use of polynomial smoothness priors and a "rule of thumb" for determining the value of the smoothing parameter. Gersovitz and MacKinnon (1978) also applied a rule of thumb in their treatment of linear models whose regression coefficients are supposed to vary smoothly with the season. Other suggestions and more discussion on how to choose smoothing parameters in polynomial distributed lag models have been provided by Ullah and Raj (1979) and Thurman *et al.* (1984); see also Judge *et al.* (1985, Chapter 9) and Trivedi (1984).

Golub *et al.* (1979; see also references therein) have addressed the problem of obtaining the value of the smoothing parameter in non-parametric regression. Their method is called generalized cross validation (GCV), and it consists essentially of finding the global minimum of a function of the smoothing parameter of the model. We shall return later to the GCV method which was applied by Engle *et al.* (1982) recently to the estimation of a non-parametric relationship between weather and electricity demand. Shiller (1984) also provided an empirical example of smoothness restrictions in non-parametric regression, but the value of the smoothing parameter seemed to have been fixed in advance.

Ridge regression constitutes a special case of smoothness priors in linear models. The smoothness assumption simply means that the coefficients of the linear model are supposed to be "small". A wide variety of techniques have been proposed for finding the smoothing parameter in ridge regression; for discussion and examples, see, for example, Dempster *et al.* (1977), Draper and Van Nostrand (1979) and Gibbons (1981). We shall discuss some of those techniques later on. Oman (1982) has studied a more general case where the estimates are shrunk towards a subspace rather than to zero. He considered the determination of the value of the smoothing parameter in that context.

This paper considers a generalization of well-known model selection criteria (MSC) such as AIC or SBIC in such a way that they can be used for determining the value of the smoothing parameter of the model in any of the above cases. The idea is to generalize the number of parameters in the penalty function of the MSC so that the generalized number is a monotonically decreasing function of the smoothing parameter. Properties of the generalized MSC which will be called smoothing criteria (SC) can be studied using asymptotic concepts. The asymptotic minimizer of the mean squared prediction error (MSPE) with respect to the smoothing parameter is unique. Shibata (1981) has defined an optimality concept for MSC. It turns out that the generalized versions of ordinary MSC optimal according to Shibata's definition have the property that their asymptotic minimizers coincide with that of the MSPE.

Another problem investigated in this paper is the sensitivity of the SC to autocorrelated errors. The conclusion is that the MSPE of the model can be expected to be insensitive to autocorrelation if the bias in the smoothness restrictions is small and/or the sample is large. In these cases, the estimated value of the smoothing parameter which is affected by autocorrelation may vary substantially with little effect on the MSPE.

This paper is organized as follows: The model and the smoothness restrictions are introduced in Section 2. Criteria for measuring the performance of different SC in determining the value of the smoothing parameter and some examples are briefly mentioned in Section 3. Section 4 comprises the optimality results. Section 5 considers a special case of smoothness restrictions, namely, ridge regression. In Section 6, the effects of autocorrelated errors on the minimization of MSPE are discussed. Section 7 points out that the small-sample properties of different smoothing criteria cannot be expected to be similar.

2. PRELIMINARIES

Consider a linear model

$$y_n = X_n\beta + \epsilon_n, \quad \epsilon_n \sim N(0, \sigma^2 I_n), \tag{2.1}$$

where y_n is a stochastic $n \times 1$ vector of the dependent variable, X_n is an $n \times p$ matrix of the finite values of the independent variables at $t = 1, \cdots, n$, $\operatorname{rank}(X_n) = p$; β is a $p \times 1$ vector of regression coefficients and ϵ_n is a stochastic $n \times 1$ error vector. Furthermore, suppose the independent variables are non-trending: $n^{-1}X_n'X_n = C_n > 0$ for $n \geq p$ and $\lim_{n\to\infty} C_n = C > 0$.

In this paper we shall consider the estimator

$$b_R(\lambda) = (X_n'X_n + \lambda R'R)^{-1}(X_n'y_n + \lambda R'r), \tag{2.2}$$

where r is a fixed $m \times 1$ vector and R a finite-valued $m \times p$ matrix of constraints, $\operatorname{rank}(R) = m < p$. Shiller (1973) derived (2.2) by assuming the existence of the following prior information about the coefficients of (2.1):

$$R\beta \sim N(r, \sigma^2\lambda^{-1}I). \tag{2.3}$$

The parameter λ in (2.3) is the ratio between the error variance of (2.1) and the prior variance of the variables of the prior distribution. Assuming that the prior is not degenerate, λ is a finite constant and independent of n. Estimator (2.2) is the posterior mean of β. Estimator (2.2) can also be obtained by minimizing, with respect to β,

$$q(\beta) = (y - X\beta)'(y - X\beta) + \lambda s's, \tag{2.4}$$

where $s = r - R\beta$. In (2.4), λ is a Lagrange multiplier not given an explicit interpretation present in the Bayesian context. However, in finite samples λ may be seen as controlling the trade-off between the sample information and the requirement that $\| s \|$ be "small".

In polynomial distributed lag models $r = 0$. In ridge regression $r = 0$ and $R = I$ so that $p = m$. Non-parametric regression also presupposes that $r = 0$ and in addition that $X_n'X_n$ is a diagonal matrix with positive diagonal elements. Of course, if the model also contains ordinary linear regressors, then the latter assertion is no longer valid.

Note that (2.2) is not a mixed estimator of Theil and Goldberger (1961). In their estimator, r is stochastic, and the stochastic properties of the mixed estimator are not the same as those of (2.2); see, for example, Judge *et al.* (1985, Chapters 3 and 22) and Teräsvirta (1981).

Finally, we shall need the residual sum of squares,

$$\begin{aligned} n\hat{\sigma}^2(\lambda) &= \{y_n - X_n b_R(\lambda)\}'\{y_n - X_n b_R(\lambda)\} \\ &= n\hat{\sigma}^2 + \hat{s}'S_n(\lambda)RU_nR'S_n(\lambda)\hat{s}, \end{aligned} \tag{2.5}$$

where $\hat{s} = r - Rb$, $b = U_n X_n' y$, $U_n = (X_n'X_n)^{-1}$, $S_n(\lambda) = (\lambda^{-1}I + RU_nR')^{-1}$ and $\hat{\sigma}^2 = n^{-1}y_n'(I - X_nU_nX_n')y_n$. In the following, $S_n(\lambda) = S_n$ for brevity.

3. OPTIMAL SMOOTHING

In the above interpretation of (2.2), λ is a fixed constant which may itself have a natural interpretation. We shall, however, concentrate on situations where λ is not known but is determined from the data. This gives rise to the problem of "optimal" selection of λ; there is a trade-off between the weight of the sample and non-sample information, and a criterion is needed to define optimal weighting. A popular way of defining optimal trade-off consists of applying a quadratic risk function

$$q_n(\lambda) = E\{b_R(\lambda) - \beta\}' A \{b_R(\lambda) - \beta\} \tag{3.1}$$

with $A \geq 0$ and choosing the minimizer of (3.1) to be the optimal λ. Note, however, that (3.1) may have several local minima.

In ridge regression, (3.1) with $A = I$ has been applied to finding the ridge parameter. If the aim of model building is forecasting, a possible choice is $A = X_n'X_n$, i.e., (3.1) becomes the MSPE. The moment matrix $X_n'X_n$ is often used in theoretical work instead of $X_h'X_h$ where X_h consists of observations not included in the actual sample. We shall choose $A = X_n'X_n$ and consider minimizing (3.1) under that assumption. The asymptotic result of this consideration is formulated in the following theorem.

Theorem 3.1. Consider the linear model (2.1) and its smoothness estimator (2.2). Assume that $s \neq 0$. Asymptotically, the minimizer of the MSPE is

$$\lambda_0 = \sigma^2 \text{ tr } (RC^{-1}R')/(s'RC^{-1}R's). \tag{3.2}$$

The proof is contained in Appendix 1.

Remark 1. Of course, $\lim_{n\to\infty} q_n(\lambda) = \sigma^2 p$ for all finite λ. However, Theorem 3.1 illuminates the situation in large samples and therefore is useful.

Remark 2. If $s = 0$, the problem is different. In that case $b_R(\lambda)$ is unbiased for any $\lambda \geq 0$. It follows from the Gauss-Markov theorem that (2.4) is minimized by the restricted least squares estimator $b_R(\infty) = b_R$ for all $n > p$, and $q_n(\infty) \equiv \sigma^2(p - m)$.

It was mentioned in the introduction that several methods of estimating λ have been proposed in the literature. From the viewpoint of this paper, a particularly interesting one is the GCV method. Wahba (1975), Craven and Wahba (1979) and Golub *et al.* (1979) have suggested selecting λ which is the minimizer of

$$\text{GCV}(\lambda) = \hat{\sigma}^2(\lambda)\{1 - k_n(\lambda)/n\}^{-2},$$

where

$$k_n(\lambda) = p - \text{tr}(S_n R U_n R') = p - m + \text{tr}(T_n) \tag{3.3}$$

with

$$T_n = (I + \lambda R U_n R')^{-1} = \lambda^{-1} S_n.$$

Expression (3.3) is called the generalized number of parameters of a restricted linear model. It is a monotonically decreasing function of λ; $k_n(0) = p$, and $k_n(\lambda) \downarrow (p - m)$ as $\lambda \to \infty$. In the latter case, $b_R(\lambda) \to b_R$ in probability.

The GCV method was originally derived for minimizing the MSPE. The concept of the generalized number of parameters may be used for a formal generalization, devoid of any further theoretical considerations, of existing MSC. The number of parameters in their penalty function is simply replaced by its generalization (3.3). Since this MSC generalization is only based on a principle of analogy, optimality properties of these generalized criteria, henceforth called smoothing criteria (SC), have to be investigated. For that purpose, we introduce the following definition.

Definition 3.1. A SC for determining λ in $b_R(\lambda)$ is called optimal if, asymptotically, given $s \neq 0$, its minimizer equals (3.2).

Teräsvirta and Mellin (1986) defined three basic types of MSC. Many well-known criteria can be shown to belong to one of these three categories, although there are important exceptions. We follow those authors by defining

$$\mathrm{SC1}(\lambda) = \ln \hat{\sigma}^2(\lambda) + k_n(\lambda) n^{-1} f(n, 0), \tag{3.4}$$

$$\mathrm{SC2}(\lambda) = \hat{\sigma}^2(\lambda) + \tilde{\sigma}^2 k_n(\lambda) n^{-1} f(n, p), \tag{3.5}$$

and

$$\mathrm{SC3}(\lambda) = \hat{\sigma}^2(\lambda) + \hat{\sigma}^2(\lambda) k_n(\lambda) n^{-1} f(n, k(\lambda)), \tag{3.6}$$

where $\tilde{\sigma}^2 = (n-p)^{-1}(y_n - X_n b)'(y_n - X_n b)$, f is a positive function of its arguments and $n^{-1} f \to 0$ as $n \to \infty$. Several SC's which are generalizations of MSC fitting into the above classifications, are listed in Table 1. Note that the GCV method is simply a SC of type 3. Functions $\hat{\sigma}^2(\lambda)$ are at least twice continuously differentiable functions of λ. The asymptotic optimality properties of minimizers (3.4)–(3.6) are characterized by the following theorem.

Theorem 3.2. All SC of types 1–3 with $\lim_{n\to\infty} f(n, \cdot) = 2$ are optimal.

The proof is given in Appendix 1.

From Table 1 it is seen that, along with GCV, there is a host of asymptotically equivalent criteria with the optimality property of Definition 3.1. On the other hand, Shibata (1981) defined an asymptotic optimality criterion for ordinary MSC. The problem was to select a finite subset of regressors when the number of regressors was either infinite or growing to infinity with the number of observations so that $p/n \to 0$. A MSC is optimal if it asymptotically minimizes the MSPE when predicting an infinite number of steps ahead. Shibata (1981) showed that a MSC of type 3 called S (see Table 1) is optimal. Consequently, all criteria asymptotically equivalent to S have the same optimality property; these criteria are those for which $\lim_{n\to\infty} f(n, \cdot) = 2$.

A well-known MSC which does not fit into the three categories of Teräsvirta and Mellin (1986) is BIC (Sawa, 1978). Nevertheless, it can in principle be generalized in the same way as the above criteria. In Appendix 2, it is seen that GBIC is optimal in the sense of Definition 3.1.

Table 1. *Smoothness Criteria Generalized from MSC and the Corresponding Limits of $f(n,\cdot)$.*

SC	Type	$f(n,\cdot)$	$\lim_{n\to\infty} f(n,\cdot)$
AIC (Akaike, 1974)	1	2	2
SBIC (Schwarz, 1978)	1	$\ln n$	∞
HQ (Hannan and Quinn, 1979)	1	$2 \ln \ln n$	∞
C_P (Mallows, 1973)	2	2	2
BEC (Geweke and Meese, 1981)	2	$(1-p/n)^{-1}\ln n$	∞
URV (Unbiased Residual Variance; Theil, 1961)	3	$\{1-k_n(\lambda)/n\}^{-1}$	1
PC (Amemiya, 1980) or FPE (Akaike, 1969)	3	$2\{1-k_n(\lambda)/n\}^{-1}$	2
GCV (Golub *et al.*, 1979)	3	$\{2-k_n(\lambda)/n\} \times\{1-k_n(\lambda)/n\}^{-2}$	2
S (Shibata, 1981)	3	2	2
S_P (Hocking, 1976)	3	$[2-\{k_n(\lambda)-1\}/(n-1)] \times\{1-k_n(\lambda)/n\}^{-1} \times\{1-k_n(\lambda)/(n-1)\}^{-1}$	2
T (Rice, 1984)	3	$2\{1-2k_n(\lambda)/n\}^{-1}$	2
BIC (Sawa, 1978)	·	·	2

4. OTHER ALTERNATIVES FOR CHOOSING THE SMOOTHING PARAMETER

The previous section discussed smoothing criteria, but other consistent methods of estimating λ are available. One way of approaching the problem is to take the minimizing equation (A1.5) and operationalize it. Since

$$E(\hat{s}'P_n\hat{s}) = s'P_ns + \sigma^2 n^{-1}\mathrm{tr}\left\{T_n^3(RC_n^{-1}R')^2\right\},$$

one may replace both sides of (A1.5) by their unbiased estimators to obtain

$$\lambda[\hat{s}'P_n\hat{s} - \tilde{\sigma}^2 n^{-1}\mathrm{tr}\left\{T_n^3(RC_n^{-1}R')^2\right\}] = \tilde{\sigma}^2\mathrm{tr}(P_n). \tag{4.1}$$

Solving (4.1) may give negative values of λ so that we may prefer a positive part variant

$$\lambda[\hat{s}'T_n^3RU_nR'\hat{s} - \tilde{\sigma}^2\mathrm{tr}\left\{T_n^3(RU_nR')^2\right\}]_+ = \tilde{\sigma}^2\mathrm{tr}(T_n^3RU_nR'), \tag{4.2}$$

where $[a]_+ = a$ if $a > 0$ and $[a]_+ = 0$ otherwise. If the l.h.s. of (4.2) equals zero then a restricted least squares estimator ($\lambda = \infty$) is selected.

A computationally simple alternative is to operationalize the asymptotic solution (3.2). A unique consistent positive part estimator corresponding to (4.2) is

$$\hat{\lambda}_+ = \frac{\tilde{\sigma}^2\mathrm{tr}(RU_nR')}{[\hat{s}'RU_nR'\hat{s} - \tilde{\sigma}^2\mathrm{tr}\left\{(RU_nR')^2\right\}]_+}.$$

Another means of determining λ is to follow a Bayesian line of argument. Although $b_R(\lambda)$ has a Bayesian interpretation, (4.2) is not an empirical Bayes solution. Assuming $R\beta \sim N(r, \sigma^2\lambda^{-1}I)$ where λ is a constant, the marginal distribution of $\hat{s} = r - Rb$ is $N(0, \sigma^2\lambda^{-1}T_n^{-1})$. Thus $\sigma^{-2}\lambda\hat{s}'T_n\hat{s} \sim \chi^2(m)$. Operationalize the χ^2 statistic by substituting $\tilde{\sigma}^2$ for σ^2 and set

$$\tilde{\sigma}^{-2}\lambda\hat{s}'T_n\hat{s} = m, \tag{4.3}$$

where m is the expectation of $\sigma^{-2}\lambda\hat{s}'T_n\hat{s}$. Solving (4.3) for λ yields an empirical Bayes estimate of this parameter. Letting $n \to \infty$, the l.h.s. of (4.3) converges in probability to $\sigma^{-2}\lambda s's$. The asymptotic solution is

$$\lambda = \sigma^2 m/(s's),$$

which coincides with (3.2) if $RC^{-1}R' = I$.

Thurman *et al.* (1984) suggested the following procedure. The necessary and sufficient condition for

$$E\left\{(b-\beta)(b-\beta)'\right\} - E[\{b_R(\lambda) - \beta\}\left\{b_R(\lambda) - \beta\right\}'] \geq 0$$

is

$$\sigma^{-2}s'(2\lambda^{-1}I + RU_nR')^{-1}s \leq 1. \tag{4.4}$$

Replace the inequality in (4.4) by an equality and substitute least squares sample values for the parameters. The value of the smoothing parameter is the solution of

$$\tilde{\sigma}^{-2}\lambda\hat{s}'\left\{2I + (\lambda/n)RC_n^{-1}R'\right\}^{-1}\hat{s} = 1. \tag{4.5}$$

The corresponding asymptotic solution is

$$\lambda = 2\sigma^2/(s's).$$

For a discussion of the rationale of (4.5), see Swa[illegible] and Mehta (1983).

5. RIDGE REGRESSION

The above results are directly applicable to ridge regression. As mentioned above, the usual ridge estimator is (2.3) with $R = I$ and $r = 0$ so that $s = -\beta$. Equation (A1.5) for the minimizer becomes

$$\lambda\beta'T_n^3C_n^{-1}\beta = \sigma^2\mathrm{tr}(T_n^3C_n^{-1}), \tag{5.1}$$

where $T_n = \{I + (\lambda/n)C_n^{-1}\}^{-1}$. This can be operationalized as above. On the other hand, Dempster *et al.* (1977) have discussed minimizing (3.1) w.r.t. λ when $A = I$. The minimizing equation is, in our notation

$$\lambda\beta'T_n^3C_n^{-2}\beta = \sigma^2\mathrm{tr}(T_n^3C_n^{-2}). \tag{5.2}$$

Dempster *et al.* (1977) suggested operationalizing (5.2) by substituting b for β and $\tilde{\sigma}^2$ for σ^2, and called the resulting ridge estimator SRIDG. If the l.h.s. of (5.2) is estimated without bias, an operational minimizing equation becomes

$$\hat{\lambda}[b'T_n^3U_n^2b - \tilde{\sigma}^2\mathrm{tr}(T_n^3U_n^3)]_+ = \tilde{\sigma}^2\mathrm{tr}(T_n^3U_n^2).$$

For $[\;]_+ = 0$, the estimate of β equals zero.

From (4.3) we have

$$\hat{\lambda}\tilde{\sigma}^{-2}b'T_nb = p. \tag{5.3}$$

The corresponding ridge estimator was called RIDGM by Dempster *et al.* (1977). Letting $n \to \infty$ in (5.2), we obtain a unique solution for λ:

$$\lambda_o = \sigma^2\mathrm{tr}(C^{-2})/(\beta'C^{-2}\beta). \tag{5.4}$$

The corresponding asymptotic solution for λ from (5.3) is

$$\lambda_o = \sigma^2 p/(\beta'\beta). \tag{5.5}$$

For $X_n'X_n = nI_n$, $n > p$, (5.1) and (5.2) equal (5.5). Solution (5.5) is also an asymptotic expression of the ridge parameter $\hat{\lambda} = \tilde{\sigma}^2 p/(b'b)$ of the HKB estimator; see, for example, Hoerl *et al.* (1975) and Judge *et al.* (1985, Chapter 22). In the orthogonal case, (5.3) can be written as

$$(1 + p/n)\hat{\lambda} = \tilde{\sigma}^2 p/(b'b).$$

Thus SRIDG, HKB and RIDGM are close to each other if $X_n'X_n$ is nearly orthogonal. HKB and RIDGM can also be expected to yield similar estimates if the sample size is large. In the extensive simulation experiment of Gibbons (1981), RIDGM and HKB together with GCV were the best estimators. The MSE ($A = I$) was used as the principal measure of performance.

SRIDG, HKB and RIDGM lead to $\hat{\lambda} = O_p(1)$. This is not true for all rules suggested in the literature. Lawless and Wang (1976) proposed the minimizer (LW)

$$\hat{\lambda} = \tilde{\sigma}^2 p/(b'X_n'X_n b) = \tilde{\sigma}^2 p/(nb'C_n b). \tag{5.6}$$

It is seen to be $O_p(n^{-1})$. This feature of the LW estimator has not been a cause of concern in Monte Carlo experiments. One reason is that in simulation studies, the sample size has traditionally been fixed. Moreover, the regressors have usually been standardized so that $X_n'X_n$ is in correlation form. This amounts to having $n \equiv 1$ in (5.6). However, in applications with non-standardized data, the difference in order is likely to have considerable influence upon the results.

6. AUTOCORRELATED ERRORS

In their study of the relation between electricity demand and weather, Engle *et al.* (1982) estimated a relationship between those two variables using non-parametric regression and GCV. They assumed that the errors of the model were an AR(1) process. The estimation results from minimizing the GCV were surprisingly insensitive to the values of the autocorrelation coefficient. In this section, possible reasons for this phenomenon will be considered. However, the starting-point will not be in the estimation of parameters with autocorrelated errors but the usual OLS estimation ignoring the autocorrelation.

Assume that in (2.1), $\epsilon_n \sim N(0, \sigma^2\Sigma_n)$ where Σ_n is an $n \times n$ positive definite matrix standardized to have ones on the main diagonal. Set $n^{-1}X_n'\Sigma_n X_n = D_n$. Suppose $D_n > 0$ for $n \geq p$ and that $\lim_{n\to\infty} D_n = D > 0$. Use $b_R(\lambda)$ as the estimator of β as in the case of independent errors. Then it turns out that, asymptotically, the value of λ that minimizes the MSPE is

$$\begin{aligned}\tilde{\lambda}_o &= \sigma^2 \mathrm{tr}(RC^{-1}DC^{-1}R')/(s'RC^{-1}R's) \\ &= \lambda_o + \sigma^2 \mathrm{tr}(RC^{-1}FC^{-1}R')/(s'RC^{-1}R's),\end{aligned} \tag{6.1}$$

where $F = \lim_{n\to\infty} n^{-1}X_n'(I - \Sigma_n)X_n = C - D$. The modifications in the proof of Theorem 3.1 needed to obtain (6.1) are in Appendix 3. The second term on the r.h.s. of (6.1) can be either positive or negative. The bias component of $q_n(\lambda)$ is not affected by the autocorrelation so that the denominator of (6.1) is the same as that of (3.2).

Expression (6.1) is only an asymptotic result, but its simplicity is an advantage when the effect of autocorrelated errors is considered. There are at least two situations where the MSPE is insensitive to autocorrelation. First, suppose $s \approx 0$ so that $\tilde{\lambda}_o$ is large. Then the minimum is usually flat and even relatively large shifts in λ have little influence on the MSPE or the coefficient estimates of the model. A large sample has the same effect: for a wide range of values of λ, the coefficient estimates will not change much.

It is possible to find upper and lower bounds for (6.1). Using a result of Rao (1973, p. 62), we have

$$\mu_n^{(n)} X_n'X_n \leq X_n'\Sigma_n X_n \leq \mu_1^{(n)} X_n'X_n, \tag{6.2}$$

where $\mu_1^{(n)} \geq \ldots \geq \mu_n^{(n)}$ are the eigenvalues of Σ_n; $\sum_{j=1}^n \mu_j^{(n)} = n$. We may obtain upper and lower bounds for the r.h.s. of (A3.3) by (6.2). If one takes those bounds and lets $n \to \infty$, one obtains:

$$\mu_L \lambda_o \leq \tilde{\lambda}_o \leq \mu_U \lambda_o, \tag{6.3}$$

where $\mu_L = \lim_{n\to\infty} \mu_1^{(n)}$ and $\mu_U = \lim_{n\to\infty} \mu_n^{(n)}$. When the error process of the model is stationary with an absolutely summable covariance function, μ_L and μ_U are the minimum and the maximum values, respectively, of the spectral density of the process, cf. Fuller (1976, section 4.2).

The preceding discussion can easily be modified to apply when the true model has $\epsilon \sim N(0, \sigma^2\Sigma)$, whereas the GLS estimator of β is based on the assumption $\epsilon \sim N(0, \sigma^2\Sigma_o)$, $\Sigma \neq \Sigma_o$. When the MSPE is flat between the lower and upper bounds (6.3), the autocorrelation does not have noteworthy influence upon the prediction accuracy of the model. In the weather and electricity example of Engle *et al.* (1982) the values of $\hat{\lambda}$ are relatively high throughout. This may be concluded from the actual generalized number of parameters which is substantially closer to $p - m$ than to p in all their estimated models. As already mentioned, the minimum of the MSPE can be expected to be fairly flat if it occurs at high values of λ.

This does not, as such, explain the fact that the coefficient estimates in the application of Engle *et al.* (1982) are insensitive to autocorrelation. Large fluctuations in the estimated coefficients are possible although the MSPE changes little. However, in the case of non-parametric regression $X_n'X_n$ is diagonal. In an extreme case, if the observations of the independent variable are fairly uniformly distributed into groups (j observations in each group) defined by the model builder, then $X_n'X_n = jI$; $pj = n$. Even then it is still possible, but less likely, that small changes in MSPE are associated with large changes in estimated coefficients. Thus a possible explanation to insensitivity noticed by Engle *et al.* (1982) is a flat minimum of MSPE. It may occur together with a large $\hat{\lambda}$ when a minimizer of types (3.4)–(3.6) is applied.

7. SMOOTHING IN SMALL SAMPLES

It is seen from Table 1 that many of the smoothness criteria generalized from the MSC are optimal. However, their small sample properties may differ widely and not necessarily uniformly. Rice (1984) has underlined this point when discussing a related topic, the choice of bandwidth for non-parametric regression. In that paper, various model selection criteria were appropriately modified to produce bandwidth estimates. It was shown that some modified criteria are optimal in the sense that the minimizers converge in probability to the MSPE minimizer as the number of observations goes to infinity.

Rice (1984) argued that many criteria in fact undersmooth in small samples. He therefore suggested another criterion (T) which penalizes undersmoothing more heavily than the other criteria he considered. Its SC modification appear in Table 1. In a Monte Carlo experiment T did perform best while some other criteria, most notably AIC, FPE (PC) and S frequently undersmoothed and performed poorly. Those findings clearly indicate that a study of smoothing in regression is not complete without consideration of the small sample properties of the various smoothing criteria and other estimation methods discussed in this paper. At the moment we do not yet have any small sample results available, but work on the problem is under way.

ACKNOWLEDGMENTS

I wish to thank Rob Engle, George Judge and John Rice for helpful discussions. They are not responsible for any errors or shortcomings of the paper. The research was supported by the Academy of Finland, the Finnish Cultural Foundation and the Yrjö Jahnsson Foundation.

APPENDIX 1: PROOFS OF THEOREM 3.1 AND 3.2

Proof of Theorem 3.1. Consider model (2.1), estimator (2.2) and its MSPE. Write MSPE as follows:

$$q_n(\lambda) = \mathrm{tr}(X_n'X_n)E[\{b_R(\lambda) - \beta\}\{b_R(\lambda) - \beta\}']. \tag{A1.1}$$

We have

$$b_R(\lambda) - \beta = (I - U_n R' S_n R)U_n' X_n' \epsilon_n - U_n R' S_n s,$$

where $U_n = (X_n'X_n)^{-1}$ and $s = r - R\beta$. The expectation in (A1.1) can be written as follows:

$$\begin{aligned} &E[\{b_R(\lambda) - \beta\}\{b_R(\lambda) - \beta\}'] \\ &\quad = \sigma^2 U_n(I - R'S_nRU_n)^2 + U_nR'S_nss'S_nRU_n \\ &\quad = \sigma^2 n^{-1}C_n^{-1}(I - n^{-1}R'S_nRC_n^{-1})^2 \\ &\qquad + n^{-2}C_n^{-1}R'S_nss'S_nRC_n^{-1}, \end{aligned} \tag{A1.2}$$

where $C_n = n^{-1}X_n'X_n$ and $S_n = (\lambda^{-1}I + n^{-1}RC_n^{-1}R')^{-1}$. Thus

$$\begin{aligned} q_n(\lambda) &= \sigma^2\mathrm{tr}(I - n^{-1}R'S_nRC_n^{-1})^2 + n^{-1}s'S_n^2RC_n^{-1}R's \\ &= \sigma^2\mathrm{tr}\left\{I - (\lambda/n)R'T_nRC_n^{-1}\right\}^2 + (\lambda^2/n)s'T_n^2RC_n^{-1}R's, \end{aligned} \tag{A1.3}$$

where $T_n = \lambda^{-1}S_n = \left\{I + (\lambda/n)RC_n^{-1}R'\right\}^{-1}$, because S_n and $RC_n^{-1}R'$ commute. The first term on the r.h.s. of (A1.3) is $O(1)$. Only if it is assumed that the minimizer $\lambda = O(n^\delta)$, $\delta \leq 1/2$, does the bias term in (A1.3) remain finite as $n \to \infty$. Thus, the case of $\delta > 1/2$ is not considered because it does not lead to a minimum of (A1.1).

If one differentiates (A1.3) w.r.t. λ, one obtains:

$$\begin{aligned} q_n'(\lambda) &= -2\sigma^2\mathrm{tr}(I - n^{-1}R'S_nRC_n^{-1})R'(\lambda^{-1}S_n)^2RC_n^{-1} \\ &\quad + 2\lambda^{-2}n^{-1}s'S_n^3RC_n^{-1}R's \\ &= -2\sigma^2n^{-1}\mathrm{tr}(T_n^3RC_n^{-1}R') + 2(\lambda/n)s'T_n^3RC_n^{-1}R's. \end{aligned} \tag{A1.4}$$

Note that $q_n'(0) < 0$ for any finite $n > p$. If one sets (A1.4) equal to zero, one obtains:

$$\lambda s' P_n s = \sigma^2 \mathrm{tr}(P_n), \tag{A1.5}$$

where $P_n = T_n^3 R C_n^{-1} R'$. For $\delta \leq 1/2$, $T_n \to I$ and $q_n(\lambda) \to \sigma^2 p$ *as* $n \to \infty$. Thus, a unique asymptotic solution for (A1.5) exists and is given by:

$$\lambda = \sigma^2 \mathrm{tr}(RC^{-1}R')/(s'RC^{-1}R's). \tag{A1.6}$$

The r.h.s. of (A1.6) is finite when $s \neq 0$, i.e., $\delta = 0$.

Proof of Theorem 3.2. From (2.5) we have

$$\hat{\sigma}^2(\lambda) = \hat{\sigma}^2 + (\lambda/n)^2 \hat{s}' T_n^2 R C_n^{-1} R' \hat{s}. \tag{A1.7}$$

If one differentiates w.r.t. λ, one obtains:

$$\hat{\sigma}^2(\lambda)' = 2(\lambda/n^2)\hat{s}' P_n \hat{s} \tag{A1.8}$$

and

$$k_n'(\lambda) = -n^{-1}\mathrm{tr}(T_n^2 R C_n^{-1} R') = -n^{-1}\left\{\mathrm{tr}(P_n) + (\lambda/n)a(\lambda/n)\right\}, \tag{A1.9}$$

where $a(\lambda/n) = \mathrm{tr}(P_n R C_n^{-1} R')$. To find the minimizer, differentiate (3.4–3.6) to obtain:

$$\frac{\partial}{\partial\lambda}\mathrm{SC1}(\lambda) = \hat{\sigma}^2(\lambda)'/\hat{\sigma}^2(\lambda) + k_n'(\lambda) n^{-1} f(n,0), \tag{A1.10}$$

$$\frac{\partial}{\partial\lambda}\mathrm{SC2}(\lambda) = \hat{\sigma}^2(\lambda)' + k_n'(\lambda)\tilde{\sigma}^2 n^{-1} f(n,p), \tag{A1.11}$$

$$\begin{aligned}\frac{\partial}{\partial\lambda}\mathrm{SC3}(\lambda) &= \hat{\sigma}^2(\lambda)'[1 + k_n(\lambda) n^{-1} f\{n, k_n(\lambda)\}] \\ &\quad + \hat{\sigma}^2(\lambda) k_n'(\lambda) n^{-1} [f\{n, k_n(\lambda)\} + k_n(\lambda) f'\{n, k_n(\lambda)\}]. \end{aligned} \tag{A1.12}$$

The minimizing equations are

$$\begin{aligned}\hat{\sigma}^2(\lambda)' &= -\hat{\sigma}^2(\lambda) k_n'(\lambda) n^{-1} f(n,0), \\ \hat{\sigma}^2(\lambda)' &= -\tilde{\sigma}^2 k_n'(\lambda) n^{-1} f(n,p), \\ \hat{\sigma}^2(\lambda)' &= -\hat{\sigma}^2(\lambda) k_n'(\lambda) n^{-1} f_1\{n, k_n(\lambda)\},\end{aligned}$$

where

$$\begin{aligned} f_1\{n, k_n(\lambda)\} &= [f\{n, k_n(\lambda)\} + k_n(\lambda) f'\{n, k_n(\lambda)\}] \\ &\quad \times [1 + n^{-1} k_n(\lambda) f\{n, k_n(\lambda)\}]^{-1}. \end{aligned}$$

It is seen from (A1.7) that if $\lambda = O_p(n^\delta)$, $\delta \geq 1$, $\hat{\sigma}^2(\lambda)$ is an inconsistent estimator of σ^2 with an asymptotic upward bias. Thus it is sufficient to consider the case $\delta < 1$. If one applies (A1.7–A1.9) one may write (A1.10–A1.12) as

$$\begin{aligned} 2\lambda\hat{s}'P_n\hat{s} &= \left\{\hat{\sigma}^2 + (\lambda/n)^2\hat{s}'T_n^2RC_n^{-1}R'\hat{s}\right\} \\ &\quad \times \left\{\mathrm{tr}(P_n) + (\lambda/n)a(\lambda/n)\right\} f(n,0), \end{aligned} \tag{A1.13}$$

$$2\lambda\hat{s}'P_n\hat{s} = \tilde{\sigma}^2\left\{\mathrm{tr}(P_n) + (\lambda/n)a(\lambda/n)\right\} f(n,p), \tag{A1.14}$$

and

$$\begin{aligned} 2\lambda\hat{s}'P_n\hat{s} &= \left\{\hat{\sigma}^2 + (\lambda/n)^2\hat{s}'T_n^2RC_n^{-1}R'\hat{s}\right\} \\ &\quad \times \left\{\mathrm{tr}(P_n) + (\lambda/n)a(\lambda/n)\right\} f_1\left\{n, k_n(\lambda)\right\} \end{aligned} \tag{A1.15}$$

respectively. Note that for $\delta < 1$, $f'\left\{n, k_n(\lambda)\right\} \to 0$ as $n \to \infty$ for all type 3 criteria considered here. Consequently,

$$\lim_{n\to\infty} f_1\left\{n, k_n(\lambda)\right\} = \lim_{n\to\infty} f\left\{n, k_n(\lambda)\right\}.$$

Letting $n \to \infty$, we have $\hat{\sigma}^2 \to \sigma^2$, $\tilde{\sigma}^2 \to \sigma^2$ and $\hat{s} \to s$ in probability. If one assumes $\delta < 1$, equations (A1.13–A1.15) become

$$2\lambda s'RC^{-1}R's = \sigma^2\mathrm{tr}\left\{(RC^{-1}R')\lim_{n\to\infty} f(n,\cdot)\right\}. \tag{A1.16}$$

From (A1.16) it is seen that SC with $\lim_{n\to\infty} f(n,\cdot) = 2$ are optimal.

APPENDIX 2: GENERALIZING SAWA'S CRITERION

Generalizing Sawa's BIC using the generalized number of parameters we have

$$\begin{aligned} \mathrm{GBIC}(\lambda) &= \ln\hat{\sigma}^2(\lambda) + 2n^{-1}\left\{k_n(\lambda) + 2\right\}\left\{\hat{\omega}^2/\hat{\sigma}^2(\lambda)\right\} \\ &\quad - 2n^{-1}\left\{\hat{\omega}^2/\hat{\sigma}^2(\lambda)\right\}^2, \end{aligned} \tag{A2.1}$$

where $\hat{\omega}^2$ is an estimator of the variance of the "pseudo-true" model, cf. Sawa (1978). If one differentiates (A2.1) w.r.t. λ, one obtains:

$$\begin{aligned} &\frac{\hat{\sigma}^2(\lambda)'}{\hat{\sigma}^2(\lambda)} + 2k_n'(\lambda)n^{-1}\frac{\hat{\omega}^2}{\hat{\sigma}^2(\lambda)} \\ &\quad - 2\left\{k_n(\lambda) + 2\right\}n^{-1}\frac{\hat{\omega}^2\hat{\sigma}^2(\lambda)'}{\hat{\sigma}^4(\lambda)} + 4n^{-1}\frac{\hat{\omega}^4\hat{\sigma}^2(\lambda)'}{\hat{\sigma}^6(\lambda)} = 0. \end{aligned} \tag{A2.2}$$

Let λ_o be the value of λ minimizing (A2.1) . Furthermore, interpret the pseudo-true model as the model where the prior and sample information is combined optimally; for the pseudo-true model, (A2.1) attains its minimum. Then $\hat{\omega}^2 = \hat{\sigma}^2(\lambda_o)$, and (A2.2) becomes

$$\left\{1 - 2n^{-1}k_n(\lambda)\right\} \frac{\hat{\sigma}^2(\lambda)'}{\hat{\sigma}^2(\lambda)} = -2k_n'(\lambda)n^{-1}$$

or

$$\hat{\sigma}^2(\lambda)' = -\hat{\sigma}^2(\lambda)k_n'(\lambda)n^{-1}f_2\left\{n, k_n(\lambda)\right\}, \tag{A2.3}$$

where

$$f_2\left\{n, k_n(\lambda)\right\} = 2\left\{1 - 2n^{-1}k_n(\lambda)\right\}^{-1}.$$

Equation (A2.3) may be used for solving λ numerically. Note that $\lim_{n\to\infty} f_2\{n, k_n(\lambda)\} = 2$ so that (A2.1) is also an optimal SC.

APPENDIX 3: MODIFIED PROOF OF THEOREM 3.1 IN THE PRESENCE OF AUTOCORRELATED ERRORS IN MODEL (2.1)

Suppose $\epsilon_n \sim N(0, \sigma^2\Sigma_n)$ in (2.1) where $\Sigma_n > 0$. Then

$$\begin{aligned} E[\{b_R(\lambda) - \beta\}\{b_R(\lambda) - \beta\}'] = \sigma^2(&U_n - U_nR'S_nRU_n) \\ &X_n'\Sigma_nX_n(U_n - U_nR'S_nRU_n) \\ &+ U_nR'S_nss'S_nRU_n, \end{aligned}$$

so that

$$\begin{aligned} q_n(\lambda) &= E[\{b_R(\lambda) - \beta\}' X_n'X_n \{b_R(\lambda) - \beta\}] \\ &= \sigma^2\{\}^4\{D_nC_n^{-1}(I - n^{-1}R'S_nRC_n^{-1})^2\} \\ &\quad + n^{-1}s'S_n^2RC_n^{-1}R's. \end{aligned} \tag{A3.1}$$

If one differentiates (A3.1) w.r.t. λ, one obtains

$$\begin{aligned} q_n'(\lambda) = 2\sigma^2n^{-1}\mathrm{tr}\left\{D_nC_n^{-1}(I - n^{-1}R'S_nRC_n^{-1})R'(\lambda^{-1}S_n)^2RC_n^{-1}\right\} \\ + 2n^{-1}s'(\lambda^{-1}S_n)^2RC_n^{-1}R'S_ns. \end{aligned} \tag{A3.2}$$

Then if one sets (A3.2) equal to zero one may show, after some manipulation, that,

$$\lambda s'P_ns = \sigma^2\mathrm{tr}(T_n^3RC_n^{-1}D_nC_n^{-1}R'). \tag{A3.3}$$

If one lets $n \to \infty$ in (A3.3), one obtains the desired result.

REFERENCES

Akaike, H. (1969), "Fitting autoregressive models for prediction". *Annals of the Institute of Statistical Mathematics* **21**, 243–247.

Akaike, H. (1974), "A new look at the statistical model identification". *IEEE Transactions on Automatic Control* **AC-19**, 716–723.

Amemiya T. (1980), "Selection of regressors". *International Economic Review* **21**, 331–354.

Craven, P., and G. Wahba (1979), "Smoothing noisy data with spline functions. Estimating the correct degree of smoothing by the method of generalized cross-validation". *Numerische Mathematik* **31**, 377–403.

Dempster, A. P., M. Schatzoff, and N. Wermuth (1977), "A simulation study of alternatives to ordinary least squares". *Journal of the American Statistical Association* **72**, 77–106 (with Discussion).

Draper, N. R., and R. C. Van Nostrand (1979), "Ridge regression and James-Stein estimation: review and comments". *Technometrics* **21**, 451–466.

Engle, R. F., C. W. J. Granger, J. Rice, and A. Weiss (1982), "Non-parametric estimates of the relation between weather and electricity demand". Department of Economics, UCSD, Discussion Paper 83–17.

Fuller, W. A. (1976), *Introduction to Statistical Time Series*. New York: Wiley and Sons.

Gersovitz, M., and J. G. MacKinnon (1978), "Seasonality in regression: an application of smoothness priors". *Journal of the American Statistical Association* **73**, 264–273.

Geweke, J., and R. Meese (1981), "Estimating regression models of finite but unknown order". *International Economic Review* **22**, 55–70.

Gibbons, D. G. (1981), "A simulation study of some ridge estimators". *Journal of the American Statistical Association* **76**, 131–139.

Golub, G. H., M. Heath, and G. Wahba (1979), "Generalized cross-validation as a method for choosing a good ridge parameter". *Technometrics* **21**, 215–223.

Hannan, E. J., and B. G. Quinn (1979), "The determination of the order of an autoregression". *Journal of the Royal Statistical Society*, Series B **41**, 190–195.

Hocking, R. R. (1976), "The analysis and selection of variables in linear regression". *Biometrics* **32**, 1–49.

Hoerl, A. E., and R. W. Kennard (1970), "Ridge regression: biased estimation for nonorthogonal problems". *Technometrics* **12**, 55–67.

Hoerl, A. E., R. W. Kennard, and K. F. Baldwin (1975), "Ridge regression: some simulations". *Communications in Statistics* **4**, 105–123.

Judge, G. G., W. E. Griffiths, R. C. Hill, H. Lütkepohl, and T.-C. Lee (1985), *The Theory and Practice of Econometrics*, 2nd edition. New York: Wiley and Sons.

Lawless, J. F., and P. Wang (1976), "A simulation of ridge and other regression estimators". *Communications in Statistics* **5**, 307–323.

Mallows, C. L. (1973), "Some comments on C_P". *Technometrics* **15**, 661–675.

Oman, S. D. (1982), "Shrinking towards subspaces in multiple linear regression". *Technometrics* **24**, 307–311.

Rao, C. R. (1973), *Linear Statistical Inference and Its Applications*, 2nd edition. New York: Wiley and Sons.

Rice, J. (1984), "Bandwidth choice for nonparametric regression". *Annals of Statistics* **12**, 1215–1230.

Sawa, T. (1978), "Information criteria for discriminating among alternative regression models". *Econometrica* **46**, 1273–1291.

Schwarz, G. (1978), "Estimating the dimension of a model". *Annals of Statistics* **6**, 461–464.

Shibata, R. (1981), "An optimal selection of regression variables". *Biometrika* **68**, 45–54.

Shiller, R. (1973), "A distributed lag estimator derived from smoothness priors". *Econometrica* **41**, 775–788.

Shiller, R. (1984), "Smoothness priors and nonlinear regression". *Journal of the American Statistical Association* **79**, 609–615.

Swamy, P. A. V. B., and J. S. Mehta (1983), "Ridge regression estimation of the Rotterdam model". *Journal of Econometrics* **22**, 365–390.

Teräsvirta, T. (1981), "A comparison of mixed and minimax estimators of linear models". *Communications in Statistics A, Theory and Methods* **10**, 1765–1778.

Teräsvirta, T., and I. Mellin (1986), "Model selection criteria and model selection tests in regression models". *Scandinavian Journal of Statistics* (in press).

Theil, H. (1961), *Economic Forecasts and Policy*, 2nd edition. Amsterdam: North-Holland.

Theil, H., and A. S. Goldberger (1961), "On pure and mixed statistical estimation in economics". *International Economic Review* **2**, 65–78.

Thurman, S. S., P. A. V. B. Swamy, and J. S. Mehta (1984), "An examination of distributed lag model coefficients estimated with smoothness priors". Federal Reserve Board, Washington, D.C., Special Studies Paper No. 185.

Trivedi, P. K. (1984), "Uncertain prior information and distributed lag analysis". In *Econometrics and Quantitative Economics*, ed. D. F. Hendry and K. F. Wallis, pp. 173–210. Oxford: Blackwells.

Ullah, A., and B. Raj (1979), "A distributed lag estimator derived from Shiller's smoothness priors. An extension". *Economics Letters* **2**, 219–223.

Wahba, G. (1975), "Smoothing noisy data with spline functions". *Numerische Mathematik* **24**, 309–317.

Benjamin Kedem [1]

A FAST GRAPHICAL GOODNESS OF FIT TEST FOR TIME SERIES MODELS

ABSTRACT

The oscillatory appearance of stationary time series is captured very economically by only a few higher order crossings which in addition contain a great deal of the spectral content of the process. A useful approximation to the variances of higher order crossings is discussed and is applied in the construction of probability limits for the hypothesized higher order crossings. From this, a graphical display of higher order crossings together with their probability limits provide a fast goodness of fit test. Examples illustrate the applicability of this device.

1. INTRODUCTION

There has been a growing interest in graphical methods in time series analysis and especially so since the popularization of electronic devices with graphics capabilities. In following this trend, the present article discusses a certain zero-crossings based graphical technique useful in testing for goodness of fit of time series models. The idea is to use plots of higher order crossings which are akin to plots of the correlogram and spectral densities or the periodogram, but which have the advantage of great simplicity. Under the Gaussian assumption, the sequence for expected higher order crossings is equivalent to the autocorrelation function and hence to the normalized spectral distribution function, but it summarizes the data differently. In this regard, the monotone property of higher order crossings plays an instrumental role since the *initial rate* of increase exhibited by higher order crossings proves to be an effective summary feature. As the higher order crossings continue to increase, their rate loses its discrimination potency

---

[1] Department of Mathematics, University of Maryland, College Park, Maryland 20742

I. B. MacNeill and G. J. Umphrey (eds.), Time Series and Econometric Modelling, 65–76.

since different processes seem to share similar rates. This is why in general very few higher crossings are used in testing goodness of fit.

The present paper gives an overview of our previous work, particularly Kedem and Reed (1986) and Kedem (1985), to which the reader is referred for mathematical details and more examples.

2. PLOTS OF HIGHER ORDER CROSSINGS

Let $\{z_t\}$, $t = 0, \pm 1, \ldots$, be a zero mean stationary Gaussian process with correlation function ρ_j and normalized spectral distribution function F, and let ∇ be the difference operator, $\nabla z_t = z_t - z_{t-1}$. It is convenient to introduce the clipped binary process

$$x_t^{(k)} = \begin{cases} 1, & \nabla^{k-1} z_t \geq 0 \\ 0, & \text{otherwise}, \end{cases} \qquad k = 1, 2 \ldots,$$

which gives rise to the indicator at time t

$$d_t^{(k)} = \begin{cases} 1, & x_t^{(k)} \neq x_{t-1}^{(k)} \\ 0, & \text{otherwise.} \end{cases}$$

The higher order crossings of order k, $D_{k,n}$, is defined by

$$D_{k,n} = d_2^{(k)} + \ldots + d_n^{(k)}.$$

It is seen that $D_{k,n}$ counts the number of axis-crossings in the $(k-1)$th differenced series $\nabla^{k-1} z_1, \ldots, \nabla^{k-1} z_n$. $D_{1,n}$ then is the usual number of zero- or axis-crossings by the original series $z_1, \ldots, z_n$.

From the point of view of the theory of stationary Gaussian processes, the sequence of higher order crossings is equivalent to the correlation and spectral structures. This is stated precisely in the following theorem.

Theorem 1. Let $\{z_t\}$ be a zero mean stationary Gaussian process with correlation function ρ_j. Then the sequence $\{\rho_j\}$ is completely determined from the sequence $\{E(D_{j,n})\}$. That is, ρ_k is determined by $E(D_{1,n}), \ldots, E(D_{k,n})$.

Proof. From Kedem and Slud (1981),

$$\cos\left(\frac{\pi E(D_{k+1,n})}{n-1}\right) = \frac{-\binom{2k}{k-1} + \rho_1\left[\binom{2k}{k} + \binom{2k}{k-2}\right] - \ldots + (-1)^k \rho_{k+1}}{\binom{2k}{k} - 2\rho_1\binom{2k}{k-1} + \ldots + (-1)^k 2\rho_k} \qquad (1)$$

and the ρ_j can be determined recursively from the $E(D_{k,n})$.

Obviously it is also true, from (1), that knowledge of $\{\rho_j\}$ is equivalent to knowledge of the sequence $\{E(D_{j,n})\}$. It follows that F is completely determined by the sequence of expected higher order crossings. This is summarized by the symbolism

$$\{E(D_{j,n})\} \Leftrightarrow \{\rho_k\} \Leftrightarrow F.$$

Thus, exactly for the same reasons that plots of ρ_k and F are extensively used in time series analysis, it is useful to observe plots of higher order crossings too.

The main thing to observe in plots of higher order crossings is the rate at which they increase and the starting point $D_{1,n}$. The fact that higher order crossings tend to increase can be attributed to the general fact that

$$D_{j,n} \leq D_{j+1,n} + 1$$

with probability one. Hence the $D_{j,n}$ tend to increase with j for fixed but large n. See also Kedem and Slud (1981).

It is instructive to observe plots of higher order crossings and thus motivate the central idea of this paper. Figure 1 displays plots of ten higher order crossings $D_{1,1000}, \ldots, D_{10,1000}$, obtained from first order autoregressive processes with different parameter values ϕ. It is seen that the initial rate of increase and starting point differ from process to process, but that as the order increases the rate is almost independent of the parameter. This same behavior has been observed in numerous cases which may be interpreted to mean that only the very first few higher crossings carry sufficient information which discriminates clearly between different processes.

Accordingly, it is suggested that plots with as few as six values of $D_{j,n}$ can be useful in goodness of fit testing. At the same time it should be noted that crossings of high order carry information also but this information is less amenable and will not be used here.

3. THE VARIANCE OF HIGHER ORDER CROSSINGS

The probability distribution of the $D_{j,n}$ is quite intractable and we shall concentrate on the more modest problem of approximating the variance of higher order crossings needed for the proposed goodness of fit test.

In general, the variance of $D_{j,n}$ is a function of the fourth order cumulant function $\kappa_x^{(j)}(r,s,t)$ of $\{X_t^{(j)}\}$ which is summable under appropriate moment conditions. Thus for $j = 1$ the following asymptotic result was proved by Kedem (1980).

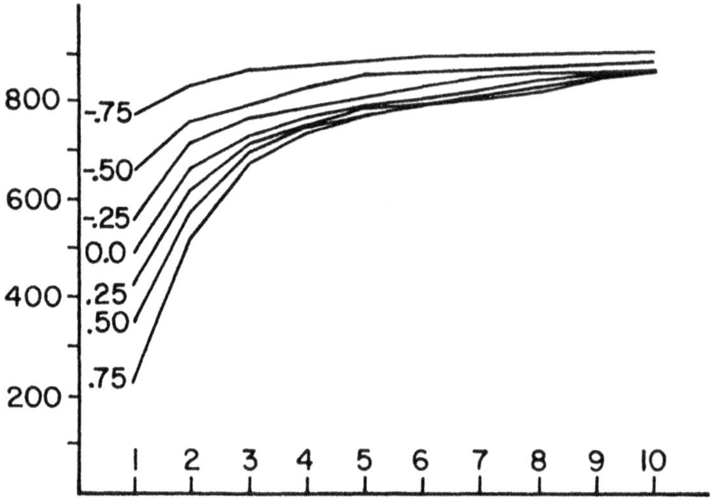

Figure 1. *Plots of* $D_{j,1000}$, $j = 1, \ldots, 10$, *from* $z_t = \phi z_{t-1} + u_t$, u_t *are independent* $N(0,1)$ *random variables and* ϕ = 0.75, 0.5, 0.25, 0, −0.25, −0.5, −0.75.

Theorem 2. If ρ_j is absolutely summable then

$$\sum_{k=-\infty}^{\infty} \mid \kappa_x^{(1)}(1,-k,1-k) \mid < \infty$$

and

$$\frac{D_{1,n} - E(D_{1,n})}{\sqrt{n}} \xrightarrow{\mathcal{L}} N\left(0, \sigma_1^2\right), \qquad n \to \infty,$$

where

$$\sigma_1^2 = \frac{1}{\pi^2} \sum_{k=-\infty}^{\infty} \left[\left(\sin^{-1}\rho_k\right)^2 + \sin^{-1}\rho_{k-1}\sin^{-1}\rho_{k+1} + 4\pi^2\kappa_x^{(1)}(1,-k,1-k)\right].$$

The same result applies to every $D_{j,n}$ provided the correlation functions of $\{\nabla^{j-1} z_t\}$ decays quickly to zero. However $\kappa_x^{(j)}$ is not known in general which makes the above result impractical.

Another approach is to hold n fixed and let j increase. In this case it is possible to obtain a useful asymptotic result under the assumption of

m-dependence. Assume that π is a point of increase for F and let

$$\lambda_j^{(k)} \equiv \Pr\left(x_t^{(k)} = 1 \mid x_{t-j}^{(k)} = 1\right).$$

Then $\lambda_1^{(k)} \to 0$ as $k \to \infty$ and it was shown by Kedem and Reed (1986) that

$$\operatorname{cov}(d_t^{(k)}, d_s^{(k)}) = o(\lambda_1^{(k)}). \tag{2}$$

The proof of this fact depends on the differential properties of $\rho_j^{(k)}$, the correlation function of $\{\nabla^k z_t\}$. (2) readily yields the following theorem.

Theorem 3. Let $\{z_t\}$ be an m-dependent stationary Gaussian process and assume that π is a point in the support of F. Then for fixed n

$$\lim_{k\to\infty} \frac{\operatorname{Var}(D_{k,n})}{(n-1)\lambda_1^{(k)}(1-\lambda_1^{(k)})} = 1.$$

This result was used in the construction of probability limits for the higher crossings under the hypothesis of white noise. However, the assumption of m-dependence cannot always be verified and thus another approximation should be used.

A rather close approximation to the variance of $D_{j,n}$ can be provided if it is assumed that the binary sequence $\{d_t^{(k)}\}$ is a Markov chain. This first order approximation has been found very satisfactory by an extensive simulation.

Define the parameters associated with the chain,

$$p^{(k)} \equiv 1 - \lambda_1^{(k)}, \qquad q^{(k)} \equiv \lambda_1^{(k)},$$

and

$$\nu^{(k)} \equiv \frac{1 - 2\lambda_1^{(k)} + \lambda_2^{(k)}}{2\left(1 - \lambda_1^{(k)}\right)}.$$

When the process is a stationary Gaussian autoregressive-moving average process with known (or hypothesized) parameters, $p^{(k)}$ and $\nu^{(k)}$ are known too explicitly. Then if $\{d_t^{(k)}\}$ is a Markov chain it can be shown (Kedem, 1985) that

$$\operatorname{Var}(D_{k,n}) = (n-1)p^{(k)}q^{(k)} + \frac{2p^{(k)}q^{(k)}\left(\nu^{(k)} - p^{(k)}\right)}{1-\nu^{(k)}} \cdot [(n-1) - V_{k,n}], \tag{3}$$

where

$$V_{k,n} = q^{(k)}\left[1 - \left(\frac{\nu^{(k)} - p^{(k)}}{q^{(k)}}\right)^{n-1}\right] \Big/ \left(1 - \nu^{(k)}\right).$$

This approximation has been compared (Kedem, 1985) with actual estimates obtained from 100 independent realizations each of length $n = 1000$. The results are given in Table 1. Although $E(D_{j,1000})$ are known explicitly when the parameters are known, these expectations are estimated too as a check of the whole simulation. It is seen that (3) agrees well with the simulation results. An algorithm for obtaining $p^{(k)}$, $\nu^{(k)}$ is given by Kedem (1985).

4. A GRAPHICAL GOODNESS OF FIT CRITERION

The proposed goodness of fit test is based on deviations of the observed path of higher crossings from the expected path where the latter is obtained under the hypothesis of an assumed model. Marked deviations of the observed path from the expected one suggest that the observed process does not *oscillate* as expected. The closeness of the two paths can be measured by appealing to (3) and to conditions under which the $D_{j,n}$ are asymptotically normal. It can be shown, using the technique of Cuzick (1976) that when $\{z_t\}$ is Gaussian the condition $\sum | \rho_k | < \infty$ implies the asymptotic normality of the $D_{k,n}$. It follows that approximate 95% probability limits for $D_{k,n}$ are, for each k and sufficiently large n,

$$(n-1)p^{(k)} \pm 1.96\{\text{Var } D_{k,n}\}^{1/2}, \tag{4}$$

where $\text{Var}(D_{k,n})$ is given by (3). When at least one observed $D_{j,n}$, $j = 1, \ldots, 6$, lies outside the limits (4) the assumed model under which (4) was derived is rejected. Before discussing the power of this test it is illustrated by a few examples.

4.1 Examples

Annual Mean Temperature. The graph of the annual mean air temperature from 1781 to 1980 at Hohenpeissenberg, Germany, is given in Figure 2. Actually the observations for 1811 and 1812 are missing and were replaced by the mean of neighboring observations. This has only a very small effect on the sequence of higher crossings.

Since annual temperature is hard to predict, we could ask, does the series oscillate as white noise? The answer is obtained from Figure 3 where it is seen that the higher order crossings are well within the bounds (4) so that at least in this sense the series resembles white noise. For comparison, the figure portrays the higher crossings of simulated white noise which also fall within the bounds as expected.

ARMA Models. Figure 4 shows the probability limits (4) under various hypotheses; these are white noise, second order autoregressive process with

Table 1. *Comparison of (3) with the standard deviation obtained from 100 independent realizations of size 1000. $E(D_{j,1000})$ and $\hat{E}(D_{j,1000})$ are rounded to the nearest integer.*

Series	j	$E(D_{j,1000})$	$\hat{E}(D_{j,1000})$ From 100 Realizations	$\{V(D_{j,1000})\}^{\frac{1}{2}}$ From (3)	$\{V(D_{j,1000})\}^{\frac{1}{2}}$ From 100 Realizations
White	1	500	497	15.81	15.96
Noise	2	666	666	13.15	13.63
	3	732	732	12.16	12.53
	4	769	770	11.57	11.49
	5	794	795	11.18	11.05
	6	813	814	10.82	10.00
AR(2)	1	424	425	9.64	9.67
$\phi_1 = 0.4$	2	484	485	9.38	9.13
$\phi_2 = -0.7$	3	536	537	10.29	10.81
	4	594	594	11.27	12.72
	5	651	652	11.87	12.02
	6	702	701	12.04	11.34
ARMA(1,1)	1	552	552	14.62	14.74
$\phi = 0.5$	2	679	679	12.96	12.87
$\phi = 0.7$	3	737	737	12.09	12.05
	4	773	772	11.27	11.52
	5	797	797	10.70	11.12
	6	814	814	10.15	10.80
ARMA(2,2)	1	884	883	10.04	10.51
$\phi_1 = -1.4$	2	897	897	9.20	9.53
$\phi_2 = -0.5$	3	903	903	8.84	9.01
$\theta_1 = 0.2$	4	908	908	8.60	8.50
$\theta_2 = 0.1$	5	911	911	8.43	8.47
	6	914	914	8.29	9.38

parameters 0.4 and −0.7, and second order autoregressive moving average process with parameters $\phi = (-1.4, -0.5)$, $\theta = (0.2, 0.1)$. The actual $D_{j,450}$ were obtained from simulated data given in an appendix by Priestley (1981).

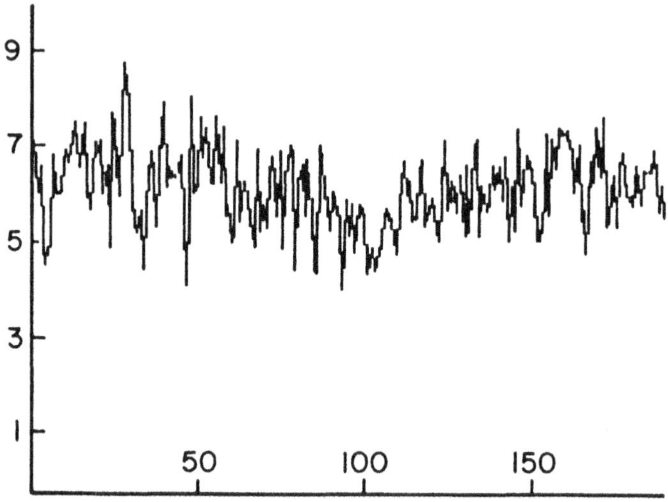

Figure 2. *Annual Temperature Series.* $n = 195$. *(Source: Report #155 of the Deutschen Wetterdienstes, West Germany (1981).)*

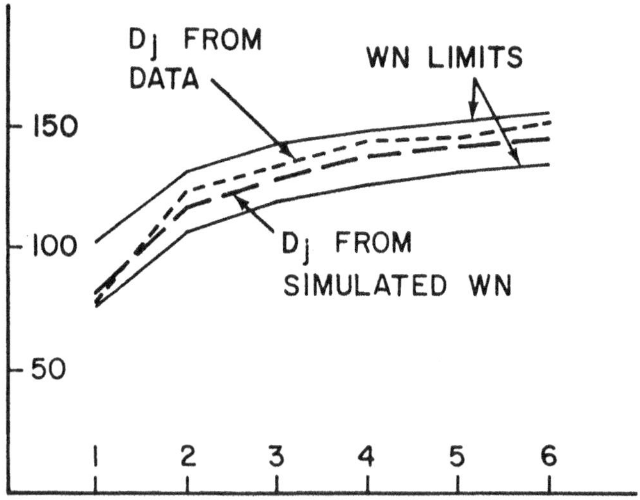

Figure 3. *Probability limits for the higher order crossings from the temperature series. The series oscillates as white noise (WN).*

The three paths fall well within their respective limits and the corresponding hypotheses are accepted.

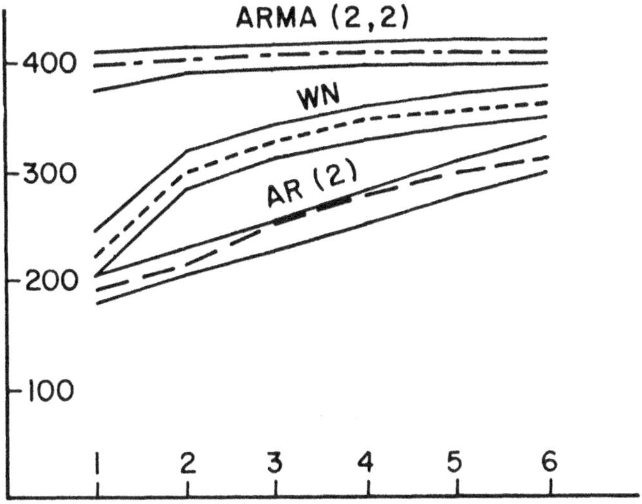

Figure 4. *Sample higher order crossings paths fall within their respective limits.*

It is seen that the three processes display different *oscillation patterns,* which are captured very economically by only six higher order crosssings. The ARMA (2,2) process is most oscillatory while the AR(2) is much smoother.

Signal Detection. Figure 5 displays two series which appear to be very similar except perhaps for scale. However their higher order crossings quickly reveal that the first one oscillates as white noise while the other oscillates roughly as a low order autoregressive process. This is illustrated in Figure 6.

Diagnostic Check. In testing the goodness of fit of a model one runs a residual analysis which usually tests whether the residual series constitutes white noise. Consider series A, D of Box and Jenkins (1976, p. 293); the fitted models are

$$\text{series A}: \quad \nabla z_t = u_t - 0.7u_{t-1}$$

and

$$\text{series D}: \quad z_t - 0.87z_{t-1} = 1.17 + u_t,$$

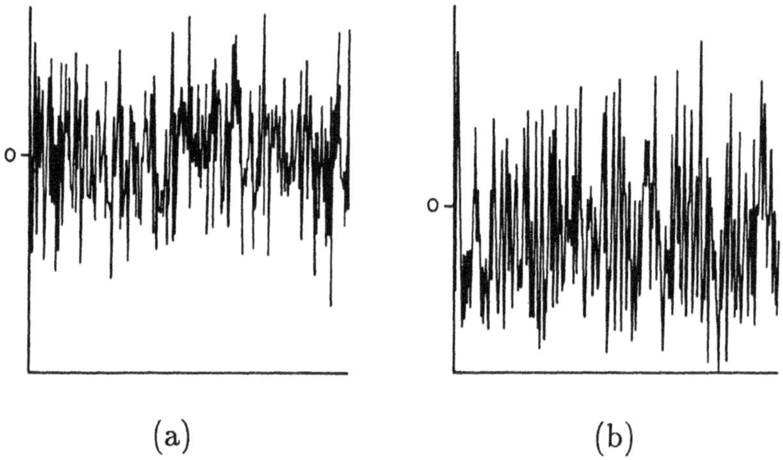

Figure 5. *Two oscillating time series.*

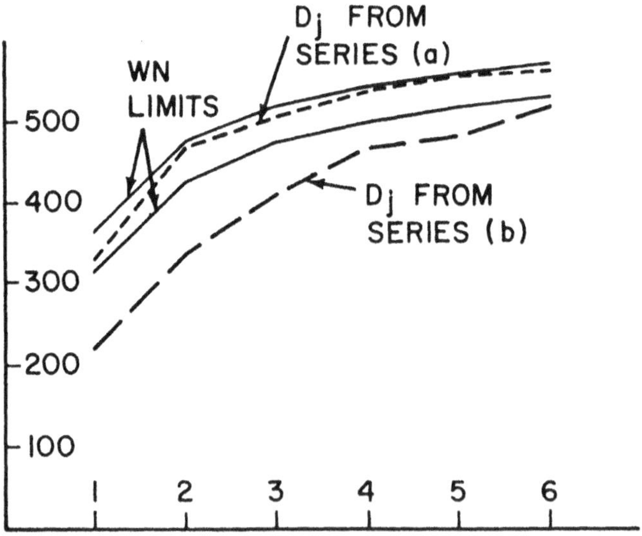

Figure 6. *The higher order crossings paths of series (a), (b). The first path is within white noise bounds.*

where $\{u_t\}$ is the residual series. Figure 7 however reveals that the two residual series are not quite white noise as signified by the axis-crossings being outside the limits (4). It is interesting to note though that the rest of the higher order crossings behave as those of white noise. Thus, except for smaller $D_{1,n}$, the two residual series oscillate as white noise.

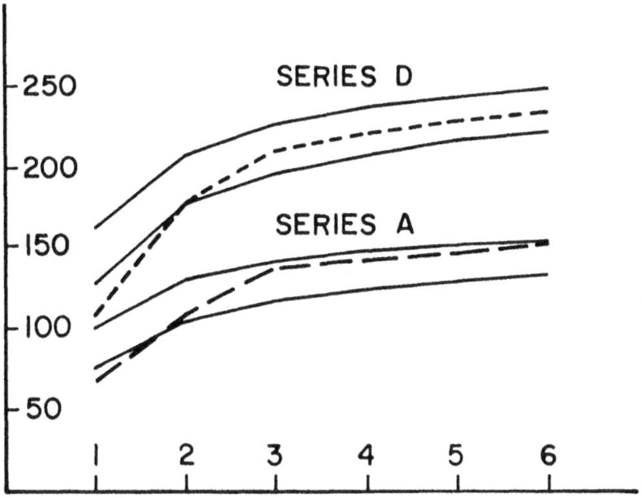

Figure 7. *Diagnostic check applied to the residuals of series A* ($n = 177$) *and series D* ($N = 290$). $D_{1,n}$ *is outside the limits (4) for white noise.*

4.2 Power Simulation

The limits (4) provide approximate 95% bounds for each value of $D_{j,n}$. However our test is based on $D_{1,n} \ldots, D_{6,n}$ simultaneously and the hypothesized model is rejected if at least one value of $D_{j,n}$ falls outside the probability bounds. It is expected that a test which is based on more than a single $D_{j,n}$ has a higher probability of rejecting a true hypothesis than 0.05 and in fact our experience indicates that with six values of $D_{j,n}$ this probability is about 0.1. The exact probability is still an open problem.

An indication of the power is provided in Table 2 which gives the power for testing the hypothesis of white noise where the alternative is the indicated process. The power is estimated from 50 independent series each of size 450. Similar results were obtained for greater series lengths.

Table 2. *Power simulation for testing white noise versus the indicated process.*

Process	Power
White Noise	.10
AR(1), $\phi = .05$	.26
MA(1), $\theta = .1$	.40
AR(1), $\phi = .2$	.90
AR(1), $\phi = .5$	1.00
AR(2), $\phi = .1,\ \phi_2 = -.15$	.88
ARMA(1,1), $\phi_1 = .1,\ \theta_1 = -.1$	.86
ARMA(2,2), $\phi_1 = .1,\ \phi_2 = -.4,\ \theta_1 = 0,\ \theta_2 = .3$	1.00
ARMA(2,2), $\phi_1 = .1,\ \phi_2 = -.2,\ \theta_1 = .2,\ \theta_2 = .1$	.88

ACKNOWLEDGMENT

The support of Grant AFOSR 82-0187 is gratefully acknowledged.

REFERENCES

Box, G. E. P., and G. M. Jenkins (1976), *Time Series Analysis, Forecasting and Control.* San Francisco: Holden-Day.

Cuzick, J. (1976), "A central limit theorem for the number of zeros of a stationary Gaussian process". *Annals of Probability* **4**, 547–556.

Kedem, B. (1980), *Binary Time Series.* New York: Marcel Dekker.

Kedem, B. (1985), "A graphical similarity measure for time series models". Report TR85-10, Mathematics Department, University of Maryland.

Kedem, B., and G. Reed (1986), "On the asymptotic variance of higher order crossings with special reference to a fast white noise test". *Biometrika* **73**, 143–149.

Kedem, B., and E. Slud (1981), "On goodness of fit of time series models: an application of higher order crossings". *Biometrika* **68**, 551–556.

Priestley, M. B. (1981), *Spectral Analysis and Time Series.* New York: Academic Press.

Bovas Abraham [1]

OUTLIERS IN TIME SERIES

1. INTRODUCTION

Every experimenter has at some time or other faced data which seem to contain some deviant or "outlying" observations. The problem of outliers in data is an old one and was one of the first to receive a statistical treatment.

One of the early references to the rejection of outliers seems to have been a remark by Bessel (see Anscombe, 1960) in a geodetic study; he remarked that he had never rejected an observation merely because of its large residual. An early attempt at developing a rejection criterion based on probability reasoning was that of Peirce (1852). He developed an outlier rejection criteron and applied it to 15 observations of the vertical semi-diameters of Venus made by Lt. Herndon, with the meridian circle at Washington in 1846. This started a lively debate which continues until today.

Major developments in the area of outlier rejection have been mainly in the estimation of a location parameter from a random sample or formulation of the parameters of a linear regression model when there is the possiblity of outliers. Approaches to this problem include:

(1) Significance tests (Dixon, 1950; Grubbs, 1950; etc.),

(2) Premium protection (Anscombe, 1960),

(3) Decision theoretic methods (Ferguson, 1961),

(4) Bayesian methods (de Finetti, 1961; Box and Tiao, 1968; Guttman, 1973; Abraham and Box, 1978),

(5) Robustness (Huber, 1964, 1981; Hampel, 1974), and

(6) Diagnostics (for example, Cook and Weisberg, 1982).

Papers in this area seem to classify the objectives of an analysis as follows: (i) making inferences about some parameters, (ii) making a general study with a view to gaining broader understanding of the problem without

[1] Department of Statistics and Actuarial Science, University of Waterloo, Waterloo, Ontario N2L 3G1

I. B. MacNeill and G. J. Umphrey (eds.), Time Series and Econometric Modelling, 77–89.

necessarily making inferences about the parameters, and (iii) discovering if outliers are present and then singling them out for further study. The approaches listed above give varying degrees of importance to these objectives. For example, the 'Robust approach' gives importance to the first objective, while the Bayesian approach addresses all three objectives.

Discussion of outliers in the context of time series is rather recent. In this paper we summarize some of the important developments and introduce some recent work in this area.

2. CHARACTERIZATION OF OUTLIERS IN TIME SERIES

Suppose z_t, $(t = 0, \pm 1, \ldots)$ is a time series observed at equally spaced intervals of time. Consider the familiar time series model

$$z_t = \Psi(B) a_t,$$

where

$$\Psi(B) = 1 - \Psi_1 B - \Psi_2 B^2 - \ldots.$$

B is a backward shift operator such that $Ba_t = a_{t-1}$ and $\{a_t,\ t = 0, \pm 1, \ldots\}$ is a sequence of independent identically distributed normal random variables with mean zero and variance σ^2. Often $\Psi(B)$ is expressed as a ratio $\theta(B)/\phi(B)$ of finite moving average and autoregressive polynomial operators.

One can characterize an aberrant observation at $t = q$ by the model $y_t = z_t + \omega_1 I_t(q)$ where y_t is the actual observation at time t, $I_t(q) = 1$ if $t = q$ (i.e., if the qth observation is aberrant) and $I_t(q) = 0$ otherwise, and ω_1 is the amount of shift. This model is referred to as the aberrant observation (AO) model (see Abraham and Box, 1979) or as the Type I model (Fox, 1972). Alternatively it might be that the aberration affects the innovation at $t = q$. This can be represented as $z_t = \Psi(B)(a_t + \omega_2 I_t(q))$ where $\Psi(B)$ and a_t are defined as before, $I_t(q) = 1$ if $t = q$ (if the qth innovation is abberant) and $I_t(q) = 0$ otherwise. This is referred to as the aberrant innovation (AI) or Type II model.

Obviously one can embed the AO and AI models in the transfer function via the intervention model approach:

$$y_t = (\omega(B)/\delta(B)) I_t(q) + \Psi(B) a_t,$$

where $\omega(B)$ and $\delta(B)$ are finite polynomial operators. If one lets $\omega(B) = \omega_1$ and $\delta(B) = 1$ one obtains the AO model; alternatively, letting $\omega(B) = \omega_2$ and $\delta(B) = \phi(B)$ yields the AI model.

The implications of the AO and AI models can be seen clearly by considering a special case $\Psi(B) = 1/[1 - \phi B]$, $|\phi| < 1$. This is the case of the well known first order autoregressive process $\{\text{AR}(1)\}$.

(i) AO: $z_t = \phi z_{t-1} + a_t$, $y_t = z_t + \omega_1 I_t(q)$; Figure 1(a) shows the plot of y_t vs. y_{t-1} where y_t represents a typical set of observations from the AR(1) model with an aberration at $t = q$. The points (y_{q-1}, y_q) and (y_q, y_{q+1}) are not consistent with the least squares line because y_q is aberrant. Although only one observation is aberrant, two points are affected. Thus the least square estimate of ϕ would not have desirable properties. In fact it can be shown that it is asymptotically biased (see, for example, Martin, 1980).

(ii) AI: $z_t = \phi z_{t-1} + a_t + \omega_2 I_t(q)$, $y_t = z_t$; Figure 1(b) shows a plot similar to that given in Figure 1(a). Here the aberration is in the innovation at q. The point (y_{q-1}, y_q) is not consistent with the least square line. However, (y_q, y_{q+1}) is consistent with the line although it is "away" from the rest. Later points (y_{q+1}, y_{q+1+t}) are also consistent with the least squares line. The least squares estimate of ϕ, in this case, is consistent. Thus the AI model does not lead to serious problems while the AO model does.

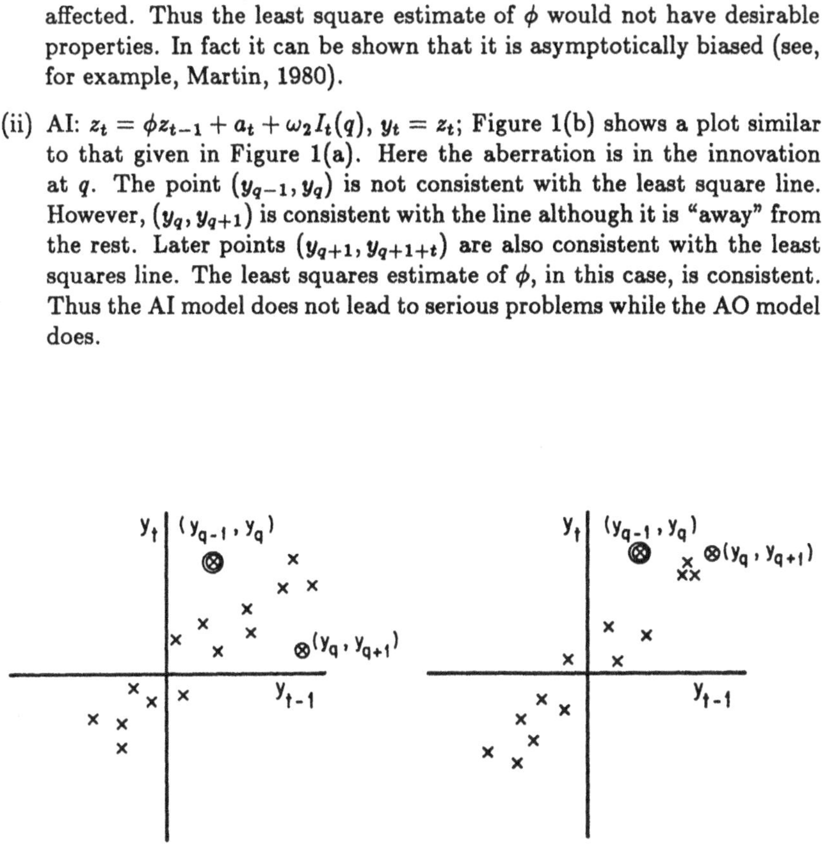

Figure 1. *(a) plot of y_t vs. y_{t-1} AR(1) with an AO outlier; (b) plot of y_t vs. y_{t-1} AR(1) with an AI outlier.*

It should be noted from the definition that an AO outlier at $t = q$ affects only y_q and consequently the residuals e_q, $e_{q+1}, \ldots$, while an AI outlier at $t = q$ affects e_q and hence $y_q, y_{q+1}, \ldots$.

3. DIFFERENT APPROACHES TO TIME SERIES ANALYSIS IN THE PRESENCE OF OUTLIERS

We now consider the major approaches to analysis of time series when outliers may be present.

3.1 Robustness

Martin (1980) discussed robust estimation of the parameter of time series models in the presence of outliers. He defined the AO and AI models in a form slightly different to that given above.. The observations $\{y_t,\ t = 0, \pm 1, \ldots\}$ are defined to be $y_t = z_t + \epsilon_t$, where z_t is the actual generating process and ϵ_t is a 'measurement error' sequence.

AO: In this case the a_t's are defined as before but ϵ_t is a random variable such that

$$P(\epsilon_t = 0) = 1 - \alpha, \quad P(\epsilon_t \neq 0) = \alpha, \quad 0 < \alpha < 1.$$

AI: Here $\epsilon_t \equiv 0$ and a_t has a heavy tailed nonnormal distribution.

With these characterizations robust estimates are obtained for the autoregressive process of order p $\{AR(p)\}$, that is:

$$z_t = \sum_{t=1}^{p} \phi_i z_{t-i} + a_t.$$

This process can also be written as

$$z_t = \mathbf{X}_t' \boldsymbol{\phi} + a_t,$$

where $\mathbf{X}_t' = (z_{t-1}, z_{t-2}, \ldots, z_{t-p})$, and $\boldsymbol{\phi}' = (\phi_1, \phi_2, \ldots, \phi_p)$. M-estimates of $\boldsymbol{\phi}$ may be obtained by minimizing $\sum_{t=p+1}^{n} \rho((z_t - \mathbf{X}_t'\boldsymbol{\phi})/\hat{\sigma})$ where ρ is a symmetric robustifying loss function, n is the number of observations and $\hat{\sigma}$ is a robust scale estimate of σ. This minimization leads to the estimating equation

$$\sum_{t=p+1}^{n} \mathbf{X}_t \psi((z_t - \mathbf{X}_t'\hat{\boldsymbol{\phi}})/\hat{\sigma}) = 0,$$

where ψ, which is the derivative of ρ, is bounded and continuous, and where $\hat{\boldsymbol{\phi}}$ denotes the M-estimate of $\boldsymbol{\phi}$.

In the AI case, since $\mathbf{X}_t \psi(\cdot)$ is bounded in $\mathbf{X}_t$, under suitable regularity conditions it can be shown that $\hat{\boldsymbol{\phi}}$ is consistent and asymptotically normal. However, in the AO case $\mathbf{X}_t \psi(\cdot)$ is not bounded and the M-estimates are "non-robust". Thus one may obtain the so-called generalized M-estimates, or bounded influence estimates, by solving the estimating equation

$$\sum_{t=p+1}^{n} \omega(\mathbf{X}_t)\mathbf{X}_t\psi((z_t - \mathbf{X}_t'\hat{\boldsymbol{\phi}})/\hat{\sigma}) = 0,$$

where $\omega(\mathbf{X}_t)$ is a weight function such that $\omega(\mathbf{X}_t)\mathbf{X}_t$ is bounded and continuous. The estimates obtained are also biased. However, this bias is smaller than that of either the M- or the least squares estimates.

When the process is autoregressive moving average (ARMA), the situation is more complicated. For the AI case, M-estimates can be obtained by the same procedure as described for the AR process. However, in the AO situation the generalized M-estimates, or bounded influence estimates, do not seem appealing. Thus Martin (1980) discussed approximate maximum likelihood type estimates which are obtained iteratively. This approach seems to be extremely cumbersome. Martin (1983) also discussed a strategy for:

(1) detecting outliers,

(2) cleaning the data with a robustly-estimated autoregressive approximation, and

(3) building a model for the cleaned-up data using the usual procedures.

3.2 Iterative Maximum Likelihood

Generalizing the procedure of Fox (1972), Chang and Tiao (1984) discussed a method of identifying and adjusting spurious observations. Their method is as follows.

Suppose that the parameters $\boldsymbol{\phi}$ and $\boldsymbol{\theta}$ of an ARMA time series model are known. Then the outlier models can be written as

$$e_t = \omega_1\pi(B)I_t(q) + a_t = \omega_1 x_{1t} + a_t \qquad \text{(AO)}$$

and

$$e_t = \omega_2 I_t(q) + a_t = \omega_2 x_{2t} + a_t, \qquad \text{(AI)}$$

where

$$\pi(B) = \phi(B)/\theta(B) = 1 - \pi_1 B - \pi_2 B^2 - \ldots,$$
$$e_t = \pi(B)z_t,$$
$$x_{1t} = \pi(B)I_t(q),$$

and

$$x_{2t} = I_t(q).$$

Approximate maximum likelihood estimates of ω_1 and ω_2 assuming q known are:

$$\hat{\omega}_1 = \sum e_t x_{1t} / \sum x_{1t}^2 = e_q - \pi_1 e_{q+1} - \ldots / \eta^2, \quad \eta^2 = \sum_{j \geq 0} \pi_j^2, \quad \pi_0 = 1,$$

and

$$\hat{\omega}_2 = \sum e_t x_{2t} / \sum x_{2t}^2 = e_q$$

and their variances are $V(\hat{\omega}_1) = \sigma^2/\eta^2$ and $V(\hat{\omega}_2) = \sigma^2$.

The likelihood ratio statistics, $\hat{\lambda}_{1q} = \hat{\omega}_1 \eta / \sigma$ and $\hat{\lambda}_{2q} = \hat{\omega}_q / \sigma$, are considered, and the statistic

$$\hat{\lambda} = \max_t \max_{1 \leq i \leq 2} \lambda_{it}$$

is used to check for outliers. When an observation is found to be an outlier, it is modified using $\hat{\omega}_1$ or $\hat{\omega}_2$ as the case may be. This procedure is repeated with the new observations until no further outliers are found. Details of the procedure are given by Chang and Tiao (1984), and by Abraham and Ledolter (1983). The distribution of $\hat{\lambda}$ is not yet well known. Also it is not clear how efficient the procedure is when more than one outlier in present.

3.3 Bayesian Approach

Abraham and Box (1979) discussed a Bayesian approach in which it is assumed that any innovation a_t has a small prior probability, α, $0 < \alpha < 1$; of being aberrant; that is, it is generated from a normal distribution with mean ω_1 and variance σ^2, $N(\omega_1, \sigma^2)$, and a complementary probability, $1 - \alpha$, that it is not aberrant or in other words is $N(0, \sigma^2)$. Under this regime it was shown that the posterior distribution of $\boldsymbol{\phi}' = (\phi_1, \ldots, \phi_p)$ in an autoregressive process of order p is given by

$$P(\boldsymbol{\phi} \mid \mathbf{y}) = \sum_r \omega_{(r)} P(\boldsymbol{\phi} \mid \textit{ a given set is aberrant}).$$

Here $\omega_{(r)}$ is the posterior probability that a particular set of r innovations are aberrant given the data. Inference about $\boldsymbol{\phi}$ can be made using $P(\boldsymbol{\phi} \mid \mathbf{y})$ and $\omega_{(r)}$ provides information about the identity of outliers.

A similar approach can be adopted in principle for the AO case .

3.4 Lagrange Multiplier (L-M) Test

Suppose $\boldsymbol{\beta}$ is a vector of parameters in some space Ω and one is interested in testing k restrictions

$$H_0: \ h_j(\boldsymbol{\beta}) = 0, \qquad j = 1, 2, \ldots, k.$$

We consider the function

$$g(\beta) = l(\beta) + \sum_{j=1}^{k} \lambda_j h_j(\beta),$$

where l is the log likelihood and λ_j $(j = 1, 2, \ldots, k)$ a set of Lagrange multipliers. Now let

$$\hat{\mathbf{d}} = \left(\frac{\partial l(\beta)}{\partial \beta_i}\right)_{\beta=\hat{\beta}}, \quad \hat{H} = \left(\frac{\partial h_i(\beta)}{\partial \beta_i}\right)_{\beta=\hat{\beta}}$$

and $S = \hat{\mathbf{d}}' I^{-1} \hat{\mathbf{d}}$ where I is the information matrix under the null hypothesis H_0. S is usually referred to as the score statistic, and it can be seen that $S = \hat{\boldsymbol{\lambda}}' \hat{H}' I^{-1} \hat{H} \hat{\boldsymbol{\lambda}}$. Under H_0, S has an asymptotic chi-square distribution with k degrees of freedom (d.f.).

We specialize by considering the familiar AR(p) process in the possible presence of an outlier, which may be written as follows:

$$z_t = \sum_{i=1}^{p} \phi_i z_{t-i} + \omega_1 \phi(B) I_t(q) + a_t + \omega_2 I_t(q).$$

When ϕ and σ^2 are known it can be shown (Yatawara, 1985) that the score statistic to test for an outlier $t = q$ is given by

$$S_q = \frac{1}{\sigma^2}\left[e_q^2 \left(\sum_{i=1}^{n-p} \phi_i e_{q+i}\right)^2 \Big/ \sum_{i=1}^{n-p} \phi_i^2\right], \quad \phi_i = 0, \quad i > p,$$

where $e_t = z_t - \sum_{i=1}^{p} \phi_i z_{t-i}$ are the recursive residuals. S_q has a chi-square distribution with 2 degrees of freedom and this may be used for tests of significance.

In practice q is unknown and it may be necessary to consider S_t, $t = p+1, \ldots, n-p$. Thus we consider the statistic

$$S_0 = \max_{(p+1) \le t \le n-p} S_t$$

and its distribution. Here we take $p = 1$ (AR(1)). The general case is considered by Yatawara (1985). When $p = 1, S_2, S_3, \ldots$ are 1 step dependent. It can be shown that when n is large S_0 has an extreme value distribution. Thus one can find the significance points, c, from the relation

$$\text{Significance level } (\alpha) = P(S_0/2 - \ln(n-2) \ge c) = 1 - \exp(-\exp(-c)).$$

Table 1. *Simulated significance points.*

n	5%	1%
50	13.04 (13.68)	15.92 (16.94)
100	14.76 (15.11)	18.20 (18.37)
200	16.12 (16.52)	19.22 (19.78)

(Values from the extreme value approximation are shown in parentheses.)

Accuracy of this approximation has been checked by a simulation experiment. We generated data from the process $z_t = .4\ z_{t-1} + a_t$ $(\sigma^2 = 1)$ and obtained S_0 for sample size 50, 100 and 200. This was repeated 500 times and the significance points obtained are shown, together with those from the extreme value approximation, in Table 1.

There seems to be good agreement between the simulated and approximate significance points when $n = 100$ or 200. For example when $n = 100$ the approximate 1% significance point is 18.37 while the simulated one is 18.20.

The effect of estimation of parameters on S_0 was studied by Abraham and Yatawara (1986). The indication is that one can consider the estimates of parameters and use S_0 as if the parameters are known.

Example: We consider the yield data reported by Abraham and Ledolter (1983), where it is shown that this series can be adequately represented by an AR(1) process. An observational outlier was introduced at $t = 45$. A plot of the data is shown in Figure 2. We analyse these data assuming that the identity of the outlier as unknown. Figure 2 shows the plot of S_t versus t $(t = 2, 3, \ldots)$. It is quite clear that $S_0 = S_{45}$ and y_{45} is an outlier. It should be noted that since the process is AR(1), S_{44} and S_{45} are large.

4. DISTINGUISHING BETWEEN OUTLIER TYPES

Suppose that we have established the existence of an outlier at $t = q$. Then it is important for further analysis to classify it as either AO or AI. From Section 3.2 for an AR(p) process it can be seen that

$$\hat{\omega}_1 = \left(e_q - \sum_{i=1}^{n-p} \phi_i e_{q+i}\right) \Big/ \left(1 + \sum_{i=1}^{n-p} \phi_i^2\right), \quad \phi_i = 0, \quad i > p,$$

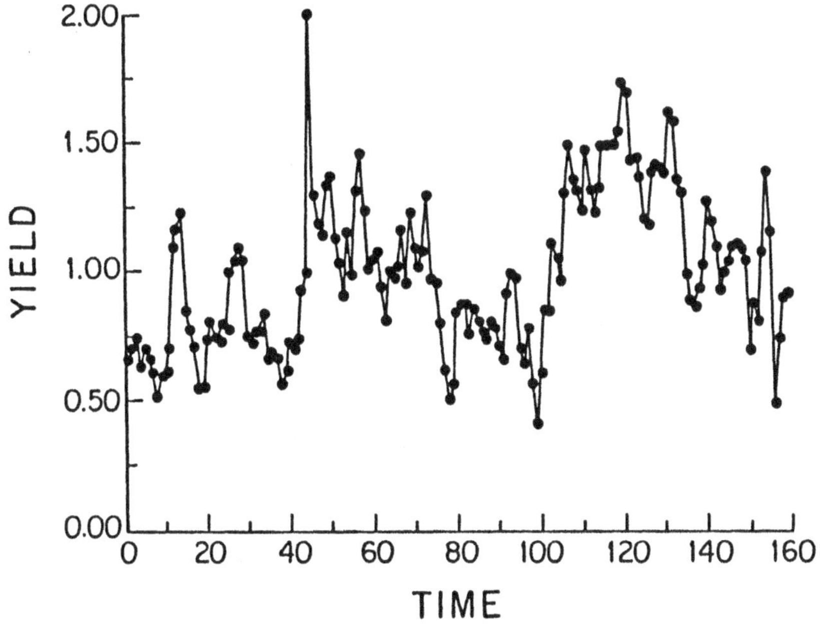

Figure 2. *Yield data with AO introduced at $t = 45$.*

and

$$\hat{\omega}_2 = e_q.$$

We now consider the sums of squares

$$SS(O) = \left(e_q - \sum \phi_i e_{q+i}\right)^2 \Big/ \left(1 + \sum \phi_i^2\right),$$
$$SS(I) = e_q^2$$

and $D = SS(O) - S(I)$. If $D_q > 0$ we take the outlier as AO and if $D_q < 0$ we consider it as AI. Now the size of $P(D_q > 0)$ depends upon ω/σ ($\omega = \omega_1$ or ω_2) and ϕ. When $\omega = 0$, $P(D_q > 0) = 0.5$. When $\omega \neq 0$ the probability calculation requires numerical integration. Here we resort to a limited simulation experiment with 500 repetitions. The results are given in Table 2 and these indicate that D_q can differentiate between the outlier types quite efficiently. For example in the AR(1) case ($\phi = .5$, $\sigma^2 = 1$, $\omega = 4.5$) when the outlier is AO, D_q declares it as AO 89% of the cases. In the AR(2) case with $\phi_1 = 1.1$, $\phi_2 = -.7$, $\sigma^2 = 1$, $\omega = 4.5$ it correctly identifies an AI outlier in 97% of the cases.

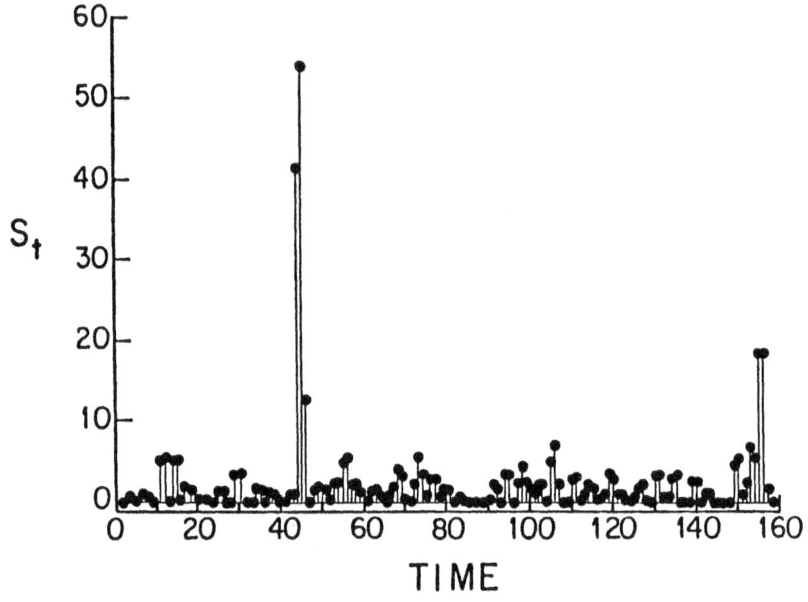

Figure 3. *Plot of S_t vs. t.*

Table 2. *Simulated* $P(D_q > 0)$

Process specifications		Outlier type	$P(D_q > 0)$
AR(1):	$\phi = .5,\ \sigma^2 = 1$ $\omega = 4.5$	AO	.89
		AI	.14
AR(2):	$\phi_1 = 1.1,\ \phi_2 = -.7$ $\sigma^2 = 1,\ \omega = 4.5$	AO	1.00
		AI	.03

5. DIAGNOSTIC CHECKS FOR OUTLIERS

There are a number of diagnostic tools available for regression models. Some of these can be adapted for time series models. In this paper we consider only autoregressive models. Suppose $(z_1, \ldots, z_n)$ is a set of observations generated by an autoregressive process of order p (AR(p)). Then we can write

$$\mathbf{Z} = X\boldsymbol{\phi} + \mathbf{a}, \tag{5.1}$$

where $\mathbf{Z}' = (z_{p+1}, \ldots, z_n)$, $\boldsymbol{\phi}' = (\phi_1, \ldots, \phi_p)$, $\mathbf{a}' = (a_{p+1}, \ldots, a_n)$ and

$$\mathbf{X} = \begin{bmatrix} z_p & z_{p-1} & \cdots & z_1 \\ \cdot & \cdot & \cdots & \cdot \\ \cdot & \cdot & \cdots & \cdot \\ \cdot & \cdot & \cdots & \cdot \\ z_n & z_{n-1} & \cdots & z_{n-p} \end{bmatrix}.$$

Then the conditional least square estimates of $\boldsymbol{\phi}$ is given by $\hat{\boldsymbol{\phi}} = (\mathbf{X}'\mathbf{X})^{-1}\mathbf{X}'\mathbf{Z}$ and $\hat{\mathbf{Z}} = \mathbf{X}(\mathbf{X}'\mathbf{X})^{-1}\mathbf{X}'\mathbf{Z} = H\mathbf{Z}$. Most of the diagnostic tools are based on the idea of deleting suspected observations and of building a measure of the change introduced by the deletion into some feature of the model. In the regression framework, deleting an observation and deleting an equation from (5.1) are equivalent. This is not true in the time series context. For deletion operations we consider the following partitioning:

$$\mathbf{X} = \begin{bmatrix} \mathbf{X}_1 \\ \mathbf{X}_2 \\ \mathbf{X}_3 \end{bmatrix} \begin{matrix} (q-p) \times p \\ k \times p \\ (n-q-k) \times p \end{matrix} \qquad \mathbf{Z} = \begin{bmatrix} \mathbf{Z}_1 \\ \mathbf{Z}_2 \\ \mathbf{Z}_3 \end{bmatrix} \begin{matrix} (q-p) \times 1 \\ k \times 1 \\ (n-q-k) \times 1 \end{matrix}$$

Then the residuals

$$\begin{aligned} \mathbf{e} = \mathbf{Z} - \hat{\mathbf{Z}} &= \begin{bmatrix} \mathbf{e}_1 \\ \mathbf{e}_2 \\ \mathbf{e}_3 \end{bmatrix} \\ &= (I - H)\mathbf{Z} \\ &= \begin{bmatrix} I - H_{11} & -H_{12} & -H_{13} \\ -H_{21} & I - H_{22} & -H_{23} \\ -H_{31} & -H_{32} & I - H_{33} \end{bmatrix} \begin{bmatrix} \mathbf{Z}_1 \\ \mathbf{Z}_2 \\ \mathbf{Z}_3 \end{bmatrix}. \end{aligned}$$

We now consider the statistics (see Draper and John, 1981; Little, 1985);

(1) $Q_{k(q)} = \mathbf{e}_2'(I - H_{22})^{-1}\mathbf{e}_2$,

(2) $R_{k(q)} = (1 - Q_{k(q)}/RSS) \mid I - H_{22} \mid$, R_{SS} = residual sum of squares,

(3) $C_{k(q)} = \mathbf{e}_2'(I - H_{22})^{-1}H_{22}(I - H_{22})^{-1}\mathbf{e}_2/p\hat{\sigma}^2$.

Monitoring these statistics for $q = p+1, \ldots, n-k-p+1$ and for $k = 1$ and $p+1$, one can spot outliers and specify their types (see Abraham and Chuang, 1986). Asymptotic distributions and approximations for efficient calculations are described in the above paper.

For illustration of the patterns, consider the Q statistic for an AR(1) process. When $k = 1$, $H_{22} = z_{q-1}^2/\sum_{t=2}^n z_{t-1}^2$, and $Q_{1(q)} = e_q^2(\sum_2^n z_{t-1}^2/\sum_{t\neq q} z_{t-1}^2)$. Note that the second factor in $Q_{1(q)}$ is approximately one. Thus Q_1 depends on e_q. If there is an AO outlier at $t = q$ then e_q and e_{q+1} are large and hence $Q_{1(q)}$ and $Q_{1(q+1)}$ are large. If the outlier is AI then only e_q is large which leads to $Q_{1(q)}$ alone being large.

If $k = 2$ then Q depends on e_q and e_{q+1}. In this case, for the AO model, $Q_{2(q-1)}$, $Q_{2(q)}$, $Q_{2(q+1)}$ are affected by the outlier. However in the AI case only $Q_{2(q-1)}$ and Q_{2q} are affected. Thus these statistics can help in specifying the outlier types.

6. CONCLUDING REMARKS

In this paper we have briefly outlined some major developments in the area of outliers in time series. It was not our intention to give an exhaustive survey. We also included some new tests based on the score statistic. Some diagnostic tools useful in detecting outliers are also introduced.

REFERENCES

Abraham, B., and G. E. P. Box (1978), "Linear models and spurious observations". *Journal of the Royal Statistical Society,* Series C **27**, 131–138.

Abraham, B., and G. E. P. Box (1979), "Bayesian analysis of some outlier problems in time series". *Biometrika* **66**, 229–236.

Abraham, B., and A. Chuang (1986), "Some diagnostic checking measures for outliers in time series". Submitted for publication.

Abraham, B., and J. Ledolter (1983), *Statistical Methods for Forecasting.* New York: Wiley and Sons.

Abraham, B., and N. Yatawara (1986), "Outliers in time series and the Lagrange multiplier test". Submitted for publication.

Anscombe, F. J. (1960), "Rejection of outliers". *Technometrics* **2**, 123–146.

Box, G. E. P., and G. C. Tiao (1968), "A Bayesian approach to some outlier problems". *Biometrika* **55**, 119–129.

Chang, I., and G. C. Tiao (1984), "Estimation of time series parameters in the presence of outliers". Technical report # 8, Statistics Research Center, Graduate School of Business, University of Chicago.

Cook, R. D., and S. Weisberg (1982), *Residuals and Influence in Regression.* New York: Chapman-Hall.

de Finetti, B. (1961), "The Bayesian approach to the rejection of outliers". In *Proceedings of the Fourth Berkeley Symposium on Mathematical Statistics and Probability,* Volume 1, ed. J. Neyman, pp. 199–210. Berkeley and Los Angeles: University of California Press.

Dixon, W. J. (1950), "Analysis of extreme values". *Annals of Mathematical Statistics* **21**, 488–506.

Draper, N. R., and J. A. John (1981), "Influential observations and outliers in regression". *Technometrics* **23**, 21–26.

Ferguson, T. S. (1961), "On the rejection of outliers". In *Proceedings of the Fourth Berkeley Symposium on Mathematical Statistics and Probability,* ed. J. Neyman, pp. 253–287. Berkeley and Los Angeles: University of California Press.

Fox, A. J. (1972), "Outliers in time series". *Journal of the Royal Statistical Society,* Series B **34**, 350–363.

Grubbs, F. E. (1950), "Sample criteria for testing outlying observations". *Annals of Mathematical Statistics* **21**, 27–58.

Guttman, I. (1973), "Care and handling of univariate or multivariate outliers in detecting spuriosity—a Bayesian approach". *Technometrics* **15**, 723–738.

Hampel, F. R. (1974), "The influence curve and its role in robust estimation". *Journal of the American Statistical Association* **69**, 383–393.

Huber, P. J. (1964), "Robust estimation of a location parameter". *Annals of Mathematical Statistics* **35**, 73–101.

Huber, P. J. (1981), *Robust Statistics.* New York: Wiley and Sons.

Little, J. K. (1985), "Influence and a quadratic form in the Andrews-Pregibon statistic". *Technometrics* **27**, 13–15.

Martin, D. (1980), "Robust methods of time series". In *Applied Time Series,* ed. D. F. Findley, pp. 683–759. New York: Academic Press.

Martin, D. (1983), "Time series outlier detection and model identification with robust filter and smoother cleaners". Paper presented at the Physical Science and Engineering Section, American Statistical Association.

Peirce, B. (1852), "Criterion for the rejection of doubtful observations". *The Astronomical Journal* **2**, 161–163.

Yatawara, N. (1985), "Detection of abrupt system changes with application to outlier analysis in time series". Unpublished manuscript, Department of Statistics and Actuarial Science, University of Waterloo.

Yer Van Hui [1] and Wai Keung Li [2]

PREDICTING DEMANDS IN A MULTI-ITEM ENVIRONMENT

ABSTRACT

Demand forecasting for each item in a multi-line inventory system is considered. Demand patterns are assumed to satisfy an integrated moving-average process, and exponential smoothing is employed in predicting demands. An empirical Bayes estimator for the smoothing parameter using pooled information from all realizations is proposed.

1. INTRODUCTION

All inventory models require information concerning demands for stocked items, and control of stock replenishment is always based upon demand forecasts. Hence, it is necessary to determine the nature of the processes generating demands in order to predict future demands.

The forecasting problem may be formulated as follows. Let $x_1, x_2, \ldots, x_{t-1}$ be the past demands of an item under study and let $\hat{x}_t$ be the one-step-ahead forecast of demand x_t at time t. Brown (1963), Kirby (1967), and Wecker (1978) have discussed the use of exponential smoothing in estimating future demand; that is:

$$\hat{x}_t = (1-\theta)x_{t-1} + \theta\hat{x}_{t-1}, \tag{1}$$

where θ is the smoothing constant and $\hat{x}_{t-1}$ is the one-step-ahead forecast of x_{t-1}. Ray (1982) also examined the use of autoregressive integrated moving-average models in demand forecasting.

Most inventory systems carry a large number of items for which individual demand forecasts are required. These items can be classified into a

[1] Department of Statistics, The Chinese University of Hong Kong, Hong Kong
[2] Department of Statistics, University of Hong Kong, Hong Kong

I. B. MacNeill and G. J. Umphrey (eds.), Time Series and Econometric Modelling, 91–99.

number of categories by item and customer characteristics, with items in the different categories treated differently. Steece and. Wood (1979) have classified items into meaningful and predictable aggregates. Their forecasting method is based on the aggregate demand series and the fractional series obtained by dividing each item demand series by the aggregate demand series.

In some cases, items can be classified into categories according to their stochastic behaviour. For example, semiconductor products with similar demand patterns are grouped in the same category. Demand patterns estimated from individual series may then be improved using pooled information from all realizations in the same category. Thisted and Wecker (1981) have examined the shrinkage estimator in exponential smoothing. Their method is based on the James-Stein (1961) estimator which assumes equal weights in the loss function.

In the sequel we assume that the demand patterns satisfy an integrated moving-average process with different moving-average parameters. In the case where the number of past observations is small, the smoothing constant estimated from individual series generally will be inaccurate. Hence, an empirical Bayes estimator is introduced, and the shrinkage estimator is reviewed. Finally, an application is discussed, and simulation results on the performance of various estimators are examined.

2. THE MODEL

Let $\{x_{i1}, x_{i2}, \ldots, x_{iT}\}$ be past demands for item i in a category of size n. We assume the demands satisfy the model

$$x_{it} = x_{i,t-1} + a_{it} - \theta_i a_{i,t-1}, \quad i = 1, 2, \ldots, n, \quad t = 1, 2, \ldots, T, \tag{2}$$

where θ_i is the moving-average parameter for item i and the a_{it}'s are independent normally distributed random errors with mean zero and variance σ_i^2. Here $\{x_{it}\}$ satisfy an integrated moving-average (IMA) process of order $(0,1,1)$. Note that the exponential smoothing predictor in (1) gives the minimum mean squared error forecast for model (2) and is widely used in demand forecasting. The problem is to estimate the smoothing constants θ_i which characterize the demand patterns.

Suppose the θ_i's are sampled from an unknown prior distribution $G(\theta)$. Let $\hat{\theta}_i$ be an asymptotically efficient estimate of θ_i based on the individual series $\{x_{it}\}$ for each item i, $i = 1, 2, \ldots, n$ (for reference, see Box and Jenkins, 1976). We wish to improve the estimate $\hat{\theta}_i$ in terms of squared error loss with additional information from the other $\hat{\theta}$'s.

3. THE SHRINKAGE ESTIMATOR

Thisted and Wecker (1981) assumed equal variances for $\hat{\theta}_i$'s and presented the following shrinkage estimator:

$$\tilde{\theta}_i = \left\{1 - (n-3)g^2 / \sum_{i=1}^{n} (\hat{\theta}_{\cdot j} - \bar{\theta})^2 \right\}^{+} (\hat{\theta}_i - \bar{\theta}) + \bar{\theta}, \quad i = 1, 2, \ldots, n, \tag{3}$$

where (i) $\bar{\theta} = \sum_{j=1}^{n} \hat{\theta}_{\cdot j}/n$, (ii) $g^2 = \exp(\sum_{j=1}^{n} \log g_{\cdot j}^2/n)$ with $g_{\cdot j}$ an estimated standard error of $\hat{\theta}_{\cdot j}$, and (iii) $\{\cdot\}^{+}$ is the positive-part rule. Efron and Morris (1973, 1975) and Morris (1983) have discussed the shrinkage estimator in a general context.

Note that the assumption of equal variances for $\hat{\theta}_i$'s is not always valid.

4. THE EMPIRICAL BAYES ESTIMATOR

Li and Hui (1983) have discussed an empirical Bayes approach in estimating random coefficient autoregressive processes. We employ the same technique here for the IMA(0, 1, 1) model in (2).

Observe that $\hat{\theta}_i$ given θ_i has an asymptotic normal distribution; that is,

$$f(\hat{\theta}_i \mid \theta_i) = (2\pi\alpha_i)^{-\frac{1}{2}} \exp\left\{-(\hat{\theta}_i - \theta_i)^2 / (2\alpha_i)\right\}, \tag{4}$$

where $\alpha_i = (1 - \theta_i^2)/T$. If one differentiates with respect to $\hat{\theta}_i$, one obtains:

$$d \log f(\hat{\theta}_i \mid \theta_i)/d\hat{\theta}_i = -(\hat{\theta}_i - \theta_i)/\alpha_i. \tag{5}$$

If $\hat{\alpha}_i$ is an estimate of α_i, then equation (5) can be written as follows:

$$\theta_i = \hat{\theta}_i + \hat{\alpha}_i \left\{d \log f(\hat{\theta}_i \mid \theta_i)/d\hat{\theta}_i\right\}. \tag{6}$$

Applications of the law of conditional probability yields an approximate Bayes estimator for θ_i with a squared error loss given by:

$$\begin{aligned} \theta_i^* &= E(\theta_i \mid \hat{\theta}_i) \\ &= \hat{\theta}_i + \hat{\alpha}_i \left\{f_G'(\hat{\theta}_i / f_G(\hat{\theta}_i))\right\}, \end{aligned} \tag{7}$$

where $f_G(\hat{\theta}_i)$ is the marginal density of $\hat{\theta}_i$ and f_G' is its derivative.

The marginal density f_G is unknown since the prior distribution is not given. Let $f_n'(\hat{\theta}_i)/f_n(\hat{\theta}_i)$ be an empirical estimate of $f_G'(\hat{\theta}_i)/f_G(\hat{\theta}_i)$ from $\hat{\theta}_1, \hat{\theta}_2, \ldots, \hat{\theta}_n$ (for reference, see Parzen, 1962). If $\hat{\alpha}_i = (1-\hat{\theta}_i^2)/T$ is an estimate of α_i, then the empirical Bayes estimate of θ_i is given as follows:

$$\theta_i^* = \hat{\theta}_i + T^{-1}(1-\hat{\theta}_i^2)\left\{f_n'(\hat{\theta}_i)/f_n(\hat{\theta}_i)\right\}, \quad i = 1, 2, \ldots, n. \tag{8}$$

The performance of the empirical Bayes estimator was studied by simulation and the results are presented in Section 6. Some theoretical results for a linear regression model were given by Singh (1985).

Next, we consider the special case where the prior distribution $G(\theta)$ is known to be normally distributed with mean θ and variance σ^2. Then $\hat{\theta}_i \sim N(\theta_i, \alpha_i)$ and $\theta_i \sim N(\theta, \sigma^2)$. The marginal density of $\hat{\theta}_i$ is given by

$$\begin{aligned} f(\hat{\theta}_i) &= \int f(\hat{\theta}_i \mid \theta_i) f(\theta_i)\, d\theta_i \\ &= \left\{2\pi\left(\alpha_i + \sigma^2\right)\right\}^{-\frac{1}{2}} \exp\left\{-(\hat{\theta}_i - \theta)^2/(2(\alpha_i+\sigma^2))\right\}. \end{aligned} \tag{9}$$

That is, $\hat{\theta}_i \sim N(\theta, \alpha_i + \sigma^2)$. The conditional density of θ_i given $\hat{\theta}_i$ is then

$$\begin{aligned} f(\theta_i \mid \hat{\theta}_i) &= f(\hat{\theta}_i \mid \theta_i) f(\theta_i) / f(\hat{\theta}_i) \\ &= (2\pi s)^{-\frac{1}{2}} \exp\left\{-\left(\theta_i - u\right)^2/(2s)\right\}, \end{aligned} \tag{10}$$

where $u = (\sigma^2\theta_i + \alpha_i\theta)/(\sigma^2+\alpha_i)$ and $s = \sigma^2\alpha_i/(\sigma^2+\alpha_i)$, or $\theta_i \mid \hat{\theta}_i \sim N(u, s)$. For squared error loss, the Bayes estimator is

$$\theta_i^B = \left(\sigma^2\theta_i + \alpha_i\theta\right) / \left(\sigma^2 + \alpha_i\right).$$

An empirical Bayes estimator for θ_i is then given by

$$\theta_i^* = \left(\sigma^2\theta_i + \hat{\alpha}_i\theta\right) / \left(\sigma^2 + \hat{\alpha}_i\right), \tag{11}$$

where α_i is estimated by $\hat{\alpha}_i$. This result can also be obtained from

$$f'(\hat{\theta}_i)/f(\hat{\theta}_i) = -(\hat{\theta}_i - \theta)/\left(\alpha_i + \sigma^2\right) \tag{12}$$

and equation (8).

5. EXAMPLE

Fifteen semiconductor products with similar stochastic behaviours in a manufacturing plant are grouped in a single category. The grouping is dependent on component characteristics and market demand. Demands $\{x_{it}\}$ for each item in the past twenty weeks were obtained from the accounting department and the exponential smoothing predictor,

$$\hat{x}_{it} = (1-\theta_i)\, x_{i,t-1} + \theta_i \hat{x}_{i,t-1}, \quad i = 1, 2, \ldots, 20, \tag{13}$$

is to be employed in demand forecasting. Precise estimates for the smoothing constants are required.

The exact likelihood estimator is chosen as the estimator of the moving-average parameter; an algorithm was given by Ansley (1979). The exact likelihood estimates $\hat{\theta}_i$'s computed for individual series $\{x_{it}\}$ are given in Table 1.

Table 1. *Exponential Smoothing Parameter Estimates for Fifteen Semiconductor Products.*

Item	$\hat{\theta}_i$	θ_i^*	$\tilde{\theta}_i$
1	.875	.855	.872
2	.800	.794	.801
3	.593	.729	.604
4	.926	.907	.920
5	.954	.939	.946
6	.739	.759	.742
7	.738	.758	.741
8	.957	.944	.950
9	.906	.885	.901
10	.920	.900	.914
11	.549	.735	.562
12	.798	.792	.799
13	.956	.942	.949
14	.677	.736	.684
15	.692	.740	.699

We assume that the moving-average parameters were sampled from an unknown prior distribution $G(\theta)$. Following the procedure discussed above, we first estimate the marginal density $f_G(\hat{\theta})$ using a kernel density estimate discussed by Parzen (1962) and Clemmer and Krutchkoff (1968). Let

$$f_n'(\hat{\theta})/f_n(\hat{\theta}) = \frac{\sum_{i=1}^{n}\left\{((\sin A_i)/A_i)^2 - ((\sin B_i)/B_i)^2\right\}}{n\sum_{i=1}^{n}((\sin B_i)/B_i)^2}, \tag{14}$$

where $\sin 0/0 = 1$, $A_i = (\hat{\theta} - \hat{\theta}_i + h)/(2h)$, and $B_i = (\hat{\theta} - \hat{\theta}_i)/(2h)$, with $h = n^{-0.7}\{\sum_{i=1}^{n}(\hat{\theta}_i - \bar{\theta})^2\}^{\frac{1}{2}}$ and $\bar{\theta} = \sum_{i=1}^{n}\hat{\theta}_i/n$. The empirical Bayes estimates, θ_i^*, are computed using equation (8) and are presented in Table 1. For reference, the shrinkage estimates, $\tilde{\theta}_i$, are also shown in Table 1 where the variance of $\hat{\theta}_i$ is estimated by $g_i^2 = (1 - \hat{\theta}_i^2)T$.

Observe that both the empirical Bayes approach and the shrinkage method adjust the exact likelihood estimates towards the centre of the marginal distribution $f_G(\hat{\theta})$. The improvements using empirical Bayes methods were examined in a simulation study, which is presented below.

6. SIMULATION STUDY

We first sampled moving-average parameters $\{\theta_1, \theta_2, \ldots, \theta_n\}$ from a known distribution $G(\theta)$. Brown (1963) noted that smoothing constants usually have a mean close to 0.8. We focused on beta distributions with means equal to 0.8 for $G(0)$. Starting values $x_{i0}, i = 1, 2, \ldots, n$ were generated independently from a lognormal distribution with parameters $\mu = 3$ and $\sigma = 2$. The error variances were chosen as $\sigma_i^2 = 2 + x_{i0}^{0.6}, i = 1, 2, \ldots, n$, with high demand items assigned larger variances. Demand data $\{x_{ij}\}$ were then generated from model (2).

The exact likelihood estimate $\hat{\theta}_i$ was obtained for each individual series. The shrinkage estimate $\tilde{\theta}_n$ and the empirical Bayes estimate θ_n^* for the nth item were then computed.

A set of simulation results is shown in Table 2 with $n = 15$, $T = 20$, a prior distribution of beta (8,2), and 10 replications. It can be seen that both the empirical Bayes estimates and the shrinkage estimates are closer to θ_n than the exact likelihood estimates. In some cases improvements through the empirical Bayes approach are very encouraging.

We fixed $n = 15$ and $T = 20$, and performed 1000 replications with various beta prior distributions. We let

$$R1(\theta^* \mid \hat{\theta}) = \sum_n (\theta_n^* - \theta_n)^2 \Big/ \sum_m (\hat{\theta}_m - \theta_m)^2 \tag{15}$$

Table 2. *Simulation Results on the Estimation of the Smoothing Parameter θ_n in 10 Replications.*

Rep.	θ_n	$\hat{\theta}_n$	θ^*_n	$\tilde{\theta}_n$
1	.830	.940	.930	.934
2	.540	.604	.671	.610
3	.658	.744	.744	.744
4	.823	.887	.855	.880
5	.738	.911	.897	.909
6	.918	.807	.836	.808
7	.843	.636	.684	.643
8	.898	.708	.756	.713
9	.879	.933	.926	.931
10	.831	.914	.904	.912

and

$$R2(\theta^* \mid \hat{\theta}) = \sum_n \mid \theta^*_n - \theta_n \mid / \sum_m \mid \hat{\theta}_m - \theta_m \mid \tag{16}$$

denote the relative mean squared errors and the relative mean absolute deviations respectively, where the summations are taken over 1000 replications. A value of less than 1 in $R1$ or $R2$ indicates that θ^* outperformed $\hat{\theta}$ in estimating θ. We also denote by $C(\theta^* \mid \hat{\theta})$ the total count that $\mid \theta^*_n - \theta_n \mid \leq \mid \hat{\theta}_n - \theta_n \mid$ in 1000 replications. With a similar notation, we define $R1(\tilde{\theta} \mid \hat{\theta})$, $R1(\theta^* \mid \tilde{\theta} \mid)$, $R2(\tilde{\theta} \mid \hat{\theta})$, $R2(\theta^* \mid \tilde{\theta})$, $C(\tilde{\theta} \mid \hat{\theta})$ and $C(\theta^* \mid \tilde{\theta})$. Simulation results are presented in Table 3. They show that both the empirical Bayes estimator and the shrinkage estimator are superior to the exact likelihood estimator. The empirical Bayes estimator also outperforms the shrinkage estimator, and the improvement is significant.

Similar simulation results were obtained for other prior distributions. Furthermore, other simulation results show that the more items in a category, the better the performance in pooling of information; that is, more information about the prior distribution is available. The empirical Bayes approach often gives significant improvement for short series.

Table 3. *Simulation Results on the Estimation of the Smoothing Parameter θ_n in 1000 Replications.*

Prior	Criteria	$\theta^* \mid \hat{\theta}$	$\tilde{\theta} \mid \hat{\theta}$	$\theta^* \mid \tilde{\theta}$
	R1	.747	.966	.773
beta (4,1)	R2	.895	.985	.909
	C	589	640	579
	R1	.680	.956	.712
beta (8,2)	R2	.839	.978	.858
	C	692	734	669
	R1	.545	.940	.580
beta (20,5)	R2	.752	.966	.778
	C	838	869	821

7. CONCLUSION

In conclusion, it can be noted that the empirical Bayes method can be extended to general autoregressive integrated moving-average demand patterns.

REFERENCES

Ansley, C. (1979), "An algorithm for the exact likelihood of a mixed autoregressive-moving average process." *Biometrika* **66**, 59–65.

Box, G. E. P., and G. M. Jenkins (1976), *Time Series Analysis, Forecasting and Control.* San Francisco: Holden-Day.

Brown, R. G. (1963), *Smoothing, Forecasting and Prediction of Discrete Time Series.* New York: Prentice-Hall.

Clemmer, B. A., and R. G. Krutchkoff (1968), "The use of empirical Bayes estimators in a linear regression model." *Biometrika* **55**, 525–534.

Efron, B., and C. Morris (1973), "Stein's estimation rule and its competitors—an empirical Bayes approach." *Journal of the American Statistical Association* **68**, 117–130.

Efron, B., and C. Morris (1975), "Data analysis using Stein's estimator and its generalizations." *Journal of the American Statistical Association* **70**, 311–319.

James, W., and C. Stein (1961), "Estimation with quadratic loss." *Proceedings of the Fourth Berkeley Symposium on Mathematical Statistics and Probability* **1**, 361–379.

Kirby, R. M. (1967), "A comparison of short and medium range statistical forecasting methods." *Management Science* **13**, 202–210.

Li, W. K., and Y. V. Hui (1983), "Estimation of random coefficient autoregressive process: an empirical Bayes approach." *Journal of Time Series Analysis* **4**, 89–94.

Morris, C. N. (1983), "Parametric empirical Bayes inference: theory and applications." *Journal of the American Statistical Association* **78**, 47–55.

Parzen, E. (1962), "On estimation of a probability density function and mode." *Annals of Mathematical Statistics* **33**, 1065–1076.

Ray, W. D. (1982), "ARIMA forecasting models in inventory control." *Journal of the Operational Research Society* **33**, 567–574.

Singh, R. S. (1985), "Empirical Bayes estimation in a multiple linear regression model." *Annals of the Institute of Statistical Mathematics* **37**, 71–86.

Steece, B. M., and S. D. Wood (1979), "An ARIMA-based methodology for forecasting in a multi-item environment." *Management Science* **25**, 167–187.

Thisted, R. A., and W. E. Wecker (1981), "Predicting a multitude of time series." *Journal of the American Statistical Association* **76**, 516–523.

Wecker, W. E. (1978), "Predicting demand from sales data in the presence of stockouts." *Management Science* **24**, 1043–1055.

Thian S. Kheoh and A. I. McLeod [1]

ON THE EFFICIENCY OF A STRONGLY CONSISTENT ESTIMATOR IN ARMA MODELS

ABSTRACT

Hannan (1975) showed that the initial estimates of the autoregressive parameters in an ARMA(p, q) model which were suggested by Box and Jenkins (1976, p. 499) are strongly consistent. In this note, the efficiency of this estimate in the ARMA(1,1) model is examined.

1. INTRODUCTION

Given the time series $\{Z_t\}$, $t = 1, 2, \ldots, n$, the mixed autoregressive-moving average model of order p and q respectively, ARMA(p, q), is defined to be

$$\phi(B)Z_t = \theta(B)a_t, \tag{1}$$

where $\phi(B) = 1 - \phi_1 B - \phi_2 B^2 - \cdots - \phi_p B^p$ and $\theta(B) = 1 - \theta_1 B - \theta_2 B^2 - \cdots - \theta_q B^q$ are polynomials in B; B is the backshift operator such that $BZ_t = Z_{t-1}$; and $\{a_t\}$ is a Gaussian white noise with zero mean and variance σ_a^2. The characteristic roots of $\phi(B) = 0$ and $\theta(B) = 0$ are assumed to lie outside the unit circle and it is further assumed that there are no common roots. Without loss of generality, it is further assumed that the time series has zero mean, i.e., $\langle Z_t \rangle = 0$, where $\langle \cdot \rangle$ denotes mathematical expectation.

Let $\boldsymbol{\phi} = (\phi_1, \ldots, \phi_p)'$ and $\boldsymbol{\theta} = (\theta_1, \ldots, \theta_q)'$. For a pure autoregression of order p, denoted by AR(p), the vector of parameters $\boldsymbol{\phi}$ can be determined by solving the Yule-Walker equation

$$P_p \boldsymbol{\phi} = \boldsymbol{\rho}_p, \tag{2}$$

[1] Department of Statistical and Actuarial Sciences, The University of Western Ontario, London, Ontario N6A 5B9 (both authors)

I. B. MacNeill and G. J. Umphrey (eds.), Time Series and Econometric Modelling, 101–105.

where

$$P_p = \begin{bmatrix} 1 & \rho_1 & \cdots & \rho_{p-1} \\ \rho_1 & 1 & \cdots & \rho_{p-2} \\ \vdots & \vdots & \ddots & \\ \rho_{p-1} & \rho_{p-2} & & 1 \end{bmatrix},$$

$$\boldsymbol{\rho}_p = (\rho_1, \ldots, \rho_p)'$$

and

$$\rho_k = \frac{\langle Z_t Z_{t-k} \rangle}{\langle Z_t^2 \rangle}, \qquad k = 1, 2, \ldots, p.$$

Then $\tilde{\boldsymbol{\phi}}$, the Yule-Walker estimate, can be obtained from (2) by simply replacing ρ_k by its estimate r_k, the sample autocorrelation function defined by

$$r_k = \frac{c_k}{c_o},$$

where $c_k = n^{-1} \sum_{t=1}^{n-k} Z_t Z_{t+k}$.

Hannan (1975) showed that in the ARMA(p, q) model the autoregressive parameters, $\phi_1, \ldots, \phi_p$, estimated by solving

$$\sum_{i=1}^{p} \tilde{\phi}_i c_{i-j} = 0, \qquad j = q+1, \ldots, q+p, \tag{3}$$

are strongly consistent. In the next section, the asymptotic efficiency of $\tilde{\phi}_1$, in the ARMA(1,1) model is derived.

2. EFFICIENCY OF $\tilde{\phi}_1$

The ARMA(p, q) model in (1) can be written as

$$Z_t - \phi_1 Z_{t-1} - \cdots - \phi_p Z_{t-p} = a_t - \theta_1 a_{t-1} - \cdots - \theta_q a_{t-1}.$$

If one multiplies by Z_{t-k} and takes expected values, one obtains the following:

$$\gamma_k - \phi_1 \gamma_{k-1} - \cdots - \phi_p \gamma_{k-p} = \gamma_{Za}(k) - \theta_1 \gamma_{Za}(k-1) - \cdots - \theta_q \gamma_{Za}(k-q), \tag{4}$$

where γ_k is the covariance function between the series Z_t and Z_{t-k} defined by $\gamma_k = \langle Z_t Z_{t-k} \rangle$, and $\gamma_{Za}(k)$ is the cross covariance function between Z_t and a_t, and is defined by $\gamma_{Za}(k) = \langle Z_{t-k} a_t \rangle$.

Upon dividing by γ_0, equation (4) can be written,

$$\rho_k - \phi_1\rho_{k-1} - \cdots - \phi_p\rho_{k-p} = 0, \qquad k \geq q+1.$$

The estimate $\tilde{\phi}$ is obtained by solving this equation and by replacing ρ_k by its sample estimate r_k.

For the ARMA(1,1) model,

$$\tilde{\phi}_1 = \frac{r_2}{r_1} = \frac{c_2}{c_1},$$

Expanding in Taylor series up to first order terms yields:

$$\tilde{\phi}_1 = \phi_1 + (c_2 - \gamma_2)\cdot\frac{1}{\gamma_1} - (c_1 - \gamma_1)\frac{\gamma_2}{\gamma_1^2}.$$

As indicated by Lomnicki and Zarembra (1957), this expansion ignores terms $O(1/n)$. Then

$$V(\tilde{\phi}_1) = \frac{1}{\gamma_1^2}V(c_2) + \frac{\phi_1^2}{\gamma_1^2}V(c_1) - \frac{2\phi_1}{\gamma_1^2}\mathrm{cov}(c_1, c_2). \tag{5}$$

It is easily shown that, apart from terms $O(1/n^2)$,

$$\begin{aligned} V(c_1) &= \frac{1}{n^2}\sum_{t=1}^{n-1}\sum_{s=1}^{n-1}\{\gamma_{t-s}^2 + \gamma_{t-s-1}\gamma_{t+1-s}\} \\ &= \frac{1}{n}\left\{\gamma_0^2 + 2\phi_1\gamma_0\gamma_1 + \gamma_1^2 + \frac{2\gamma_1^2}{1-\phi_1^2}(1-\phi_1^2)\right\}, \end{aligned} \tag{6}$$

$$\begin{aligned} V(c_2) &= \frac{1}{n^2}\sum_{t=1}^{n-2}\sum_{s=1}^{n-2}\{\gamma_{t-s}^2 + \gamma_{t-s-2}\gamma_{t+1-s}\} \\ &= \frac{1}{n}\left\{\gamma_0^2 + 3\phi_1^2\gamma_1^2 + 2\phi_1^3\gamma_0\gamma_1 + \frac{2\gamma_1^2}{1-\phi_1^2}(1+\phi_1^4)\right\}, \end{aligned} \tag{7}$$

and

$$\begin{aligned} \mathrm{cov}(c_1, c_2) &= \frac{1}{n^2}\sum_{t=1}^{n-1}\sum_{s=1}^{n-2}\{\gamma_{t-s}\gamma_{t-s-1} + \gamma_{t-s-2}\gamma_{t+1-s}\} \\ &= \frac{2}{n}\left\{\phi_1\gamma_1^2 + \gamma_1(1+\phi_1^2)\left(\gamma_0 + \frac{\phi_1\gamma_1}{1-\phi_1^2}\right)\right\}. \end{aligned} \tag{8}$$

If one substitutes equations (6), (7) and (8) into (5) one may obtain the following:

$$V(\tilde{\phi}_1) = \frac{1}{n}\left\{\frac{1+\phi_1^2}{\rho_1^2} - \frac{4\phi_1}{\rho_1} + 2\right\},$$

where

$$\rho_1 = \frac{(1-\phi_1\theta_1)(\phi_1-\theta_1)}{1+\theta_1^2-2\phi_1\theta_1}.$$

Let $\hat{\phi}_1$ denote the maximum likelihood estimate of ϕ_1. Box and Jenkins (1976, p. 242) give the asymptotic variance of $\hat{\phi}_1$ as follows:

$$V(\hat{\phi}_1) = \frac{(1-\phi_1\theta_1)^2(1-\phi_1^2)}{n(\phi_1-\theta_1)^2}.$$

Hence the asymptotic efficiency of $\tilde{\phi}_1$ relative to $\hat{\phi}_1$ is

$$\begin{aligned}\text{Eff} &= \frac{V(\hat{\phi}_1)}{V(\tilde{\phi}_1)} \\ &= \frac{\rho_1^2(1-\phi_1\theta_1)^2(1-\phi_1^2)}{(\phi_1-\theta_1)^2(2\rho_1^2-4\rho_1\phi_1+\phi_1^2+1)}.\end{aligned}$$

It follows that,

$$\lim_{\substack{\theta_1=0\\ \phi_1\to 0}} \text{Eff} = 1 \text{ and } \lim_{\substack{\phi_1=0\\ \theta_1\to 0}} \text{Eff} = 1.$$

Hence $\tilde{\phi}_1$ is as efficient as $\hat{\phi}_1$ when θ_1 is close to zero.

Remark: The result of Section 2 also applies to the estimator of θ_1 in an ARMA(1,1) model given by $\tilde{\theta}_1 = ri(2)/ri(1)$, where $ri(\cdot)$ denotes the inverse autocorrelation function.

The asymptotic efficiency for various models is presented in Table 1. This table shows that $\tilde{\phi}_1$ is indeed as efficient as the maximum likelihood estimator when θ_1 is near zero, but that the efficiency is very poor when both ϕ_1 and θ_1 are close to negative one or when both ϕ_1 and θ_1 are close to positive one.

Table 1. *Asymptotic Efficiency of* $\tilde{\phi}_1$ *Relative to* $\hat{\phi}_1$

ϕ_1	θ_1: −0.90	−0.60	−0.30	0.00	0.30	0.60	0.90
−0.95	0.003	0.288	0.908	1.000	0.983	0.961	0.944
−0.75	0.012	0.174	0.739	1.000	0.919	0.815	0.736
−0.50	0.045	0.215	0.688	1.000	0.846	0.652	0.516
−0.25	0.107	0.290	0.696	1.000	0.783	0.510	0.340
0.25	0.340	0.510	0.783	1.000	0.696	0.290	0.107
0.50	0.516	0.652	0.846	1.000	0.688	0.215	0.045
0.75	0.736	0.815	0.919	1.000	0.739	0.174	0.012
0.95	0.944	0.961	0.983	1.000	0.908	0.288	0.003

REFERENCES

Box, G. E. P., and G. M. Jenkins (1976), *Time Series Analysis, Forecasting and Control.* San Francisco: Holden-Day.

Hannan, E. J. (1975), "The estimation of ARMA models". *Annals of Statistics* **3**, 975–981.

Lomnicki, Z. A., and S. K. Zarembra (1957), "On the estimation of autocorrelation in times series". *Annals of Mathematical Statistics* **28**, 140–158.

Leo A. Aroian [1]

RECENT RESULTS FOR TIME SERIES IN M DIMENSIONS

ABSTRACT

Recent results for time series in m dimensions are reviewed and a brief summary is given of previous fundamental theory. Many important fields which have been unexplored so far, are outlined briefly.

1. INTRODUCTION

In this paper, established results for time series in m dimensions are reviewed and further needed researches are considered. Other systems of spatial-time formulations are compared to those of time series in m dimensions. Excellent examples of time series in m dimensions are concerned with the characteristics of rivers, lakes, and oceans. Other examples include: the gulf stream; atmospheric characteristics such as the jet stream; pollution of streams; social problems; geographical problems; as well as others in science, industry, economics and business.

2. TWO INTERESTING LONG-MEMORY PROBLEMS

Example 1. N. G. Pisias and T. C. Moore, Jr. (1981) stated: "In the latter part of the Pleistocene, variations in global ice volume have been dominated by an approximate 100,000-year cycle. Analysis of a 2-Myr-long oxygen isotope record (${}_8O^{16}/{}_8O^{18}$) from an equatorial Pacific core indicates this is only true for the last 900,000 years." They state also that, besides this cycle, there are two shorter cycles with periods of 41,000 and 21,000 years. It can be shown that these last two periods are associated with periodic changes in

[1] Institute of Administration and Management, Union College, Schenectady, New York 12308

I. B. MacNeill and G. J. Umphrey (eds.), Time Series and Econometric Modelling, 107–112.

the 41,000 and 21,000-year components of the tilt of the Earth's axis, and in the precession of the equinoxes as predicted by the astronomical theory of the ice ages. The eccentricity of the Earth's orbit varies with a period of 413,000 and 100,000 years. The 100,000-year component in the spectrum of global ice-volume changes is not predicted by simple linear forcing of the Earth's orbital variations; thus the origin of this phenomena is still being investigated. Presently many new cores have been obtained in the north polar and south polar regions. This allows a spatial-time series approach.

Example 2. Another interesting example is concerned with the ages of bristle-cone pines *(Pincus longaeva)* in Arizona, Utah, and California, the oldest living trees—4,500 years. A bristlecone pine, in East Central Nevada, cut down in 1964, exceeded 5,000 years in age while trees now dead had lives exceeding 9,000 years (Hitch, 1982). Their ages, found by counting tree rings, vary according to their locations, thus providing an interesting example to model by spatial time series. The ages of these trees have been used to calibrate carbon dating.

3. GENERAL REMARKS

The first articles discussing stationary spatial time processes (STM) involved moving average (MA) processes, autoregressive (AR) processes and their combination, ARMA processes. Ordinary time series have dimension $m = 0$ since they deal with the time parameter only. Their analysis is important either as time increases (into the future) or as it decreases, thus delving into the past. STM processes are more complicated because, while t has two directions, each spatial dimension may change also in two ways; for $m = 1$, this means 4 possibilities, although some are essentially the same. For $m = 2$ this means 8 possibilities. Models such as

$$z_t = a_t - \theta_1 a_{t-1} - \theta_2 a_{t+1}$$

make very little sense but spatial time models, such as the following, do:

$$z_{x,t} = a_{x,t} - \theta_1 a_{x+1,t-1} - \theta_2 a_{x-1,t-1},$$

or

$$z_{x,t} = a_{x,t} - \theta_1 a_{x+1,t-1} - \theta_2 a_{x,t-1} - \theta_3 a_{x-1,t-1}.$$

A simple moving average model, such as the following,

$$z_{x,t} = -\theta_1 a_{x,t-1} - \theta_2 a_{x-1,t-1} + a_{x,t}, \quad -\infty < t < t_0$$

may be written with backward shift operators, as follows:

$$z_{x,t} = (1 - \theta_1 B_t - \theta_2 B_x B_t)\, a_{x,t} = \theta\,(B_x, B_t)\, a_{x,t}, \tag{3.1}$$

where $\theta(B_x, B_t)$ represents the generating function. Then, if $|\ \theta_1\ | + |\ \theta_2\ | < 1$, it can be shown that

$$\sigma_z^2 = \sigma_a^2(1 + \theta_1^2 + \theta_2^2),$$

$$\gamma_{01} = -\theta_1 \sigma_a^2,$$

$$\gamma_{10} = \theta_1 \theta_2 \sigma_a^2,$$

and

$$\gamma_{11} = -\theta_2 \sigma_a^2,$$

where γ_{rs} is the covariance function. In all models considered here it is assumed for convenience that $E(z_{x,t}) = 0$. Also, it can be noted that

$$\theta_2 = \{-\rho_{11} \mp \rho_{11}[1 - 4(\rho_{01}^2 + \rho_{11}^2)]^{1/2}\}[2(\rho_{01}^2 + \rho_{11}^2)]^{-1}.$$

Thus, the requirement that

$$1 - 4(\rho_{01}^2 + \rho_{11}^2) > 0,$$

proves the restriction

$$\rho_{01}^2 + \rho_{11}^2 < 1/4$$

and

$$-\rho_{10}^{-1} < \rho_{11}^{-1} + \rho_{01}^{-1} < \rho_{10}^{-1} \qquad \text{for } 0 < \rho_{10} < 1,$$

or

$$\rho_{10}^{-1} < \rho_{11}^{-1} + \rho_{01}^{-1} < -\rho_{10}^{-1} \qquad \text{for } -1 < \rho_{10} < 0.$$

The only possible values of $\rho_{01}, \rho_{01}, \rho_{11}$ are those in the intersection of the above restrictions.

The coefficients of the corresponding AR model are found by inversion of $a_{x,t} = \{\theta^{-1}(B_x B_t)\} z_{x,t}$. Estimation, simulation, confidence limits for forecasting θ_i, and the power spectrum are given by Aroian (1985) and Voss *et al.* (1980).

The autocovariance generating function for the process

$$z_{x,t} = \varepsilon_{x,t} - \theta_1 \varepsilon_{x,t+1} - \theta_2 \varepsilon_{x+1,t+1},$$

or

$$z_{x,t} = (1 - \theta_1 F_t - \theta_2 F_x F_t)\varepsilon_{x,t},$$

is

$$\Gamma(B_x, B_t) = \sigma_\varepsilon^2 \theta(B_x, B_t)\theta(F_x, F_t).$$

This is the backward model for time and space with the same values of θ_1 and θ_2 as given by (3.1).

Now, consider a backward-forward model in x:

$$z_{x,t} = a_{x,t} - \theta_1 a_{x+1,t-1} - \theta_2 a_{x-1,t-1},$$

with

$$\gamma_{\ell,k} = E(z_{x,t} z_{x-\ell,t-k})$$

representing the autocovariance between $z_{x,t}$ and $z_{x-\ell,t-k}$. The autocovariance generating function $\Gamma(B_x, B_t)$ of such an MA process equals $\sigma_a^2 \theta(B_x, B_t)\theta(F_x, F_t)$, and $\gamma_{\ell,k}$ is the coefficient of both $B_x^\ell B_t^k$ and $B_x^\ell B_t^{-k}$. This result holds also for models where the characteristic function is $\theta(F_x, B_t)$ with autocovariance generating function given by $\sigma_a^2 \theta(F_x, B_t)\theta(B_x, F_x)$; $\Gamma_{\ell,k}$ is the coefficient of both $F_x^\ell B_t^k$ and $B_x^\ell F_t^k$.

Hence

$$\Gamma(F_x, B_t) = \sigma_a^2(1 - \theta_1 F_x B_t - \theta_2 B_x B_t)(1 - \theta_1 B_x F_t - \theta_2 F_x F_t),$$

$$\sigma_z^2 = \sigma_a^2(1 + \theta_1^2 + \theta_2^2),$$

$$\gamma_{20} = \theta_1\theta_2, \gamma_{11} = -\theta_2$$

and

$$\gamma_{1-1} = -\theta_1.$$

For details regarding the ARMA model see the preceding references.

Multivariate models have been proposed by Aroian (1985), but results are quite limited; this is a wonderful area for future research. The same thing is true for forecasting when proceeding with the vector x backward and forward in all directions and combining this with time either backward or forward. Another problem area is non-stationary series. A regression approach appears feasible both for AR, MA, and ARMA models whether processes are stationary or non-stationary. However, the model must be correctly specified for least squares approximations to be effective. More work should be done on maximum likelihood, although approximate results have been obtained (Aroian, 1985). Exact results are possible but the expenditure of effort surely should be directed to other more fruitful aspects. The applications field is wide open, and it is hoped that this area may go forward. For present and past applications see Aroian (1985). What about the work of others in this field? Haugh (1985) has analyzed the connections between these results and those of Pfeifer and Deutsch (1980), and shows that their model is a subset of time series in m dimensions. Bennett's models (1979) are essentially the same as those of Aroian (1980, 1985) with a quite different mathematical approach. D. S. Stoffer (1985) generalized the work of Pfeifer and Deutsch (1980) using modified Kalman smoothed estimates and mentioned also the work of Larimore (1977).

ACKNOWLEDGMENT

The author wishes to thank Roger Yetzer and Professor Herman Zimmerman for providing him with information in this example.

REFERENCES

Aroian, L. A. (1980), "Time series in m dimensions, definitions, problems, and prospects." *Communications in Statistics B, Simulation and Computation* **5**, 453–465.

Aroian, L. A. (1985), "Time series in M dimensions: past, present, and future." In *Time Series Analysis: Theory and Practice 6*, ed. O. D. Anderson, J. K. Ord, and E. A. Robinson, pp. 241–261. Amsterdam: Elsevier Science.

Bennett, R. J. (1979), *Spatial Time Series*. London: Pion.

Haugh, L. D. (1984), "An introductory overview of some recent approaches to modeling spatial time series." In *Time Series Analysis: Theory and Practice 5*, ed. O. D. Anderson, pp. 287–301. Amsterdam: Elsevier Science.

Hitch, C. J. (1982), "Dendrochronology and serendipity." *American Scientist* **70**, 300–305.

Larimore, W. E. (1977), "Statistical inference on stationary random fields." *Proceedings of the IEEE* **65**, 961–970.

Perry, R. J., and L. A. Aroian (1985), "Autoregressive models, M= 1, theory and examples." In *Time Series Analysis: Theory and Practice 6*, ed. O. D. Anderson, J. K. Ord, and E. A. Robinson, pp. 263–271. Amsterdam: Elsevier Science.

Pfeifer, P. E., and S. Deutsch (1980), "A three-stage iterative procedure for space-time modeling." *Technometrics* **22**, 35–47.

Pisias, N. G., and T. C. Moore, Jr. (1981), "The evolution of Pleistocene climate: a time series approach." *Earth and Planetary Science Letters* **52**, 450–458.

Stoffer, D. S. (1985), "Maximum likelihood fitting of starmax models to incomplete space-time series data." In *Time Series Analysis: Theory and Practice 6*, ed. O. D. Anderson, J. K. Ord, and E. A. Robinson, pp. 283–296. Amsterdam: Elsevier Science.

Voss, D. A., C. A. Oprian, and L. A. Aroian (1980), "Moving average models, time series in m dimensions." *Communications in Statistics B, Simulation and Computation* **5**, 467–489.

Yetzer, Roger (1985), Ph.D. thesis in progress, Union College, Schenectady, New York.

B. C. Sutradhar, [1] I. B. MacNeill [2] and H. F. Sahrmann [3]

TIME SERIES VALUED EXPERIMENTAL DESIGNS: ONE-WAY ANALYSIS OF VARIANCE WITH AUTOCORRELATED ERRORS

ABSTRACT

A methodology is developed for analysing factorial designs when the observations at a particular treatment combination form a time series. Maximum likelihood estimators of treatment effects and of time series parameters are found. Analogues of the standard F-ratios are proposed for testing treatment effects. A detailed discussion is given for the one-way classification with error variables generated by AR(1) processes. A simulation study for the case of two treatments is presented.

1. INTRODUCTION

Several models related to, but different from, that considered in the sequel have been discussed in the literature. Azzalini (1981) considered the model, $y_i(t) - \mu = \phi\{y_i(t-1) - \mu\} + a_i(t)$, where: $y_i(t)$ is the tth observation $t = 1, 2, \ldots, n_i$ of the ith time series $i = 1, \ldots, k$; ϕ is the AR(1) parameter for each time series; and $a_i(t)$ are the independent error variables having identical normal distributions with zero mean and variance σ^2. Azzalini dealt mainly with the estimation of the parameters μ, ϕ, and σ^2 of the above model with special emphasis on the asymptotic results when $k \to \infty$ and n_i is fixed, say $n_i = n$. Azzalini (1984) extended this model as follows:

$$y_i(t) = \mu_i + \xi_t + z_i(t), \qquad i = 1, \ldots, k, \quad t = 1, \ldots, n, \tag{1.1}$$

[1] Department of Mathematics and Statistics, Memorial University of Newfoundland, St. John's, Newfoundland A1C 5S7

[2] Department of Statistical and Actuarial Sciences, The University of Western Ontario, London, Ontario N6A 5B9

[3] Math Services Laboratory, 3M Canada Inc., P.O. Box 5757, London, Ontario N6A 4T1

I. B. MacNeill and G. J. Umphrey (eds.), Time Series and Econometric Modelling, 113–129.

with $z_i(t) = \phi z_i(t-1) + a_i(t)$. In equation (1.1): $y_i(t)$ is as before; μ_i is the random effect of the ith subject or the effect due to the ith time series such that $\mu_i \sim N(x_i'\beta, \sigma_\mu^2)$, where x_i is the p-dimensional vector of covariates and β is a $p \times 1$ vector of unknown parameters; ξ_t is the time effect due to non-stationarity of the series; and $z_i(t)$ follows an AR(1) process. If $p = 1$, Azzalini's (1984) scheme is a random effects two-way classification with autocorrelated errors. Andersen *et al.* (1981) studied a model similar to (1.1) but dealt mainly with a two-way fixed effects model with correlated errors. In Azzalini's notation, Andersen *et al.* (1981) considered μ_i as the ith row effect and ξ_t as the tth column or time effect. Box (1954b) also studied a model similar to that of Andersen *et al.*

In each of the above articles, time has been considered as a specific factor. More specifically, Andersen *et al.* (1981) applied their theoretical results to clinical trial data. They analyzed data on the plasma citrate concentration of $k = 10$ subjects measured at $n = 14$ equally spaced time points, to detect whether the plasma citrate concentration changes during the day. In this application one notes that time is a specific factor. Azzalini (1984) also considered the same numerical example.

However, in many situations where an investigation is repeated over time on physically independent material, and where external conditions can be treated as random, it may be sensible to treat time as a non-specific factor (see Cox, 1984). For example, the use of automated data acquisition equipment may make it possible to obtain many observations on the same treatment combination, but with only a small time interval between consecutive observations. The motivation for the formulation of the model considered below and for the concomitant analysis was a process control problem in which it was expensive to change the process parameters, but in which it was possible to make many observations in a short period of time for a fixed set of parameters. These observations formed a time series characterized by a high degree of correlation among contiguous observations. In other cases where data are collected over time it may not be clear that the white noise assumption is valid; one suspects for the most part these cases are analyzed routinely by ANOVA methods without challenging the independence assumption.

At this point, it may be useful to comment on the use of the regression approach to analysis of variance with ARMA (p,q) error structure. Berndt and Savin (1975) have discussed Wald, likelihood ratio, Lagrange multiplier and max-root tests for testing the linear hypothesis in the multivariate linear regression model. However, they have shown that these tests based on exact distributions conflict with each other when applied to a given data set. In a later paper, Berndt and Savin (1977) showed that even in the asymptotic case, the Wald, likelihood ratio and Lagrange multiplier tests yield conflicting inferences. Rothenberg (1984) has suggested that, under many regularity conditions on the behaviour of the error covariance matrix and the coefficient matrix of the linear regression model, Edgeworth-corrected critical values may be used for the above three tests. He claims that his size-adjusted tests do not conflict in the case of a one-dimensional hypothesis; for example, in testing $\alpha_a = \alpha_2$, where for $k = 2$, α_1 and α_2 are the treatment effects. This is because, in such cases, all three size-adjusted tests appear to have the same approximate power function. However, these tests fail to give a unique inference for multidimensional hypotheses; for example, in testing $\alpha_1 = \alpha_2 = \alpha_3$ or $\alpha_1 = \alpha_2 = \alpha_3 = \alpha_4$, where α's are the treatment effects. Unlike the regression procedures discussed above, the methods given in the sequel provide an exact analysis for testing hypotheses involving treatment effects. The test in the present approach is computationally simpler than the tests used in the regression approach, and the proposed method of analysis is suitable for testing one-dimensional as well as multidimensional hypotheses. Furthermore, in general, one is also concerned about the inference from the size-adjusted test as it is based on approximations of different types at many stages (see Rothenberg, 1984, for details). However, a detailed comparison between the present approach and the regression approach discussed in the econometric literature is beyond the scope of this paper.

The sequel develops the classical, as opposed to the regression, approach to ANOVA for time series valued experimental designs in which time is a non-specific factor. The analysis is carried out in the context of a one-way classification with error variables forming an AR(1) time series, and with autoregressive parameters of all series assumed equal. The methodology may be generalized to more complex factorial designs with ARMA(p, q) error structure, where the autoregressive and/or moving average parameters of all series may or may not be equal. Distribution theory is discussed in Section 4 for the case where the ARMA parameters are known. In the final section, tables of the 5% and 1% values of the proposed test statistic are provided by a simulation study for AR(1) residuals and two treatments.

2. MODEL FOR TIME SERIES VALUED EXPERIMENTAL DESIGNS

The fixed-effects model with time series valued error variables is considered below; more precisely,

$$Y_i(t) = \mu + \alpha_i + z_i(t), \tag{2.1}$$

where

$$\phi_i(B)\{z_i(t)\} = \theta_i(B)\{a_i(t)\}, \quad i = 1, 2, \ldots, k, t = 1, 2, \ldots, n,$$

and where $Y_i(t)$ is the observation at time t due to the ith treatment, μ is the overall mean effect, α_i is the effect of the ith treatment, and $z_i(t)$ is a component of an ARMA(p,q) process. The notation of Box and Jenkins (1976) will be used; that is:

$$\phi_i(B) = (1 - \phi_{1i}B - \phi_{2i}B^2 - \cdots - \phi_{pi}B^p)$$

and

$$\theta_i(B) = (1 - \theta_{1i}B - \theta_{2i}B^2 - \cdots - \theta_{qi}B^q),$$

where B denotes the backshift operator and where $\phi(B)$ and $\theta(B)$ have their zeros outside the unit circle. The $a_i(t)$ are independently and identically distributed as $N(0, \sigma_a^2)$.

We consider in detail the AR(1) case, although, as indicated above, the methodology may be extended to more complex ARMA models. Hence

$$Y_i(t) = \mu + \alpha_i + z_i(t), \tag{2.2}$$

with

$$\phi_i(B)\{z_i(t)\} = a_i(t),$$

where $\phi_i(B) = (1 - \phi_{1i}B)$, and $|\ \phi_{1i}\ | < 1$. In this model ϕ_{1i} denotes the AR(1) parameter for observations derived from the ith treatment. Data collected in accordance with model (2.2) may be graphed as in Figure 1.

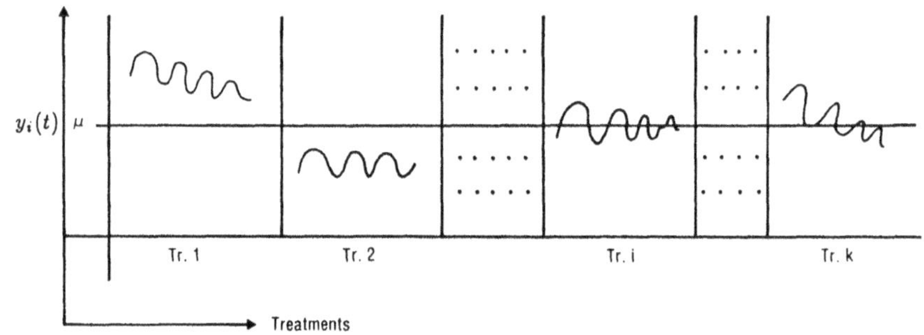

Figure 1. *Example of treatment time series traces.*

Analogous to, and in generalization of standard ANOVA definitions, we define a particular weighted sum of treatment effects to be zero; that is,

$$\frac{\sum_{i=1}^{k}(1 - \phi_{1i})^2\alpha_i}{\sum_{i=1}^{k}(1 - \phi_{1i})^2} = 0. \tag{2.3}$$

Although with this general definition one may estimate the parameters of model (2.2), in the following discussion for the sake of simplicity, we shall restrict ourselves to the case where $\phi_{1i} = \phi$ for all i. In many practical cases, it may be reasonable to assume that the autoregressive parameters are equal for all the series (see Azzalini, 1981, 1984; Andersen *et al.*, 1981).

From model (2.2) we note that

$$\begin{aligned} a_i(t) &= z_i(t) - \phi_{1i} z_i(t-1) \\ &= \{Y_i(t) - \mu - \alpha_i\} - \phi_{1i}\{Y_i(t-1) - \mu - \alpha_i\}. \end{aligned}$$

Parameters may then be estimated by minimizing the conditional sum of squares, S, where

$$\begin{aligned} S &= \sum_{i=1}^{k}\sum_{t=2}^{n} a_i^2(t) \\ &= \sum_{i=1}^{k}\sum_{t=2}^{n} [(Y_i(t) - m_i) - \phi\{Y_i(t-1) - m_i\}]^2, \end{aligned}$$

with $m_i = \mu + \alpha_i$. One may now derive the normal equations and invoke (2.3), to obtain the following equations for estimation of $m_1, \ldots, m_k$ and ϕ:

$$m_i = \frac{\sum_{t=2}^{n}\{Y_i(t) - \phi Y_i(t-1)\}}{(n-1)(1-\phi)}, \qquad i = 1, \ldots, k,$$

and

$$\phi = \frac{\sum_{i=1}^{k}\sum_{t=2}^{n}\{Y_i(t) - m_i\}\{Y_i(t-1) - m_i\}}{\sum_{i=1}^{k}\sum_{t=2}^{n}\{Y_i(t) - m_i\}^2}.$$

The Newton-Raphson iterative formula for obtaining an explicit solution is as follows:

$$\begin{pmatrix} \hat{m}_1 \\ \vdots \\ \hat{m}_i \\ \vdots \\ \hat{m}_k \\ \hat{\phi}_1 \end{pmatrix} = \begin{pmatrix} \hat{m}_1 \\ \vdots \\ \hat{m}_i \\ \vdots \\ \hat{m}_k \\ \hat{\phi}_1 \end{pmatrix} - \begin{pmatrix} F_{11} & \cdots & F_{1i} & \cdots & F_{1k} & F_{1\phi} \\ \vdots & & \vdots & & \vdots & \vdots \\ F_{i1} & \cdots & F_{ii} & \cdots & F_{ik} & F_{i\phi} \\ \vdots & & \vdots & & \vdots & \vdots \\ F_{k1} & \cdots & F_{ki} & \cdots & F_{kk} & F_{k\phi} \\ G_1 & \cdots & G_i & \cdots & G_k & G_\phi \end{pmatrix}^{-1} \begin{pmatrix} F_1 \\ \vdots \\ F_i \\ \vdots \\ F_k \\ G \end{pmatrix}, \quad (2.4)$$

where

$$F_i = (n-1)m_i(1-\phi) - \sum_{t=2}^{n}\{Y_i(t) - \phi Y_i(t-1)\},$$

$$G = \phi\sum_{i=1}^{k}\sum_{t=2}^{n}\{Y_i(t) - m_i\}^2 - \sum_{i=1}^{k}\sum_{t=2}^{n}\{(Y_i(t) - m_i\}\{Y_i(t-1) - m_i)\},$$

$$F_{ii} = \frac{\partial F_i}{\partial m_i} = (n-1)(1-\phi),$$

$$F_{ij} = \frac{\partial F_i}{\partial m_j} = 0, \quad i \neq j,$$

$$F_{i\phi} = \frac{\partial F_i}{\partial \phi} = -m_i(n-1) + \sum_{t=2}^{n} Y_i(t-1),$$

$$G_i = \frac{\partial G}{\partial m_i} = 2(n-1)m_i(\phi-1) + \sum_{t=2}^{n}\{Y_i(t) + (1-2\phi)Y_i(t-1)\},$$

and

$$G_\phi = \frac{\partial G}{\partial \phi} = \sum_{i=1}^{k}\sum_{t=2}^{n}\{Y_i(t-1) - m_i\}^2.$$

We remark here that, for the case $\phi_{1i} = \phi$, equation (2.3) reduces to $\sum \alpha_i = 0$, which is the relationship among treatment effects assumed in standard ANOVA; this condition is used in solving the normal equations.

Since the error variables have been assumed to be independently and identically distributed as $N(0, \sigma_a^2)$, σ_a^2 can be estimated by conditional maximum likelihood as follows:

$$\hat{\sigma}_a^2 = \frac{1}{k(n-1)}\sum_{i=1}^{k}\sum_{t=2}^{n}[(Y_i(t) - \hat{m}) - \hat{\phi}\{Y_i(t-1) - \hat{m}\}]^2. \tag{2.5}$$

3. TESTS FOR TREATMENT EFFECTS

For the usual ANOVA model, viz, $Y_i(t) = \mu + \alpha_i + a_i(t)$ with $a_i(t)$ are i.i.d. $N(0, \sigma_a^2)$, one tests the null hypothesis, $H_0: \ \alpha_1 = \alpha_2 = \cdots = \alpha_k = 0$, against the alternative $H_A: \ \alpha_i \neq 0$, for some i, by using the classical F-statistic; namely,

$$F = \frac{kn(n-1)\sum_{i=1}^{k}(\overline{Y}_{i\cdot} - \overline{Y}_{\cdot\cdot})^2}{(k-1)\sum_{i=1}^{k}\sum_{t=1}^{n}(Y_i(t) - \overline{Y}_{i\cdot})^2}, \tag{3.1}$$

where $\overline{Y}_{i\cdot} = \sum_{t=1}^{n} Y_i(t)/n$ and $\overline{Y}_{\cdot\cdot} = \sum_{i=1}^{k}\sum_{t=1}^{n} Y_i(t)/(nk)$. For model (2.2) this statistic is inappropriate, since the dependence among observations implied by the model alters the amount of information provided by the observations. In the case of high positive correlation, as would be the case for ϕ near unity, a new observation will provide very little new information. This radically alters the expected values of the components of the F-ratio.

Also, in contrast to standard ANOVA, the treatment sum of squares and the error sum of squares are not independent, except for the case $\phi_{1i} = 0$. For these reasons the usual critical points obtained from the F-tables are invalid. The following analysis characterizes the degree of invalidity and indicates that in cases of large correlation between contiguous observations, the standard F-test is in substantial error.

The purpose of this paper is to provide analogous statistics which fit into the standard ANOVA paradigm, but which take into account the effects of correlation on the expected values of the sums of squares, particularly when the errors form an AR(1) process. In the sequel, we assume the autoregressive parameter is the same value for all series, and adopt the usual ANOVA notation.

Let $Y_i(t)$ be the observation at time t which is due to the ith treatment and which is generated as follows:

$$Y_i(t) = \mu + \alpha_i + z_i(t), \tag{3.2}$$

with

$$z_i(t) = \phi z_i(t-1) + a_i(t),$$

where $a_i(t)$ are i.i.d. $N(0, \sigma_a^2)$. Then the following statistic is proposed for testing H_0 versus H_A:

$$F^* = \frac{kn(n-1)}{(k-1)} \frac{\sum_{i=1}^{k}(\overline{Y}_{i\cdot} - \overline{Y}_{\cdot\cdot})^2}{\sum_{i=1}^{k}\sum_{t=1}^{n}(Y_i(t) - \overline{Y}_{i\cdot})^2} \frac{c_2(\hat{\phi})}{c_1(\hat{\phi})}, \tag{3.3}$$

where

$$c_1(\phi) = \frac{1}{(1-\phi)^2}\left[1 - \frac{2\phi(1-\phi^n)}{n(1-\phi^2)}\right]$$

and

$$c_2(\phi) = \frac{1}{n-1}\left[\frac{n}{1-\phi^2} - c_1(\phi)\right].$$

H_0 is rejected for large values of F^*.

We now examine the effects of autocorrelation on the sums of squares. Since $\sum \alpha_i = 0$ and

$$\overline{Y}_{i\cdot} = \mu + \alpha_i + \sum_{t=1}^{n} z_i(t)/n,$$

then

$$\sum_{i=1}^{k} \overline{Y}_{i\cdot} = k\mu + \sum_{i=1}^{k}\sum_{t=1}^{n} z_i(t)/n.$$

Because the $a_i(t)$'s are independently and identically distributed as $N(0, \sigma_a^2)$, it follows that

$$k^{-1}E\left[\left(\sum_{i=1}^{k}\overline{Y}_{i\cdot}\right)^2\right] = k\mu^2 + \frac{1}{n^2k}\sum_{i=1}^{k}\sum_{t=1}^{n}\sum_{t'=1}^{n}E(z_i(t)z_i(t'))$$
$$= k\mu^2 + \sigma_a^2 c_1(\phi)/n, \qquad (3.4)$$

where $c_1(\phi)$ is given by (3.3). Also, similar calculations show that

$$E\left[\sum_{i=1}^{k}\overline{Y}_{i\cdot}^2\right] = k\mu^2 + \sum_{i=1}^{k}\alpha_i^2 + \sigma_a^2 c_1(\phi)/n. \qquad (3.5)$$

Hence, from (3.4) and (3.5), one obtains for the treatment sum of squares, $TrSS$,

$$E(TrSS) = E\left\{n\sum_{i=1}^{k}(\overline{Y}_{i\cdot} - \overline{Y}_{\cdot\cdot})^2\right\}$$
$$= n\sum_{i=1}^{k}\alpha_i^2 + \sigma_a^2(k-1)c_1(\phi).$$

Therefore, under H_0,

$$E\left\{n\sum_{i=1}^{k}(\overline{Y}_{i\cdot} - \overline{Y}_{\cdot\cdot})^2/\sigma_a^2\right\} = (k-1)c_1(\phi). \qquad (3.6)$$

The expression $(k-1)c_1(\phi)$ will be referred to as the "degrees of freedom" (d.f.) for $TrSS$.

Now consider the error sum of squares, RSS. Equation (3.2) implies that

$$E\left[\sum_{i=1}^{k}\sum_{t=1}^{n}\{Y_i(t)\}^2\right] = kn\mu^2 + n\sum_{i=1}^{k}\alpha_i^2 + kn\sigma_a^2/(1-\phi^2). \qquad (3.7)$$

Hence (3.5) and (3.7) yield

$$E(RSS) = E\left\{\sum_{i=1}^{k}\sum_{t=1}^{n}\{Y_i(t) - \overline{Y}_{i\cdot}\}^2\right\}$$
$$= k(n-1)\sigma_a^2 c_2(\phi), \qquad (3.8)$$

where $c_2(\phi)$ is given by (3.3). Hence, under both H_0 and H_A, $E\{RSS/\sigma_a^2\} = k(n-1)c_2(\phi)$. The expression $k(n-1)c_2(\phi)$ will be referred to as the "degrees of freedom" for the error sum of squares.

Thus,

$$E[RSS/\{k(n-1)c_2(\phi)\}] = \sigma_a^2$$

and

$$E[TrSS/\{(k-1)c_1(\phi)\}] = \sigma_a^2 + \frac{n}{(k-1)c_1(\phi)} \sum_{i=1}^{k} \alpha_i^2.$$

These expected "mean squares" suggest that H_0 be rejected when F^* is "large", where "large" is determined from distributional results.

It follows from (3.3) that the effect produced by autocorrelated errors is embodied in the ratio $c_2(\phi)/c_1(\phi)$ which may be estimated as follows:

$$\frac{c_2(\hat{\phi})}{c_1(\hat{\phi})} = \frac{n(1-\hat{\phi})\{1-\hat{\phi}(n+1)/(n-1)\} + 2\hat{\phi}(1-\hat{\phi}^n)/(n-1)}{\{n(1-\hat{\phi}^2) - 2\hat{\phi}(1-\hat{\phi}^n)\}}.$$

For $\phi = -0.9(0.1)0.9$; $c_1(\phi)$, $c_2(\phi)$ and $c_2(\phi)/c_1(\phi)$ are graphed in Figures 2 and 3 for $n = 25$ and 75 respectively. The case of independent observations is represented by $\phi = 0$. The functions $c_1(\phi)$ and $c_2(\phi)$ show the effects of autocorrelated errors on the "degrees of freedom" for the treatment sum of squares and for the error sum of squares respectively. The graph of $c_2(\phi)/c_1(\phi)$ shows how the usual F statistic (for the independent case) is affected by autocorrelated errors. Although $c_2(\phi)/c_1(\phi)$ depends upon sample size, for a wide range of n, MacNeill *et al.* (1985) indicated that the ratio may be approximated by $(1-\phi)/(1+\phi)$ for $|\phi| < 0.9$.

4. DISTRIBUTION OF THE TEST STATISTIC

The test statistic F^* may be written in the form

$$F^* = d\sum_{i=1}^{k}(\overline{Y}_{i\cdot} - \overline{Y}_{\cdot\cdot})^2 \Big/ \left[\sum_{i=1}^{k}\sum_{t=1}^{n}\{Y_i(t) - \overline{Y}_{i\cdot}\}^2\right], \tag{4.1}$$

where

$$d = \{kn(n-1)/(k-1)\}\{c_2(\phi)/c_1(\phi)\}.$$

We now examine the distribution of F^* where d is assumed known. Hence, we consider

$$\begin{aligned}\Pr(F^* \le f) &= \Pr\left[\left\{\sum_{i=1}^{k}(\overline{Y}_{i\cdot} - \overline{Y}_{\cdot\cdot})^2 - (f/d)\sum_{i=1}^{k}\sum_{t=1}^{n}(Y_i(t) - \overline{Y}_{i\cdot})^2\right\} \le 0\right] \\ &= \Pr[\{Q_1 - (f/d)Q_2\} \le 0], \end{aligned} \tag{4.2}$$

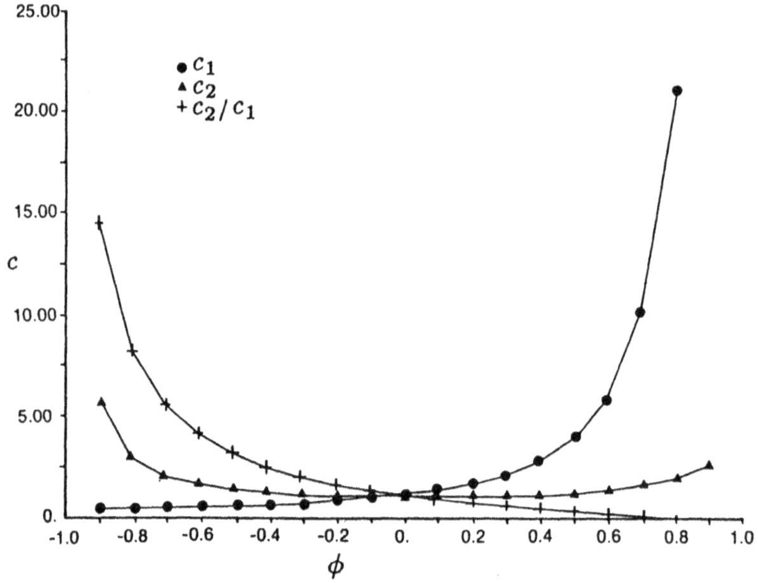

Figure 2. *c_1, c_2, and c_2/c_1 for $n = 25$. One point outside range.*

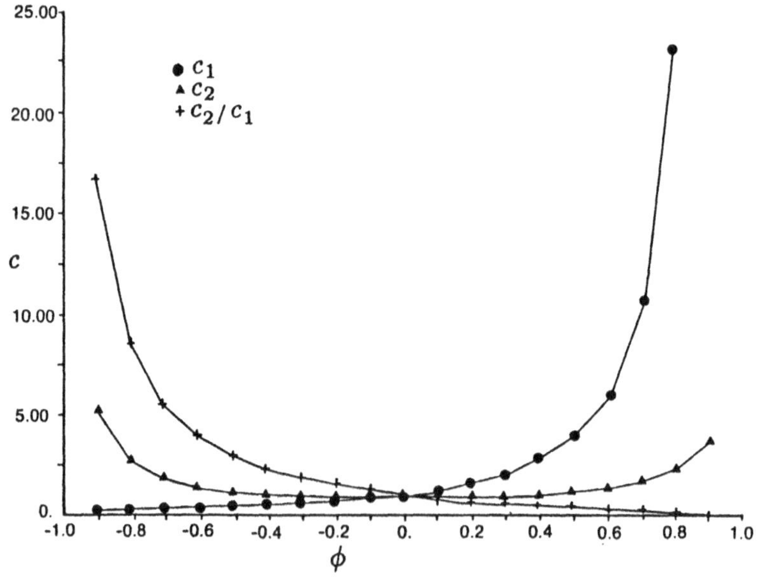

Figure 3. *c_1, c_2, and c_2/c_1 for $n = 75$. One point outside range.*

where

$$Q_1 = \sum_{i=1}^{k} (\overline{Y}_{i\cdot} - \overline{Y}_{\cdot\cdot})^2$$

and

$$Q_2 = \sum_{i=1}^{k} \sum_{t=1}^{n} \{Y_i(t) - \overline{Y}_{i\cdot}\}^2.$$

Thus the problem is one of determining the distribution of the difference of two quadratic forms in the same variables.

We consider equivalent formulations for Q_1 and Q_2 in terms of these same variables. The quadratic form Q_1 may be written as follows:

$$Q_1 = \overline{\mathbf{Y}}'_{i\cdot} \mathbf{A} \overline{\mathbf{Y}}_{i\cdot}, \tag{4.3}$$

where

$$\overline{\mathbf{Y}}'_{i\cdot} = (\overline{Y}_{1\cdot}, \overline{Y}_{2\cdot}, \ldots, \overline{Y}_{k\cdot})$$

and

$$\mathbf{A} = \begin{bmatrix} (1 - 1/k) & -1/k & \cdots & -1/k \\ -1/k & (1 - 1/k) & & -1/k \\ \vdots & & \ddots & \\ -1/k & -1/k & & (1 - 1/k) \end{bmatrix}.$$

We rewrite Q_1 as follows:

$$Q_1 = \mathbf{Y}^{*\prime} \left(\mathbf{I}_k \otimes \frac{1}{n} \mathbf{1}_n \right) \mathbf{A} \left(\mathbf{I}_k \otimes \frac{1}{n} \mathbf{1}'_n \right) \mathbf{Y}^*,$$

where

$$\mathbf{Y}^{*\prime} = (Y_{11}, \ldots, Y_{1n}, Y_{21}, \ldots, Y_{2n}, \ldots, Y_{k1}, \ldots, Y_{kn}),$$
$$\mathbf{1}'_n = (1, 1, \ldots, 1)_{1 \times n},$$

$\mathbf{I}_k$ is an identity matrix of order k, and $\otimes$ denotes the Kronecker product. Q_1 can be written in more compact form as follows:

$$Q_1 = \mathbf{Y}^{*\prime} \{ (\mathbf{I}_k - k^{-1} \mathbf{U}_k) \otimes \mathbf{U}_n \} \mathbf{Y}^* / n^2, \tag{4.4}$$

where $\mathbf{U} = \mathbf{1}\mathbf{1}'$.

We next consider the quadratic form in the denominator, Q_2. First we observe that

$$Q_2 = \sum_{i=1}^{k} \mathbf{Y}'_i \mathbf{B}\, \mathbf{Y}_i,$$

where $\mathbf{Y}_i' = (Y_{i1}, \ldots, Y_{in})$ and $\mathbf{B}$ is the $n \times n$ matrix given by

$$\mathbf{B} = \begin{bmatrix} (1-1/n) & -1/n & \cdots & -1/n \\ -1/n & (1-1/n) & \cdots & -1/n \\ \vdots & \vdots & & \vdots \\ -1/n & -1/n & \cdots & (1-1/n) \end{bmatrix}.$$

Q_2 can be written in compact form as follows:

$$Q_2 = \mathbf{Y}^{*\prime}\{\mathbf{I}_k \otimes (\mathbf{I}_n - n^{-1}\mathbf{U}_n)\}\mathbf{Y}^*, \tag{4.5}$$

where $\mathbf{Y}^*$ and $\mathbf{U}_n$ are defined as in (4.4).

Now, we rewrite (4.2) using the results embodied in expressions (4.5) and (4.6):

$$\begin{aligned} &\Pr[\{Q_1 - (f/d)Q_2\} \le 0] \\ &\quad = \Pr[\{\mathbf{Y}^*((\mathbf{I}_k - k^{-1}\mathbf{U}_k) \otimes \mathbf{U}_n)\mathbf{Y}^*/n^2 \\ &\qquad - \mathbf{Y}^{*\prime}(\mathbf{I}_k \otimes (\mathbf{I}_n - n^{-1}\mathbf{U}_n))(f/d)\mathbf{Y}^*\} \le 0] \\ &\quad = \Pr[\mathbf{Y}^{*\prime}\mathbf{R}\mathbf{Y}^* \ge 0], \end{aligned} \tag{4.6}$$

where

$$\mathbf{R} = \mathbf{M}_f - \mathbf{P},$$

with

$$\mathbf{M}_f = (f/d)\{\mathbf{I}_k \otimes (\mathbf{I}_n - n^{-1}\mathbf{U}_n)\}$$

and

$$\mathbf{P} = \{(\mathbf{I}_k - k^{-1}\mathbf{U}_k) \otimes \mathbf{U}_n\}/n^2,$$

where d is given in (4.1). It may be noted that the vector $\mathbf{Y}^*$ in (4.6) has the following distribution:

$$\mathbf{Y}^* = (Y_{11}, \ldots, Y_{1n}, \ldots, Y_{k1}, \ldots, Y_{kn})' \sim N_{kn}(\mathbf{m}, \mathbf{I}_k \otimes \mathbf{\Lambda}),$$

where

$$\mathbf{m} = (m_1\mathbf{1}_n', \ldots, m_i\mathbf{1}_n', \ldots, m_k\mathbf{1}_n')',$$

with $m_i = \mu + \alpha_i$ for $i = 1, \ldots, k$, and

$$\mathbf{\Lambda} = \frac{\sigma_a^2}{1-\phi^2}\begin{bmatrix} 1 & \phi & \phi^2 & \cdots & \phi^{n-1} \\ \phi & 1 & \phi & \cdots & \phi^{n-2} \\ \vdots & \vdots & \vdots & & \vdots \\ \phi^{n-1} & \phi^{n-2} & \phi^{n-3} & \cdots & 1 \end{bmatrix}.$$

Now, consider the transformation $\mathbf{\Lambda}^{-\frac{1}{2}}\mathbf{Y}_i = \mathbf{Z}_i$, where $\mathbf{Z}_i' = (Z_{i1}, \ldots, Z_{in})$. Then the distribution function given by (4.6) reduces to

$$\Pr(\mathbf{Z}^{*\prime}\mathbf{R}_\sigma\mathbf{Z}^* \geq 0), \tag{4.7}$$

where

$$\mathbf{Z}^* \sim N\left[\begin{pmatrix} \mathbf{\Lambda}^{-\frac{1}{2}}(m_1 \otimes \mathbf{1}_n) \\ \vdots \\ \mathbf{\Lambda}^{-\frac{1}{2}}(m_k \otimes \mathbf{1}_n) \end{pmatrix}, \mathbf{I}_{kn}\right],$$

and $\mathbf{R}_\sigma$ is the $kn \times kn$ matrix given by

$$\mathbf{R}_\sigma = \mathbf{M}_{f\sigma} - \mathbf{P}_\sigma,$$

with

$$\mathbf{P}_\sigma = \{(\mathbf{I}_k - k^{-1}\mathbf{U}_k) \otimes \mathbf{\Lambda}^{\frac{1}{2}}\mathbf{U}_n\mathbf{\Lambda}^{\frac{1}{2}}\}/n^2$$

and

$$\mathbf{M}_{f\sigma} = \{\mathbf{I}_k \otimes \mathbf{\Lambda}^{\frac{1}{2}}(\mathbf{1}_n - n^{-1}\mathbf{U}_n)\mathbf{\Lambda}^{\frac{1}{2}}\}(f/d).$$

For a given f, our objective is to derive the distribution of $\mathbf{Z}^{*\prime}\mathbf{R}_\sigma\mathbf{Z}^*$ both under H_0 and H_A. Since $\mathbf{P}_\sigma' = \mathbf{P}_\sigma$ and $\mathbf{M}_{f\sigma}' = \mathbf{M}_{f\sigma}$, it may be observed that $\mathbf{R}_\sigma$ is a symmetric matrix. Suppose the rank of $\mathbf{R}_\sigma$ is $m \leq nk$. Then there exists an orthogonal matrix $\mathbf{\Gamma}$ such that

$$\mathbf{R}_\sigma = \mathbf{\Gamma}\mathbf{\Lambda}_\lambda\mathbf{\Gamma}', \tag{4.8}$$

where

$$\mathbf{\Lambda}_\lambda = \text{diag}(\lambda_1, \lambda_2, \ldots, \lambda_j, \ldots, \lambda_m), \qquad m \leq nk.$$

Therefore, the probability in (4.7) is given by:

$$\begin{aligned} \Pr(\mathbf{Z}^{*\prime}\mathbf{R}_\sigma\mathbf{Z}^* \geq 0) &= \Pr(\mathbf{U}^{*\prime}\mathbf{\Gamma}'\mathbf{R}_\sigma\mathbf{\Gamma}\mathbf{U}^* \geq 0) \\ &= \Pr(\mathbf{U}^{*\prime}\mathbf{\Lambda}_\lambda\mathbf{U}^* \geq 0), \end{aligned} \tag{4.9}$$

where $\mathbf{U}^* \sim N(\boldsymbol{\delta}, \mathbf{I}_{nk})$, and where

$$\boldsymbol{\delta} = \begin{bmatrix} \delta_1 \\ \delta_2 \\ \vdots \\ \delta_{nk} \end{bmatrix} = \mathbf{\Gamma}'\boldsymbol{\theta},$$

with

$$\boldsymbol{\theta} = \begin{bmatrix} \theta_1 \\ \theta_2 \\ \vdots \\ \theta_{nk} \end{bmatrix} = \begin{bmatrix} \mathbf{\Lambda}^{-\frac{1}{2}}(m_1 \otimes \mathbf{1}_n) \\ \vdots \\ \mathbf{\Lambda}^{-\frac{1}{2}}(m_k \otimes \mathbf{1}_n) \end{bmatrix}.$$

Define $Q_f = \sum_{j=1}^{m} \lambda_j u_j^2$. Then the probability given by (4.9) can be written as follows:

$$\Pr\left(Q_f = \sum_{j=1}^{m} \lambda_j u_j^2 \geq 0\right), \qquad m \leq nk, \tag{4.10}$$

where λ_j, $j = 1, \ldots, m$, are the eigenvalues of the matrix $\mathbf{R}_\sigma$ defined by (4.7). Note that no λ_j in (4.10) is zero; however, some are positive and the remainder negative. Moreover, some may be multiple roots. Also note in (4.10) that the u_j^2's are independent and u_j^2 has a non-central chi-square distribution with 1 degree of freedom and non-centrality parameter δ_j^2, where δ_j is given in (4.9). Thus, Q_f in (4.10) can be written as a difference of two linear combinations of independent non-central chi-square distributions, where each linear combination contains only positive coefficients. For the case when U and V are two linear combinations in independent non-central chi-square variables, the theoretical distribution of $T = (U - V)$ has been studied by Press (1966), among others. In view of Press's result, the distribution of Q_f is known and $\Pr\{Q_f \geq 0\}$ in (4.10) can, in theory, be calculated. However, it should be recalled that ϕ has been assumed known, and it should be noted that the theory is awkward to apply for purposes of calculation.

5. SIMULATION STUDY FOR THE CASE OF TWO TREATMENTS WITH AR(1) ERROR VARIABLES

Consider the model given by (2.2). For the case when $\phi_{1i} = \phi$, $i = 1, \ldots, k$, it has been suggested in Section 3 that for testing $H_0 : \alpha_1 = \alpha_2 = \cdots = \alpha_k = 0$ versus $H_A : m_i \neq 0$ for some i, an appropriate test statistic is

$$F^* = \frac{kn(n-1)}{k-1} \frac{c_2(\hat{\phi})}{c_1(\hat{\phi})} \frac{\sum_{i=1}^{k}(\overline{Y}_{i\cdot} - \overline{Y}_{\cdot\cdot})^2}{\sum_{i=1}^{k}\sum_{t=1}^{n}(Y_{it} - \overline{Y}_{i\cdot})^2}.$$

The theoretical probability distribution of this test statistic when ϕ is known has been given in Section 4. For significance testing one needs to know the values of f such that $P(F^* \leq f) = 0.95$ or $P(F^* \leq f) = 0.99$. In the special case, $\phi = 0$ and not estimated, one may use the usual F tables to determine f. But in the present case, calculations of exact probabilities through (4.6) to (4.9) of Section 4, and hence the determination of f for $\phi = -0.9(-.1)0.9$, is cumbersome. Furthermore, ϕ is generally not known.

Therefore, a simulation study was conducted for the two treatment case ($k = 1$), and tables were constructed for the 5% and 1% values of F^* for $\phi = -0.9(0.1)0.9$ for $n = 75$ and $n = 100$. For each set of parameters, 1000 simulations were carried out. The series were simulated, and α_1, α_2,

ϕ and σ_a^2 were estimated by the Newton-Raphson iteration procedure as described in Section 2. The F^* statistic was calculated for each simulation and the 5% and 1% points were estimated. Generally speaking, the estimated percentage points were found to increase in magnitude as $|\phi|$ increased. Hence, percentage values were smoothed over the range of values of the autoregressive parameter. The results are shown in Tables 1 and 2.

Table 1. *Simulation estimates of 5% and 1% points of the distribution of the F^* statistic for two treatments and $AR(1)$ errors: sample size of 75.*

ϕ	$\hat{\phi}$	$F^*_{m(0.05)}$	$F^*_{m(0.01)}$
–0.9	–0.890	4.53	7.82
–0.8	–0.795	4.32	7.40
–0.7	–0.691	4.15	7.06
–0.6	–0.599	4.02	6.82
–0.5	–0.506	3.94	6.66
–0.4	–0.405	3.89	6.59
–0.3	–0.305	3.88	6.61
–0.2	–0.209	3.91	6.72
–0.1	–0.111	3.98	6.91
–0.0	–0.011	4.09	7.19
0.1	0.084	4.23	7.56
0.2	0.179	4.42	8.02
0.3	0.279	4.65	8.56
0.4	0.377	4.92	9.20
0.5	0.472	5.23	9.92
0.6	0.571	5.57	10.72
0.7	0.667	5.96	11.62
0.8	0.763	6.38	12.60
0.9	0.855	6.85	13.67

It should be noted that the effect of large correlation is to reduce the number of "degrees of freedom". It can also be noted that the requirement to estimate ϕ also increases the critical point.

Table 2. *Simulation estimates of 5% and 1% points of the distribution of the F^* statistic for two treatments for AR(1) errors: sample size of 100.*

ϕ	$\hat{\phi}$	$F^*_{m(0.05)}$	$F^*_{m(0.01)}$
-0.9	-0.893	4.13	8.24
-0.8	-0.791	4.00	7.73
-0.7	-0.698	3.89	7.31
-0.6	-0.599	3.81	6.97
-0.5	-0.501	3.76	6.73
-0.4	-0.402	3.74	6.57
-0.3	-0.306	3.74	6.50
-0.2	-0.204	3.77	6.52
-0.1	-0.106	3.83	6.64
0.0	-0.011	3.92	6.83
0.1	0.088	4.04	7.12
0.2	0.188	4.19	7.50
0.3	0.287	4.36	7.97
0.4	0.385	4.56	8.52
0.5	0.481	4.79	9.16
0.6	0.576	5.05	9.90
0.7	0.677	5.34	10.72
0.8	0.770	5.65	11.63
0.9	0.866	6.00	12.63

CONCLUSIONS

Highly automated data acquisition systems make possible the rapid gathering of large numbers of observations in industrial processes. These observations may be densely packed over time with the consequence that they may be highly positively correlated. Tables 1 and 2 and Figures 2 and 3 indicate the dramatic effect of the presence of such autocorrelations. Hence the straightforward use of the standard F-test may be highly misleading. The modifications proposed to the F-test account for autocorrelation in the replications.

ACKNOWLEDGMENTS

This research was partially supported by the Natural Sciences and Engineering Research Council of Canada. The authors would like to thank a referee for suggestions.

REFERENCES

Andersen, A. H., E. B. Jensen, and G. Schou (1981), "Two-way analysis of variance with correlated errors". *International Statistical Review* **49**, 153–167.

Azzalini, A. (1981), "Replicated observations of low order autoregression time series". *Journal of Time Series Analysis* **2**, 63–70.

Azzalini, A. (1984), "Estimation and hypothesis testing for collections of autoregressive time series". *Biometrika* **71**, 85–90.

Berndt, E. R., and N. E. Savin (1975), "Conflict among criteria for testing hypotheses in the multivariate linear regression model". Discussion Paper 75-21 (revised), Department of Economics, University of British Columbia.

Berndt, E. R., and N. E. Savin (1977), "Conflict among criteria for testing hypotheses in the multivariate linear regression model". *Econometrica* **45**, 1263–1277.

Box, G. E. P., and G. M. Jenkins (1976), *Time Series Analysis, Forecasting and Control.* San Francisco: Holden-Day.

Cox, D. R. (1984), "Interaction". *International Statistical Review* **52**, 1–31.

MacNeill, I. B., H. Sahrmann, and B. C. Sutradhar (1985), "Time series valued experimental designs". *International Statistical Institute 45th Session Conference Volume*, 469–470.

Press, S. J. (1966), "Linear combinations of non-central chi-square variates". *Annals of Mathematical Statistics* **37**, 480–487.

Rothenberg, T. J. (1984), "Hypothesis testing in linear models when the error covariance matrix is nonscalar". *Econometrica* **52**, 827–842.

Estela Bee Dagum [1]

MONTHLY VERSUS ANNUAL REVISIONS OF CONCURRENT SEASONALLY ADJUSTED SERIES

1. INTRODUCTION

Statistics Canada's official policy of using concurrent seasonal adjustment was established in 1975; gradually, other foreign statistical agencies followed it. The old practice for seasonally adjusting a current (monthly or quarterly) observation was to apply year-ahead seasonal factors generated from a series that ended in the month of December of the previous year. Since these projected factors were calculated ahead of the actual time they were applied, they didn't take into account the most recent information incorporated into the series. On the other hand, the use of a concurrent seasonal factor to produce a current seasonally adjusted datum implies the use of all the data in the series up to and including the current month's observation.

The main reason for using concurrent instead of seasonal factor forecasts is that the former are subject to smaller revisions as new observations are added to the series. This important result has been confirmed in several empirical studies; see among others, Dagum (1978), Bayer and Wilcox (1981), Kenny and Durbin (1982), McKenzie (1982, 1984) and Dagum and Morry (1984).

There are two sources influencing the size of the revisions of current seasonally adjusted data: (1) differences in the moving averages or linear filters applied to the same observation as later data become available; and (2) the innovations that enter the series with new observations. Ideally one would like to minimize revisions due to filter discrepancies. Two studies by Dagum (1982a,b) have shown that if the current observation is seasonally adjusted using a concurrent seasonal factor instead of a year-ahead factor, the corresponding concurrent linear filter is subject to smaller revisions than any of the year-ahead seasonal filters. The same conclusions have been

[1] Director Time Series Research and Analysis Division, Statistics Canada, Ottawa, Ontario K1A 0T6

I. B. MacNeill and G. J. Umphrey (eds.), Time Series and Econometric Modelling, 131–146.

reached in a recent study by Pierce and McKenzie (1985) from a time series analysis viewpoint.

The use of concurrent seasonal factors for current seasonal adjustment poses the problem of how often should the series be revised. In this regard, Kenny and Durbin (1982) recommended that revisions should be made after one month and thereafter each calendar year. Dagum (1982b) supported these conclusions and furthermore, recommended an additional revision at six months if the seasonal adjustment method is the X-11-ARIMA without ARIMA extrapolation. In this case, the X-11-ARIMA (Dagum, 1980) closely approximates the Census Method II-X-11 version (Shiskin *et al.,* 1967) except for changes in the treatment of outliers and the use of more accurate end weights for the seasonal moving averages.

Recently, Burridge and Wallis (1984) showed that the X-11 filters are not internally consistent in a signal extraction sense. In fact, they observed that the transfer function of the first year revised concurrent filter differed more from that of the symmetric filter (to which it should converge) than the transfer function of the concurrent filter itself. It would appear as if the transition from asymmetric to symmetric filters was not gradual for all frequencies.

This paper deals with the problem of consistency between successive filters in relation with the revision pattern of the concurrent linear filters of X-11-ARIMA and X-11. Section 2 introduces two measures of filter revisions given by the root mean square differences between the frequency response functions of the analysed filters. Section 3 estimates and discusses the time paths of the revision of the concurrent seasonal adjustment filters of X-11 and X-11-ARIMA for consecutive month-spans. Section 4 estimates and analyses the time paths of the monthly and annual revisions of the concurrent and remaining asymmetric filters, and Section 5 gives the main conclusions of this study.

2. MEASURES OF FILTER REVISIONS

Under the assumption of an additive decomposition model and no replacement of extreme values, the seasonally adjusted estimates from X-11-ARIMA with and without ARIMA extrapolations are obtained by the application of a set of moving averages or linear filters. For central or middle observations, say $n+1 \leq t \leq T-n$, the filter is always the *same* and *symmetric* whereas for the remaining n observations on both ends of the series, the filters are *asymmetric* and *different* for each observation.

We can express the seasonally adjusted value, for recent years, from

X-11-ARIMA and X-11 by

$$y_t^{(m)} = \sum_{j=-m}^{n} h_{m,j} x_{t-j} = h^{(m)}(B) x_t, \tag{2.1}$$

where $y_t^{(m)}$ is the seasonally adjusted estimate from a series x_{t-n}, x_{t-n+1}, $\ldots$, $x_t, \ldots, x_{t+m}$; $h_{m,j}$ denotes the moving average weights to be applied to the series and $h^{(m)}(B)$ denotes the corresponding linear filter using the backshift operator B, such that $B^n x_t = x_{t-n}$.

For $m = 0$, $y_t^{(0)}$ is the concurrent seasonally adjusted value and $h^{(0)}(B)$ the corresponding concurrent filter; for $m = 1$, $y_t^{(1)}$ is the first-period (month, quarter) revised seasonally adjusted figure and for $m = n$, $y_t^{(n)}$ is the final seasonally adjusted value in the sense that it is estimated with a symmetric filter $h^{(n)}(B)$, where $h_{n,j} = h_{n,-j}$ for all j. For any two points in time $t + k$, $t + \ell$ $(k < \ell)$, the revision of the seasonally adjusted value is given by

$$r_t(\ell, k) = y_t^{(\ell)} - y_t^{(k)}, \qquad k < \ell. \tag{2.2}$$

This revision reflects: (1) the innovations introduced by the new observations $x_{t+k+1}, x_{t+k+2}, \ldots, x_{t+\ell}$; and (2) the differences between the two asymmetric filters $h^{(\ell)}(B)$ and $h^{(k)}(B)$. If one fixes $k = 0$ and lets ℓ vary from 1 to n, then (2.2) gives a sequence of revisions of the concurrent seasonally adjusted value for different time spans or lags. The *total revision* of the concurrent estimate is obtained for $\ell = n$. If one fixes $\ell = k + 1$ and lets k take values from 0 to $n - 1$, then equation (2.2) gives the sequence of *single period revisions* of each estimated seasonally adjusted value, and in particular, if one starts at $k = 0$ one obtains the $n - 1$ successive single-period revisions of each estimated seasonally adjusted value before it becomes final. If one fixes $\ell = k + 12$ and lets k take values from 0 to $n - 12$ then equation (2.2) gives the sequence of annual revisions. The revisions in which we are interested here are those introduced by filter discrepancies, and these will be studied by looking at the frequency response functions of the corresponding filters.

Equation (2.1) represents a linear system where $y_t^{(m)}$ is the convolution of the input x_t and a sequence of weights $h_{m,j}$ called the *impulse response function* of the filter. The properties of this function can be explored using its Fourier transform which is called the *frequency response function,* defined by,

$$H^{(m)}(\omega) = \sum_{j=-m}^{n} h_{m,j} e^{-i2\pi\omega j}, \qquad 0 \leq \omega \leq 1/2, \tag{2.3}$$

where ω is the frequency in cycles per unit time. Theoretically the frequency response function of a (discrete) filter exists over the interval $[-1/2,\ 1/2]$

but, in practice, it is sufficient to describe it for $\omega \in [0,\ 1/2]$ so that it is completely described in the frequency domain. $H(\omega)$ fully describes the effects of the linear filter on the given input. In general, the frequency response function may be expressed in polar form as follows:

$$H(\omega) = A(\omega) + iB(\omega) = G(\omega)e^{i\phi(\omega)}, \tag{2.4}$$

where $G(\omega) = [A(\omega) + B(\omega)]^{1/2}$ is called the gain of the filter, and $\phi(\omega) = \arctan[B(\omega)/A(\omega)]$ is called the phase angle of the filter and is expressed in radians. The gain and the phase angle vary with the frequency ω. For symmetric filters, the phase angle is zero or $\pm\pi$ and for asymmetric filters it can take any value between $\pm\pi$; it is undefined at those frequencies where the gain is zero.

Following Dagum (1982a,b) we introduce next three measures of filter revision based on the root mean square revision of different filters over all the frequencies.

The first measure is:

$$R^{(\ell,0)} = \left[2\int_0^{1/2} \| H^{(\ell)}(\omega) - H^{(0)}(\omega) \|^2 \, d\omega\right]^{1/2},$$
$$0 \le \omega \le 1/2, \quad \ell = 1,2,3,\ldots,n, \tag{2.5}$$

where $H^{(0)}(\omega)$ is the frequency response function of the concurrent seasonal adjustment filter and $H^{(\ell)}(\omega)$ is the frequency response function of a filter shifted ℓ periods with respect to the concurrent. Taking into consideration that for monthly series the symmetric seasonal adjustment filter of X-11 can be well approximated with 7 years of data plus one (see Young, 1968; Wallis, 1974) and similarly for X-11-ARIMA (see Dagum, 1983); the $H^{(0)}(\omega)$ corresponds to the filter applied to the last observation of a series consisting of *at least* 85 data points. This filter becomes central or symmetric after the series is extended with forty-two more observations and thus $H^{(42)}(\omega)$ denotes the frequency response function of the symmetric filter.

Equation (2.5) gives the time path of the concurrent filter as it approaches the symmetric filter for $\ell = 1,2,\ldots,42$.

The second measure of filter revision to be used in this study refers to the differences between consecutive filters defined by

$$R^{(k+1,k)} = \left[2\int_0^{1/2} \| H^{(k+1)}(\omega) - H^{(k)}(\omega) \|^2 \, d\omega\right]^{1/2},$$
$$0 \le \omega \le 1/2, \quad k = 0,1,2,\ldots,n-1. \tag{2.6}$$

Equation (2.6) gives the time path of single-period revisions of the filters as new observations enter into the series. In the case of monthly data to be discussed in this study, $R^{(k+1,k)}$ gives the time path of the *monthly revisions*.

A third measure gives the time path of annual revisions and is defined by:

$$R^{(k+12,k)} = \left[2 \int_0^{1/2} \| H^{(k+12)}(\omega) - H^{(k)}(\omega) \|^2 \, d\omega \right]^{1/2},$$
$$0 \leq \omega \leq 1/2, \quad k = 0, 1, 2, 3, \ldots, n - 12. \tag{2.7}$$

Equations (2.6) and (2.7) are useful for assessing the frequency of revisions of the concurrent seasonal adjustment filter as new observations enter into the series.

3. TIME PATH OF THE CONCURRENT SEASONAL ADJUSTMENT FILTERS OF X-11 AND X-11-ARIMA

The $R^{(\ell,0)}$, $\ell = 1, 2, \ldots, 42$, measure given in equation (2.5) has been calculated for the X-11 and X-11-ARIMA concurrent filters. The ARIMA extrapolation model applied is the classical $(0,1,1)(0,1,1)_{12}$ IMA type (Box and Jenkins, 1970) of the following form:

$$(1 - B)(1 - B^{12})X_t = (1 - \theta B)(1 - \Theta B^{12})a_t. \tag{3.1}$$

Since the extrapolations affect significantly the concurrent filter depending on the parameter values of θ and Θ, we selected some combinations of values often found when modelling economic time series. These are:

$\theta = 0.40$	$\Theta = 0.40$	$\theta = 0.60$	$\Theta = 0.40$	$\theta = 0.80$	$\Theta = 0.40$
$\theta = 0.40$	$\Theta = 0.60$	$\theta = 0.60$	$\Theta = 0.60$	$\theta = 0.80$	$\Theta = 0.60$.
$\theta = 0.40$	$\Theta = 0.80$	$\theta = 0.60$	$\Theta = 0.80$	$\theta = 0.80$	$\Theta = 0.80$

The smaller the value of θ, the more flexible or changing the trend-cycle component is assumed to be. Similarly, the smaller the value of Θ, the more flexible or moving the seasonal component is assumed to be.

Table 1 gives a summary of the values of: (1) the total revisions of the concurrent filters $R^{(42,0)}$; (2) the revisions of the concurrent filters after a year of observations has been incorporated into the series, $R^{(12,0)}$, and (3) the revisions after 13 months, i.e., $R^{(13,0)}$.

We note from Table 1 that the total revisions of the concurrent filters are always smaller if ARIMA extrapolations are used. These observations

Table 1. *Root Mean Square Revisions Over all Frequencies of the Concurrent Seasonal Adjustment Filter for Selected Month-Spans*

X-11-ARIMA Method		Total Revisions $R^{(42,0)}$	First Year Revision $R^{(12,0)}$	13 month-span Revision $R^{(13,0)}$
Without Extrapolations (X-11)		.36	.29	.30
With Extrapolations from Model (0,1,1)(0,1,1) (Parameter Values)				
$\theta = .40$	$\Theta = .40$	.34	.30	.34
$\theta = .40$	$\Theta = .60$	.32	.27	.31
$\theta = .40$	$\Theta = .80$	.30	.25	.27
$\theta = .60$	$\Theta = .40$	.34	.32	.34
$\theta = .60$	$\Theta = .60$	.32	.28	.30
$\theta = .60$	$\Theta = .80$	.30	.25	.26
$\theta = .80$	$\Theta = .40$	.34	.33	.34
$\theta = .80$	$\Theta = .60$	.32	.29	.30
$\theta = .80$	$\Theta = .80$	.30	.26	.26

conform to those given by Dagum (1982a,b) although these earlier studies referred only to the revisions of the seasonal frequency bands whereas here we are analysing the revisions over all the frequencies. The root mean square total revision reduction ranges from 20% for $\Theta = .80$ to 6% for $\Theta = .40$.

Second, the speed of convergence of the concurrent seasonal adjustment filter to the symmetric filter is faster for X-11-ARIMA with extrapolations than without extrapolations (X-11). After 13 months, $R^{(13,0)}$ represents between 88% to 100% of the total revision $R^{(42,0)}$, depending on the values of Θ. For $\Theta = .40$, which implies a fast moving seasonality, the total revision is completed after the first year whereas for $\Theta = .80$, which corresponds to a more rigid or stable seasonal pattern, only 88% of the total revision is corrected during the same period.

It is important to point out here that the revisions from the concurrent filter to the 12 month lag filter are not monotonic for each frequency ω. In fact, although $R^{(12,0)} < R^{(42,0)}$ *over all* the ω's, the reverse occurs for the revisions associated with the frequencies ω that fall between 0 and 0.050 which generally are attributed to trend and cyclical variations. This remark agrees with that of Burridge and Wallis (1984), who showed the presence of inconsistencies between the concurrent and the 12 month lag filters. Table 2 shows the revision measures $R^{(12,0)}$, $R^{(24,0)}$ and $R^{(42,0)}$ for two frequency bands, namely, $0 \leq \omega \leq 0.050$ and $0.050 < \omega < 0.50$ which are often attributed to the trend-cycle and seasonal plus irregular variations respectively.

Table 2. *Root Mean Square Revisions of the Concurrent Seasonal Adjustment Filter for Selected Frequency Bands and Selected Month-Spans*

X-11-ARIMA Method	12 month-span Revisions $R^{(12,0)}$		24 month-span Revisions $R^{(24,0)}$		Total Revisions $R^{(42,0)}$	
	Tr.	Ir.	Tr.	Ir.	Tr.	Ir.
Without Extrapolations (X-11)	.09	.31	.05	.39	.05	.38
With Extrapolations from Model (0,1,1)(0,1,1) (Parameter Values)						
$\theta = .40$ $\Theta = .40$	.24	.31	.14	.36	.15	.36
$\theta = .40$ $\Theta = .60$	.21	.27	.12	.34	.12	.34
$\theta = .40$ $\Theta = .80$	.18	.24	.11	.32	.11	.32
$\theta = .60$ $\Theta = .40$	.28	.32	.17	.37	.17	.36
$\theta = .60$ $\Theta = .60$	.24	.28	.15	.35	.15	.34
$\theta = .60$ $\Theta = .80$	.21	.25	.13	.32	.13	.32
$\theta = .80$ $\Theta = .40$	.34	.33	.20	.36	.20	.35
$\theta = .80$ $\Theta = .60$	.29	.29	.18	.34	.18	.34
$\theta = .80$ $\Theta = .80$	.26	.26	.16	.32	.16	.32

Tr. = Trend-cycle ($0 \leq \omega \leq .05$)

Ir. = Irregular ($.05 \leq \omega \leq .5$)

We can see that the revisions of the low frequency band are larger after 12 months than when 24 or 42 months have been added to the series. Although not shown, smaller discrepancies are also observed for $R^{(13,0)}$ and $R^{(14,0)}$. These discrepancies disappear for most cases after 16 months and, in all cases, after 24 months where the concurrent filter revisions are equivalent to those obtained from the final filter. These results imply that second year revisions would suffice from the viewpoint of filter changes.

We also observe larger revisions of the frequencies attributed to trend and cyclical variations when the ARIMA extrapolations are used. In practice, however, these revisions would not be observed for they correspond mainly to short cycles (3 years or less) which would not be present if the series is well represented by an IMA model of the $(0,1,1)(0,1,1)_s$ type used for the extrapolations.

Summarizing the above observations, the time path of the various concurrent filters shows that the use of ARIMA extrapolations is highly beneficial from the viewpoints of: (1) the size of the total revisions, which is significantly decreased; and (2) the period of time required for the concurrent filter to converge to the final symmetric filter, which is also significantly decreased.

4. TIME PATH OF MONTHLY AND ANNUAL REVISIONS OF THE SEASONAL ADJUSTMENT ASYMMETRIC FILTERS OF X-11 and X-11-ARIMA

4.1 Monthly Revisions

The $R^{(k+1,k)}$ measure of equation (2.6) has been calculated for the X-11-ARIMA concurrent filters with extrapolations from the ARIMA models discussed in the previous section and for X-11-ARIMA without extrapolations (X-11). The monthly revisions $R^{(k+1,k)}$ also measure the distance between consecutive asymmetric filters.

One important set of single-period revisions is that corresponding to the first year; that is, for time lags $\ell = k+1 = 1,2,\ldots,11$. These eleven monthly revisions should improve the seasonal adjustment filter because of the improvement in the weight system of the 13-term Henderson trend-cycle filter which becomes symmetric after six observations have been added to the series. We will discuss later whether it is advisable or not to revise eleven times the concurrent filter in order to improve the seasonally adjusted estimates.

Another set of consecutive filter revisions of interest includes $\ell = 12$ and $\ell = 24$. These revisions reflect the improvement in the 3×5 (7-term) seasonal moving average weights which change from year to year (being constant within the year) until they become symmetric (after three years). Finally, the revisions at $\ell = 13$ and $\ell = 25$ are important because they are due to the fact that in the X-11-ARIMA the seasonal estimates are forced to sum to zero (12) over each calendar year if an additive (multiplicative) decomposition model is applied.

The observations drawn from Table 3 can be summarized as follows. First, the pattern of monthly revisions is the same whether ARIMA extrapolations are used or not. This pattern is characterized by a rapid decrease in the monthy revisions for $\ell = 1, 2$, and 3; and slow thereafter till $\ell = 11$; then a large increase (reversal of direction) occurs at $\ell = 12$ followed by a rapid decrease for $\ell = 13$ then another large increase at $\ell = 24$ followed by a rapid decrease at $\ell = 25$.

Table 3. *Monthly Root Mean Square Revisions, $R^{(k+1,k)}$, Over All Frequencies of the Concurrent and Asymmetric Filters of X-11-ARIMA With and Without Extrapolations (X-11)*

Monthly Revisions $\ell = k+1$	Without Extrapolations	With Extrapolations, from Model (0,1,1)(0,1,1) and selected parameter values								
		$\theta = .40$ $\Theta = .40$	$\theta = .40$ $\Theta = .60$	$\theta = .40$ $\Theta = .80$	$\theta = .60$ $\Theta = .40$	$\theta = .60$ $\Theta = .60$	$\theta = .60$ $\Theta = .80$	$\theta = .80$ $\Theta = .40$	$\theta = .80$ $\Theta = .60$	$\theta = .80$ $\Theta = .80$
1	0.122	0.176	0.148	0.123	0.136	0.115	0.090	0.100	0.079	0.063
2	0.066	0.130	0.113	0.098	0.102	0.087	0.074	0.078	0.065	0.054
3	0.024	0.089	0.081	0.073	0.071	0.065	0.058	0.056	0.051	0.045
4	0.022	0.056	0.054	0.051	0.048	0.046	0.044	0.041	0.039	0.038
5	0.037	0.034	0.033	0.033	0.033	0.033	0.033	0.033	0.032	0.033
6	0.041	0.022	0.018	0.018	0.024	0.021	0.021	0.026	0.024	0.024
7	0.033	0.019	0.015	0.014	0.016	0.013	0.012	0.018	0.015	0.015
8	0.018	0.022	0.019	0.018	0.014	0.009	0.008	0.009	0.009	0.008
9	0.014	0.029	0.027	0.026	0.012	0.011	0.010	0.007	0.007	0.006
10	0.025	0.041	0.038	0.037	0.019	0.017	0.016	0.013	0.011	0.010
11	0.030	0.057	0.053	0.051	0.026	0.024	0.023	0.015	0.012	0.012
12	0.210	0.259	0.228	0.199	0.277	0.245	0.215	0.293	0.262	0.232
13	0.108	0.136	0.118	0.101	0.104	0.088	0.073	0.075	0.062	0.050
14	0.059	0.103	0.091	0.080	0.080	0.069	0.060	0.060	0.051	0.043
15	0.022	0.071	0.065	0.059	0.057	0.052	0.047	0.045	0.040	0.036
16	0.019	0.046	0.043	0.041	0.038	0.037	0.035	0.033	0.031	0.025
⋮	⋮	⋮	⋮	⋮	⋮	⋮	⋮	⋮	⋮	⋮
24	0.122	0.150	0.134	0.119	0.161	0.145	0.129	0.172	0.155	0.139
25	0.066	0.079	0.069	0.060	0.058	0.050	0.042	0.040	0.034	0.028
26	0.036	0.063	0.057	0.050	0.048	0.042	0.037	0.030	0.030	0.026
⋮	⋮	⋮	⋮	⋮	⋮	⋮	⋮	⋮	⋮	⋮
36	0.054	0.062	0.058	0.054	0.067	0.062	0.059	0.072	0.067	0.063
37	0.031	0.032	0.029	0.027	0.021	0.020	0.019	0.013	0.012	0.011

The significant decreases for the first three consecutive revisions are due to the improvement of the Henderson filter weights. What looks like a reversal of direction in the size of the filter revisions at $\ell = 12$ and $\ell = 13$, is due to the improvement in the seasonal weights which become less asymmetric from year to year until three full years are added to the series.

Second, the effect of the ARIMA extrapolations can be observed in the monthly revisions during the first year, particularly at $\ell = 1, 2$ and 3 where the revisions tend to be larger for small θ and Θ whereas the opposite occurs for large θ and Θ.

Third, we note that the consecutive single-period revisions *do not decrease monotonically* within the year. Although the revision values are very small for $\ell \geq 4$; there are reversals of direction at $\ell = 5, 6$ and 10 when no extrapolations are used. There is only one reversal and at a later lag if ARIMA extrapolations are used; this occurs at $\ell = 7$ for $\theta = .40$ and at $\ell = 10$ for $\theta = .60$ and $\Theta = .80$. This pattern repeats for the second year after a large jump at $\ell = 12$ and again during the third year after lag $\ell = 24$.

Since the monthly revisions during the first year are not monotonically decreasing, *it is not advisable to revise every time a new observation enters into the series.* Revising the concurrent filter eleven times will introduce unwanted revisions because *the distance between consecutive asymmetric filters does not decrease monotonically* as the lags for filters approach to $\ell = 12$. Although not shown here, this *inconsistency* of the distance between asymmetric filters is mainly due to the phase angle of the filters and affects particularly the high frequencies ω associated with the irregulars. This type of *inconsistency between the distance of consecutive asymmetric filters* does not imply however that the time path of the total distance of each asymmetric filter with respect to the final is inconsistent. In fact, the total distances of each asymmetric filter, i.e., $R^{(42,k)}$, $k = 0, 1, 2, \ldots 41$, decrease monotonically with increasing k.

Finally, we observe that the two largest single period revisions occur at $\ell = 1$ and $\ell = 12$.

4.2 Annual Revisions

The $R^{(k+12,k)}$ measure of equation (2.7) has been calculated for the X-11-ARIMA asymmetric filters with and without ARIMA extrapolations.

Table 4 shows that for each asymmetric filter, $k = 0, 1, 2, \ldots 30$, its corresponding annual revision converges monotonically and very fast to zero. For example, the revision of the concurrent filter of X-11 after 12 months (first annual revision) is 0.29, its second annual revision is 0.20 and its third annual revision is 0.11. A similar pattern is followed by the remaining asymmetric filters. This monotonic convergence holds for the root mean square revisions over all frequencies. For the band of low frequencies associated with cyclical

Table 4. *Annual Root Mean Square Revisions, $R^{(k+12,k)}$, Over All Frequencies of the Concurrent and Asymmetric Filters of X-11-ARIMA With and Without ARIMA Extrapolations (X-11)*

Annual Revisions $\ell = k + 12$	Without Extrapolations	With Extrapolations, from Model (0,1,1)(0,1,1) and selected parameter values								
		$\theta = .40$ $\Theta = .40$	$\theta = .40$ $\Theta = .60$	$\theta = .40$ $\Theta = .80$	$\theta = .60$ $\Theta = .40$	$\theta = .60$ $\Theta = .60$	$\theta = .60$ $\Theta = .80$	$\theta = .80$ $\Theta = .40$	$\theta = .80$ $\Theta = .60$	$\theta = .80$ $\Theta = .80$
12	0.29	0.30	0.27	0.23	0.31	0.28	0.25	0.33	0.29	0.26
13	0.27	0.35	0.31	0.27	0.34	0.30	0.26	0.34	0.30	0.26
14	0.26	0.35	0.31	0.26	0.34	0.30	0.26	0.34	0.30	0.26
15	0.26	0.34	0.30	0.26	0.34	0.30	0.26	0.34	0.30	0.26
16	0.26	0.33	0.30	0.26	0.33	0.29	0.26	0.33	0.29	0.26
17	0.26	0.33	0.29	0.26	0.33	0.29	0.26	0.33	0.29	0.26
18	0.28	0.34	0.30	0.26	0.34	0.30	0.26	0.34	0.30	0.26
19	0.27	0.34	0.30	0.26	0.34	0.30	0.26	0.34	0.30	0.26
20	0.27	0.34	0.30	0.26	0.34	0.30	0.26	0.34	0.30	0.26
21	0.27	0.34	0.30	0.26	0.34	0.30	0.26	0.34	0.30	0.26
22	0.27	0.34	0.30	0.26	0.34	0.30	0.26	0.34	0.30	0.26
23	0.27	0.34	0.31	0.26	0.34	0.30	0.26	0.34	0.30	0.26
24	0.20	0.19	0.17	0.15	0.20	0.18	0.16	0.20	0.18	0.16
25	0.18	0.22	0.19	0.17	0.20	0.18	0.16	0.20	0.18	0.16
26	0.16	0.21	0.19	0.17	0.20	0.18	0.16	0.20	0.18	0.16
27	0.16	0.20	0.18	0.16	0.20	0.18	0.16	0.19	0.17	0.16
28	0.16	0.20	0.17	0.16	0.19	0.18	0.16	0.20	0.17	0.16
29	0.16	0.19	0.17	0.16	0.19	0.18	0.16	0.20	0.17	0.16
30	0.16	0.20	0.17	0.16	0.20	0.18	0.15	0.20	0.17	0.15
31	0.17	0.20	0.17	0.16	0.20	0.18	0.15	0.20	0.17	0.15
32	0.16	0.20	0.17	0.16	0.20	0.18	0.15	0.20	0.17	0.15
33	0.16	0.20	0.18	0.16	0.20	0.18	0.16	0.20	0.17	0.15
34	0.16	0.20	0.18	0.16	0.20	0.18	0.16	0.20	0.17	0.15
35	0.16	0.20	0.18	0.16	0.20	0.18	0.16	0.20	0.17	0.15
36	0.11	0.10	0.09	0.08	0.09	0.09	0.08	0.09	0.09	0.08
37	0.09	0.10	0.09	0.08	0.09	0.09	0.08	0.09	0.08	0.08
38	0.08	0.10	0.09	0.08	0.09	0.09	0.08	0.09	0.08	0.08
39	0.07	0.09	0.08	0.08	0.09	0.09	0.08	0.09	0.08	0.08
40	0.07	0.08	0.07	0.07	0.08	0.08	0.07	0.08	0.07	0.07
41	0.07	0.08	0.07	0.07	0.08	0.08	0.07	0.08	0.07	0.07
42	0.07	0.08	0.07	0.07	0.08	0.08	0.07	0.08	0.07	0.07

variations this monotonic convergence is not observed at $\ell = 12$ as discussed in Section 3.

Table 4 also shows that the size of the annual revisions are rather constant for consecutive filters within each year; that is, from $\ell = 12$ to $\ell = 23$, $\ell = 24$ to $\ell = 35$ and $\ell = 37$ to $\ell = 42$. This pattern of annual revisions implies that all changes observed in month-to-month comparisons within the same year are attributed mainly to the innovations entering into the series.

However, the most common practise of revising current seasonally adjusted data consists of keeping constant the concurrent estimate from the time it appears until the end of the year and then revising annually the current and earliest years, generally up to three. Consequently, first-year revisions are given by $R^{(0,0)}$, $R^{(1,0)}$, $R^{(2,0)}, \ldots, R^{(11,0)}$, second-year revisions by $R^{(12,0)}$, $R^{(13,1)}$, $R^{(14,2)}, \ldots, R^{(23,11)}$, and third-year revisions by $R^{(24,12)}$, $R^{(25,13)}$, $R^{(26,14)}, \ldots, R^{(35,23)}$. Table 4 shows the second and third-year revisions and Table 5 shows first-year revisions. Given that these latter are generally the most relevant for decision making, the revisions are shown for two frequency bands and for all frequencies. Since the pattern is similar when using ARIMA extrapolation only two of the cases are shown.

Table 5. *First Year Root Mean Square Revisions, $R^{(\ell,0)}$, of the Concurrent Seasonal Adjustment Filter for Selected Frequency Bands and Over All Frequencies*

	Without Extrapolation			With Extrapolations $\theta = .40$ $\Theta = .80$			With Extrapolations $\theta = .80$ $\Theta = .80$		
ℓ	Tr.	Ir.	Tot.	Tr.	Ir.	Tot.	Tr.	Ir.	Tot.
1	.02	.13	.12	.04	.13	.12	.05	.06	.06
2	.03	.14	.13	.07	.13	.13	.08	.08	.08
3	.04	.14	.13	.09	.13	.13	.11	.08	.09
4	.04	.14	.13	.10	.13	.13	.13	.08	.09
5	.04	.16	.15	.11	.13	.13	.14	.08	.09
6	.04	.18	.17	.11	.13	.13	.15	.08	.09
7	.05	.17	.16	.11	.13	.13	.15	.08	.09
8	.05	.17	.16	.11	.13	.13	.15	.08	.09
9	.05	.17	.16	.11	.13	.13	.15	.08	.09
10	.05	.17	.16	.11	.14	.14	.15	.08	.09
11	.05	.17	.16	.11	.14	.14	.15	.08	.09

Tr. = Trend-cycle $(0 \leq \omega \leq .05)$

Ir. = Irregular $(.05 < \omega \leq .5)$

Tot. = Total $(0 \leq \omega \leq .5)$

We can observe that there is a mononotic increase in the revisions of the low frequencies from $\ell = 1$ to $\ell = 3$ or 4 and thereafter they remain rather constant. On the other hand, the revision of the high frequencies seems to be rather constant for $\ell \geq 2$ if ARIMA extrapolations are used and for $\ell \geq 7$ if no extrapolations are made.

The advantage of this common scheme of revisions is that when doing month to month comparisons, all changes within the first year are due mainly to the innovations since the filter applied during the current year is always the same, i.e., the concurrent seasonal adjustment filter. Furthermore, the filters of the previous years are modified by almost the same amount within each year with the exceptions being $\ell = 1, 2, 12$ and 13 in most cases.

4.3 Combining Monthly and Annual Revisions

The results discussed in Sections 4.1 and 4.2 suggest that a better scheme of revisions than the common practice should include monthly as well as annual revisions since the largest single period revisions occur at $\ell = 1$ and 12. It is expected that: (1) adjusting concurrently each month, say from January to November and revising only once when the next month is available, and (2) adjusting concurrently December when it first appears and then revising the first year and earlier years when January is added, should improve the reliability of the filter applied during the current year while maintaining simultaneously the filter's homogeneity for month to month comparisons.

The first-year revisions would then be $R^{(1,1)}$, $R^{(2,1)}$, $R^{(3,1)}, \ldots, R^{(10,1)}$ and $R^{(11,1)}$. These revisions are shown in Table 6.

We can see that the pattern is very similar to that for the concurrent filter but the size of the revisions are smaller in all cases which agrees with our expectations.

This scheme of combining monthly and annual revisions has been shown to produce smoother seasonally adjusted series and smaller revisions when applied to real data (Kenny and Durbin, 1982; Dagum and Morry, 1984).

5. CONCLUSIONS

This study has addressed the problem of how often the concurrent seasonal adjustment filter of X-11-ARIMA with and without extrapolations should be revised. It is shown that:

(I) The time path of the concurrent filter for consecutive month-spans, $\ell = 1, 2, \ldots, 41$ approaches nearly monotonically to the final symmetric filter ($\ell = 42$). The use of the ARIMA extrapolation option decreases the size of the total revision while increasing the concurrent filter's speed of convergence to the final symmetric filter. However, it has been ob-

Table 6. *First-Year Root Mean Square Revisions, $R^{(\ell,1)}$, of the First-Month Revised Seasonal Adjustment Filter for Selected Bands and Over All Frequencies*

	Without Extrapolation			With Extrapolations $\theta = .40$ $\Theta = .80$			With Extrapolations $\theta = .80$ $\Theta = .80$		
ℓ	Tr.	Ir.	Tot.	Tr.	Ir.	Tot.	Tr.	Ir.	Tot.
2	.01	.07	.07	.03	.10	.10	.04	.06	.06
3	.02	.07	.07	.05	.10	.10	.07	.07	.07
4	.02	.07	.07	.07	.10	.10	.09	.07	.07
5	.02	.08	.08	.07	.10	.10	.11	.07	.08
6	.03	.10	.10	.08	.11	.11	.11	.07	.08
7	.03	.11	.11	.08	.11	.11	.12	.07	.08
8	.03	.11	.11	.08	.11	.11	.12	.07	.08
9	.03	.11	.11	.08	.11	.11	.12	.07	.08
10	.03	.12	.11	.08	.11	.11	.12	.07	.08
11	.03	.12	.12	.09	.12	.12	.12	.07	.08

Tr. = Trend-cycle ($0 \leq \omega \leq .05$)
Ir. = Irregular ($.05 < \omega \leq .5$)
Tot. = Total ($0 \leq \omega \leq .5$)

served that the revisions of the low frequencies are larger after 12 months than when 24 or 42 have been added to the series. This inconsistency was already noted by Burridge and Wallis (1984) when fitting ARIMA models to the transfer functions of the concurrent and first-year revised filters. This inconsistency disappears for most cases after 16 months and, in all cases, the revisions of the concurrent filter are equivalent to those obtained from the final filter after 24 months have been added to the series. These results imply that second-year revisions should suffice from the viewpoint of filter changes.

(II) The monthly revisions of the concurrent filter do not approach monotonically either to the annual or to the final filters. The larger one-single period revisions occur at $\ell = 1, 2, 3, 12, 13, 24$, and 25. There are significant decreases for the first three consecutive revisions due to the improvement of the end weights of the Henderson trend-cycle filter. There is a reversal of direction in the size of the filter revision at $\ell = 12$ and $\ell = 24$ due to an improvement in the seasonal weights which become less asymmetric from year to year until three full years are added to the series. There are two large decreases at $\ell = 13$ and $\ell = 25$ which are due to the fact that, the seasonal estimates are forced to add to zero (12) over each calendar year if an additive (multiplicative) decomposition model is assumed.

The annual revisions of the concurrent filter and remaining monthly asymmetric filters of the first year approach monotonically to the final filtering root mean square over all the frequencies but not for each frequency; in particular, those frequencies associated with the trend-cycle component are revised more as compared to the total revision (distance between concurent and final filter). It is also observed that the annual revisions are rather constant for each filter within the same year.

Taking into consideration the patterns of monthly and annual revisions, the *best combination of frequency of revision of the concurrent filter* would be to revise when a new month appears, keep the estimate constant for the remainder of the year and then, revise annually when the first month of the next year is available. This scheme offers the following advantages: (1) by revising each month once, the reliability of the concurrent filter increases significantly and since the revised filter is kept constant during the first year, changes in the month-to-month comparisons are due only to the innovations; and (2) by revising annually, the reliability of the filters improves, while the comparability of consecutive filters is maintained since they are all revised by almost a constant amount (without introducing frequency distortions) within each year. This scheme has shown to produce smoother seasonally adjusted series and smaller revisions when applied to real data by Kenny and Durbin (1982) and Dagum and Morry (1984).

REFERENCES

Box, G. E. P., and G. M. Jenkins (1970), *Time Series Analysis, Forecasting and Control.* San Francisco: Holden-Day.

Bayer, A., and D. Wilcox (1981), "An evaluation of concurrent seasonal adjustment". Technical Report, Washington, D.C. Board of Governors of the Federal Reserve System, Special Studies Section.

Burridge, P., and K. F. Wallis (1984), "Unobserved-components models for seasonal adjustment filters". *Journal of Business and Economic Statistics* **2**, 350–359.

Dagum, E. B. (1978), *Comparison and Assessment of Seasonal Adjustment Methods for Labor Force Series.* Washington, D.C.: U.S. Government Printing

Office, Stock No. 052-003-00603-1.

Dagum, E. B. (1980), *The X-11-ARIMA Seasonal Adjustment Method.* Ottawa: Statistics Canada Catalogue No. 12-564E.

Dagum, E. B. (1982a), "The effects of asymmetric filters on seasonal factor revisions". *Journal of the American Statistical Association* **77**, 732–738.

Dagum, E. B. (1982b), "Revisions of seasonally adjusted data due to filter changes". *Proceedings of the Business and Economic Statistics Section, American Statistical Association*, pp. 39–45.

Dagum, E. B. (1983), "Spectral properties of the concurrent and forecasting linear filters of the X-11-ARIMA method". *Canadian Journal of Statistics* **2**, 73–90.

Dagum, E. B., and M. Morry (1984), "Basic issues on the seasonal adjustment of the Canadian consumer price index". *Journal of Business and Economic Statistics* **2**, 250–259.

Kenny, P., and J. Durbin (1982), "Local trend estimation and seasonal adjustment of economic time series". *Journal of the Royal Statistical Society*, Series A **145**, 1–41.

McKenzie, S. (1982), "An evaluation of concurrent adjustment on Census Bureau time series". *Proceedings of the Business and Economic Statistics Section, American Statistical Association*, pp. 46–55.

McKenzie, S. (1984), "Concurrent seasonal adjustment with Census X-11". *Journal of Business and Economic Statistics* **2**, 235–249.

Pierce, D., and S. McKenzie (1985), "On concurrent seasonal adjustment". Special Studies No. 164, U.S. Federal Reserve Board.

Shiskin, J., A. H. Young, and J. C. Musgrave (1967), "The X-11 variant of census method II seasonal adjustment program". Technical Paper 15, Washington, D.C.: U.S. Bureau of Census.

Wallis, K. F. (1974), "Seasonal adjustment and relations between variables". *Journal of the American Statistical Association* **69**, 18–31.

Young, A. H. (1968), "Linear approximations to the census and BLS seasonal adjustment methods". *Journal of the American Statistical Association* **63**, 445–457.

D. S. Stoffer and T. Panchalingam [1]

A WALSH-FOURIER APPROACH TO THE ANALYSIS OF BINARY TIME SERIES

ABSTRACT

A nonparametric approach to analyzing a stationary binary time series $\{X(n), n = 0, \pm 1, \pm 2, \ldots\}$ taking values in $\{0, 1\}$ is discussed. The analysis is accomplished in the spectral domain using the Walsh-Fourier transform which is based on Walsh functions. This seems to be a natural alternative to the trigonometric functions used in the usual spectral analysis since the Walsh functions take on only two values, +1 or −1, (or "on" and "off", as does the series $X(n)$ itself). This approach enables the investigator to analyze a binary series in terms of square-waves and sequency (switches or changes per unit time) rather than sine-waves and frequency (cycles per unit time). We discuss (1) the basic theory of Walsh-Fourier analysis, (2) the computational aspects involved in calculating the discrete Walsh-Fourier transform, and (3) the analysis of simulated and real binary data in the sequency domain. We suggest that these methods would enhance the analysis of time series which take values in a discrete finite set.

1. INTRODUCTION

Implicit in the spectral (Fourier) analysis of time series is one of two "extreme" assumptions about the process: (a) the very long stretch of the time series is the only time series we want to consider and consists of the superposition of not too many sinusoidal terms of substantially different frequencies; (b) the time series is to be regarded as a realization of an ergodic Gaussian process; it is one of many possible time series and the analyses are directed toward the properties of the ensemble of the series, not toward those of a specific realization (Brillinger and Tukey, 1982).

[1] Department of Mathematics and Statistics, University of Pittsburgh, Pittsburgh, Pennsylvania 15260 (both authors)

I. B. MacNeill and G. J. Umphrey (eds.), Time Series and Econometric Modelling, 147–163.

However, there are many situations in which time series are patently non-normal. Similarly, there are processes, such as those that take values in a discrete finite set, which cannot be thought of as the superpositions of well separated sinusoids. For the case of continuous valued non-normal time series it is perhaps still reasonable, in some cases, to do spectral analysis using Fourier (trigonometric) methods. However, in the cases where the time series takes values in a discrete finite set, it makes little sense to correlate the data with sines and cosines. As an alternative, we suggest that the spectral analysis of discrete-valued time series be accomplished in the "sequency" domain via the Walsh-Fourier transform (Ahmed and Rao, 1975; Kohn 1980a, 1980b; Morettin, 1981, 1983; Stoffer, 1985). This seems to be a natural alternative to the usual Fourier analysis since the Walsh-Fourier transform is based on the "square-wave" Walsh functions (Ahmed and Rao, 1975; Fine, 1949, 1950, 1957; Morettin, 1974b, 1981, to mention a few). This approach would enable investigators to analyze a discrete-valued time series (which we may think of as a square-waveform) in terms of square-waves and sequency (switches per unit time) rather than sine-waves and frequency. As empirically demonstrated by Beauchamp (1975), "the respective roles of Walsh and Fourier spectral analysis for discontinuous and smooth-varying signals are clear. Where the signal is derived from sinusoidally-based waveforms, Fourier analysis is relevant. Where the signal contains sharp discontinuities and a limited number of levels, Walsh analysis is appropriate".

The Walsh functions, which are defined via the Rademacher functions (see Ahmed and Rao, 1975; Kohn, 1980a), form a complete orthonormal sequence on $[0,1)$ and take on only two values, $+1$ and -1 (or "on" and "off"). They are ordered by the number of zero-crossings which is called *sequency*. If $W(n,\gamma)$, $n = 0,1,2,\ldots,$ $0 \leq \gamma < 1$, denotes the nth Walsh function, then $W(n,\cdot)$ makes n zero-crossings in $[0,1)$.

The first eight sequency ordered discrete Walsh functions $W(n, m/N)$, $n, m = 0,1,\ldots,7$, corresponding to a sample of size $N = 2^3$ are shown in Figure 1 in an 8×8 matrix. We note that other types of orderings exist, for example Paley order and Hadamard order (Ahmed and Rao, 1975). However, sequency ordering is more natural in that it is comparable to the frequency ordering of sines and cosines. We will discuss methods of generating the Walsh functions in Section 3.

Walsh spectral analysis has been used for several purposes, primarily in the Engineering sciences, such as speech processing, word recognition, image coding and transmission, filtering and multiplexing. See, for example, the 1971 and 1973 Proceedings on the Applications of Walsh Functions, Beauchamp (1975) and Harmuth (1972) to mention a few. Applications of the Walsh-Fourier transform in statistics are scarce and we discuss a few here. Robinson (1972) compared the Walsh-Fourier transform using a Parzen ker-

$$\begin{array}{l} W(0,m/N) \rightarrow \\ W(1,m/N) \rightarrow \\ W(2,m/N) \rightarrow \\ W(3,m/N) \rightarrow \\ W(4,m/N) \rightarrow \\ W(5,m/N) \rightarrow \\ W(6,m/N) \rightarrow \\ W(7,m/N) \rightarrow \end{array} \begin{bmatrix} 1 & 1 & 1 & 1 & 1 & 1 & 1 & 1 \\ 1 & 1 & 1 & 1 & -1 & -1 & -1 & -1 \\ 1 & 1 & -1 & -1 & -1 & -1 & 1 & 1 \\ 1 & 1 & -1 & -1 & 1 & 1 & -1 & -1 \\ 1 & -1 & -1 & 1 & 1 & -1 & -1 & 1 \\ 1 & -1 & -1 & 1 & -1 & 1 & 1 & -1 \\ 1 & -1 & 1 & -1 & -1 & 1 & -1 & 1 \\ 1 & -1 & 1 & -1 & 1 & -1 & 1 & -1 \end{bmatrix}$$

Figure 1. *Sequency-ordered discrete Walsh functions for a sample of size* $N = 8$.

nel with the usual Fourier transform for first order Markov processes. Ott and Kronmal (1976) used the Walsh transform in classification and prediction problems for strictly stationary binary time series. Panchalingam (1985) analyzed simulated and real binary time series in the sequency domain.

Theoretical results concerning the statistical application of Walsh spectral analysis to stationary time series are relatively new and limited to the works of Kohn (1980a,b), Morettin (1974a, 1981, 1983) and Stoffer (1985). We note here that some work has been done by others on the statistical aspects of Walsh spectral analysis for "dyadic" stationary time series (see Morettin, 1981, for discussions and references). Although dyadic time has some theoretical appeal, it is of little practical use. We, therefore, concentrate on real time stationary processes.

2. THE WALSH-FOURIER TRANSFORM

Our discussion will be restricted to univariate time series; the multivariate versions follow in an obvious way (see Kohn, 1980b, Section 3).

Throughout this section, we suppose that $X(0), X(1), \ldots, X(N-1)$ is a sample of length $N = 2^p$ (p a positive integer) from a zero-mean, weakly stationary time series, $X(n)$, with absolutely summable autocovariance function, $\gamma(h)$, $h = 0, \pm1, \pm2, \ldots$. Let

$$d_N(\lambda) = N^{-\frac{1}{2}} \sum_{n=0}^{N-1} X(n)W(n,\lambda), \qquad 0 \leq \lambda < 1$$

be the finite Walsh-Fourier transform of the data. The logical covariance (Kohn, 1980a) is defined to be

$$\tau(j) = N^{-1} \sum_{k=0}^{N-1} \gamma(j \oplus k - k),$$

where $j \oplus k$ means the dyadic addition of j and k. It can then be shown that the variance of $d_N(\lambda)$ is given by

$$\operatorname{Var}\{d_N(\lambda)\} = \sum_{j=0}^{N-1} \tau(j)W(j,\lambda). \tag{1}$$

Taking the limit $(N \to \infty)$ in (1) we see that $\operatorname{Var}\{d_N(\lambda)\} \to f(\lambda)$, where

$$f(\lambda) = \sum_{j=0}^{\infty} \tau(j)W(j,\lambda), \qquad 0 \le \lambda < 1, \tag{2}$$

is called the Walsh-Fourier spectrum of $X(n)$. We note that $f(\lambda)$ exists since the absolute summability of $\gamma(h)$ implies the absolute summability of $\tau(j)$ (Kohn, 1980a, Lemma 3). Specifically, Kohn has shown the existence of $f(\lambda)$ under the condition that

$$\lim_{p\to\infty} \sum_{|j|<2^p} \left(1 - \frac{|j|}{2^p}\right) |\gamma(j)| < \infty.$$

If $X(0),\ldots,X(N-1)$ is a sample of length $N = 2^p$, the transform $d_N(\lambda)$ is calculated for $\lambda = m/N$, $m = 0,\ldots,N-1$. It can be shown that

$$W(n, m/N) = W(m, n/N) \quad m,n = 0,1,\ldots,N-1 \tag{3}$$

and hence the value of λ in the discrete Walsh-Fourier transform corresponds to sequency. As with the usual Fourier analysis, if the mean of the series is unknown, the only sequency of the form m/N at which the transform cannot be evaluated is at the zero sequency $(m = 0)$. To see this, let $\mu = E[X(n)]$, all n, and note that for $m = 0,1,\ldots,N-1$,

$$N^{-1}\sum_{n=0}^{N-1} W(n, m/N) = \begin{cases} 1 & \text{if } m = 0 \\ 0 & \text{if } m \neq 0. \end{cases} \tag{4}$$

Relationships (3) and (4) are given by Kohn (1980a, Lemma 1). It is clear from (4) that the mean centered transform will be the uncentered transform except at $m = 0$, and in particular

$$E\{d_N(m/N)\} = N^{-\frac{1}{2}} \sum_{n=0}^{N-1} \mu W(n, m/N) = \begin{cases} N^{\frac{1}{2}}\mu & \text{if } m = 0 \\ 0 & \text{if } m \neq 0, \end{cases}$$

$m = 0,1,\ldots,N-1$.

Various authors have established central limit theorems for the finite Walsh-Fourier transform, $d_N(m/N)$, under a wide range of conditions (Kohn, 1980a; Morettin, 1983; Stoffer, 1985). We state the version given by Morettin (1983).

Theorem 2.1. Let $X(n)$ be a zero-mean strictly stationary time series with finite moments and let $c_r(j_1,\ldots,j_r) = \text{cum}\{X(j_1),\ldots,X(j_r)\}$ be the rth cumulant of $X(n)$, $j_1,\ldots,j_r = 0,\pm1,\pm2,\ldots$. Further, suppose $\sum_{j_1=0}^{\infty}\cdots\sum_{j_{r-1}=0}^{\infty} | c_r(j_1,\ldots,j_{r-1}) | < \infty$. Then $d_N(\lambda)$ converges in distribution to a normal variate with zero mean and variance $f(\lambda)$ given by (2).

In order to be able to estimate consistently the Walsh-Fourier spectral density, we shall need asymptotic results for smoothing the transform $d_N(\cdot)$. The following lemma is given by Kohn (1980a, Corollary 3).

Lemma 2.1. Let $X(n)$ be a strictly stationary zero-mean time series with absolutely summable autocovariance function, and suppose λ_N and μ_N are dyadically rational (i.e., their binary representations are finite).

(1) If $\lambda = \mu$, $| \lambda_N - \mu_N | \geq N^{-1}$, $\lambda_N \oplus \lambda \to 0$ and $\mu_N \oplus \mu \to 0$ as $N = 2^p \to \infty$, then $E\{d_N(\lambda_N)d_N(\mu_N)\} \to 0$.

(2) If $\lambda_N \oplus \lambda \to 0$, then $E\{d_N^2(\lambda_N)\} \to f(\lambda)$ as $N \to \infty$.

In particular, Theorem 2.1 and Lemma 2.1 gives us the following useful result for estimating the Walsh-Fourier spectral density.

Corollary 2.1. Let $\lambda_{j,N} = j/N$, $1 \leq j \leq N-1$ and suppose for $\{\lambda_{j(1),N},\ldots,\lambda_{j(M),N}\}$, $\lambda_{j(k),N} \oplus \lambda \to 0$ as $N \to \infty$, $k = 1,\ldots,M$, and

$$| \lambda_{j(\ell),N} - \lambda_{j(k),N} | \geq N^{-1} \text{ for } \ell \neq k,\ \ell,k = 1,\ldots,M.$$

Then $\mathbf{d}_N \xrightarrow{L} N(0,\Delta)$ where $\mathbf{d}_N = (d_N(\lambda_{j(1),N}),\ldots,d_N(\lambda_{j(M),N}))'$ and Δ is an $M \times M$ diagonal matrix with $f(\lambda)$ along the diagonal. Also, $\mathbf{d}_N'\mathbf{d}_N \xrightarrow{L} f(\lambda)\chi_M^2$ so that $M^{-1}\mathbf{d}_N'\mathbf{d}_N$ is an estimate of $f(\lambda)$ having variance $2f^2(\lambda)/M$.

In particular, if we let $M \to \infty$ as $N \to \infty$ with $M/N \to 0$ in Corollary 2.1, the smoothed periodogram $M^{-1}\mathbf{d}_N'\mathbf{d}_N$ is a mean square consistent estimate of the Walsh-Fourier spectrum $f(\lambda)$.

3. COMPUTATION OF THE WALSH-FOURIER TRANSFORM

The Walsh functions are usually calculated via the Hadamard matrix which, for a sample of size $N = 2^p$ (p a positive integer) is defined to be a symmetric orthogonal $N \times N$ matrix whose (u, v)th element, $u, v = 0, 1, \ldots, N-1$, is equal to $\phi(u, v) = \prod_{j=1}^{p}(-1)^{u_j v_j}$ where the binary representations of u and v are given by $(u_1, \ldots, u_p)$ and $(v_1, \ldots, v_p)$, respectively, with $u_j, v_j = 0$ or 1. The Hadamard matrix can be generated recursively by $H(0) = 1$, and

$$H(k+1_) = \begin{bmatrix} H(k) & H(k) \\ H(k) & -H(k) \end{bmatrix}, k = 0, 1, 2, \ldots.$$

For example, for a sample of size $N = 4$ we would calculate

$$H(1) = \begin{bmatrix} 1 & 1 \\ 1 & -1 \end{bmatrix}, \text{ and } H(2) = \begin{bmatrix} 1 & 1 & 1 & 1 \\ 1 & -1 & 1 & -1 \\ 1 & 1 & -1 & -1 \\ 1 & -1 & -1 & 1 \end{bmatrix}.$$

The Hadamard matrix gives the Walsh functions as rows (or columns) in what is called natural or Hadamard ordering. To obtain the Walsh functions in sequency ordering we simply reorder the rows of $H(\cdot)$ according to the number of sign changes. Another method used "bit-reversal Gray code" (Ahmed and Rao, 1975) to rearrange the rows of $H(\cdot)$. These methods, however, are not very efficient.

If $H(2) = [\mathbf{H}_0(2), \mathbf{H}_1(2), \mathbf{H}_2(2), \mathbf{H}_3(2)]$ where $\mathbf{H}_i(2), i = 0, 1, 2, 3$ is the ith column of $H(2)$, then the corresponding Walsh-ordered Hadamard matrix is $H_w(2) = [\mathbf{H}_0(2), \mathbf{H}_2(2), \mathbf{H}_3(2), \mathbf{H}_1(2)]$. The procedure of obtaining the Walsh-ordered Hadamard transform from its definition either requires storage of the Hadamard matrix, or recomputation whenever the elements of $H_w(p)$ are needed. Hence, either the sample length is restricted to about $p = 10$ or 20, or the procedure is extremely slow. There are, however, fast methods which can reduce the number of computations (additions and subtractions) by a factor of approximately $2^{p-1}/p$ from the number using the definition. The method we discuss here and a computer subroutine are given by Ahmed and Rao (1975, Chapter 6). The Walsh-Hadamard matrix can be computed as

$$H_w(p) = \prod_{i=1}^{p} H_i(p) \cdot B \tag{5}$$

where

$$H_i(p) = \begin{bmatrix} F_s & & & & & & & 0 \\ & G_s & & & & & & \\ & & \ddots & & & & & \\ & & & F_s & & & & \\ & & & & G_s & & & \\ & & & & & \ddots & & \\ & & & & & & F_s & \\ 0 & & & & & & & G_s \end{bmatrix}, \quad s = 2^{i-1},$$

with

$$F_s = \begin{bmatrix} I_s & I_s \\ I_s & -I_s \end{bmatrix},$$
$$G_s = \begin{bmatrix} I_s & -I_s \\ I_s & I_s \end{bmatrix},$$

and I_s being the $s \times s$ identity matrix. The matrix B in (5) is a matrix which bit-reverses the order of the data. For example, with $N = 2^3$ the bit-reversal of $1 = (0,0,1)$ is $4 = (1,0,0)$, and the bit-reversal of $3 = (0,1,1)$ is $6 = (1,1,0)$ so that $X(1)$ is exchanged with $X(4)$ and $X(3)$ is exchanged with $X(6)$ in the data vector. If $\mathbf{X} = (X(0), \ldots, X(N-1))'$ is the data vector ($N = 2^p$), the fast finite Walsh-Fourier transform is computed as

$$\mathbf{d}_N(\boldsymbol{\lambda}_N) = N^{-\frac{1}{2}} H_w(p)\mathbf{X} = N^{-\frac{1}{2}} \prod_{i=1}^{p} H_i(p) \cdot B\mathbf{X},$$

where $\boldsymbol{\lambda}_N = (0/N, 1/N, \ldots, (N-1)/N)'$. For example, if $N = 2^3$, the Walsh-ordered Hadamard matrix can be decomposed as

$$H_w(3) = \begin{bmatrix} F_1 & & & 0 \\ & G_1 & & \\ & & F_1 & \\ 0 & & & G_1 \end{bmatrix} \begin{bmatrix} F_2 & 0 \\ 0 & G_2 \end{bmatrix} F_4 B,$$

where

$$B = \left[\begin{array}{cccc|cccc} 1 & 0 & 0 & 0 & 0 & 0 & 0 & 0 \\ 0 & 0 & 0 & 0 & 1 & 0 & 0 & 0 \\ 0 & 0 & 1 & 0 & 0 & 0 & 0 & 0 \\ 0 & 0 & 0 & 0 & 0 & 0 & 1 & 0 \\ \hline 0 & 1 & 0 & 0 & 0 & 0 & 0 & 0 \\ 0 & 0 & 0 & 0 & 0 & 1 & 0 & 0 \\ 0 & 0 & 0 & 1 & 0 & 0 & 0 & 0 \\ 0 & 0 & 0 & 0 & 0 & 0 & 0 & 1 \end{array}\right].$$

The vector of periodogram ordinates is obtained by squaring each element of $\mathbf{d}_N(\boldsymbol{\lambda}_N)$. Let $I_N(m) = d_N^2(m/N)$ denote the mth periodogram ordinate, $m = 0, 1, \ldots, N-1$. It is seen that $I_N(m) = \sum_{j=0}^{N-1} \hat{r}(j)W(j, m/N)$, where $\hat{r}(j) = N^{-1}\sum_{k=0}^{N-1} X(k)X(j \oplus k)$. Employing relationships (3) and (4) we may write

$$\hat{r}(j) = N^{-1} \sum_{m=0}^{N-1} I_N(m)W(m, j/N). \tag{6}$$

Thus, for large N, the quickest way to compute $\hat{r}(j)$ is to use the fast Walsh-Fourier transform twice, once to compute $I_N(m)$ and once to compute the right-hand side of (6).

4. SEQUENCY ANALYSIS OF BINARY TIME SERIES

First, we discuss a possible model for a binary-valued time series where Walsh-Fourier analysis is desirable as well as superior to trigonometric analysis. Consider a binary version of the signal-plus-noise models used for Gaussian processes. In general, write the binary time series $\{X(n), n = 0, 1, 2, \ldots,\}$, $X(n) \in \{0, 1\}$, as

$$X(n) = S(n) + \varepsilon(n), \qquad n = 0, 1, 2, \ldots, \tag{7}$$

where $S(n)$ is a random stationary binary (not necessarily 0 or 1) signal which possibly depends on parameters $\boldsymbol{\theta} = (\theta_1, \ldots, \theta_q)'$, $q > 0$, and $\varepsilon(n)$ is a zero-mean binary (not necessarily 0 or 1) white noise process which is uncorrelated with the signal $S(n)$. Let $X(n)$ satisfy the conditions of Theorem 2.1.

For a specific example of such a model consider a two state (0 or 1) Markov process. Let $X(n)$ be the value of the process at time n, let θ be the probability of being in state 1 at any given time and let p_{ij}, $i, j = 0, 1$ denote the transition probabilities. Then we may write

$$E[X(n)] = E[X(n-1)]p_{11} + E[1 - X(n-1)]p_{01}$$

from which we obtain the signal-plus-noise model

$$X(n) = X(n-1)p_{11} + (1 - X(n-1))p_{01} + \varepsilon(n), \tag{8}$$

where $\varepsilon(n)$, $n = 0, 1, 2, \ldots$, is a zero mean Bernoulli type white noise process whose value depends on the value of $X(n-1)$, but is uncorrelated with the signal $S(n) = X(n-1)p_{11} + (1 - X(n-1))p_{01}$. To see this note that

$$\begin{aligned}\mathrm{Cov}\,(\varepsilon(n), X(n-1)) &= E\{[X(n) - p_{01} - X(n-1)(p_{11} - p_{01})]\, X(n-1)\} \\ &= \theta p_{11} - \theta p_{01} - \theta(p_{11} - p_{01}) = 0.\end{aligned}$$

Hence (8) is of the binary signal-plus-noise form given in (7), where the binary signal $S(n)$ is a function of two parameters p_{01} and p_{11}.

For the binary signal-plus-noise models given by (7), Walsh-Fourier analysis would be useful for detecting whether or not a binary signal exists in the time series (as opposed to the series being white noise), and if so, determine the cyclic behavior (in terms of sequency) of the binary signal.

For example, if we denote the autocovariance functions of $S(n)$ and $\varepsilon(n)$ by $\gamma_S(h)$ and $\gamma_\varepsilon(h)$, $h = 0, \pm 1, \pm 2, \ldots$, respectively, where $\gamma_\varepsilon(h) = 0$ for $h \neq 0$, and assume that $\gamma_S(h)$ is absolutely summable, then the Walsh-Fourier spectral densities of $S(n)$ and $\varepsilon(n)$ exist. Denote $f_S(\lambda) = \sum_{j=0}^{\infty} \tau_S(j) W(j, \lambda)$ and $f_\varepsilon(\lambda) = \sum_{j=0}^{\infty} \tau_\varepsilon(j) W(j, \lambda)$ as the Walsh-Fourier spectra of $S(n)$ and $\varepsilon(n)$, respectively, where $\tau_S(j)$ and $\tau_\varepsilon(j)$ are the logical covariances of $S(n)$ and $\varepsilon(n)$, respectively. Clearly then, by the model assumptions, the Walsh-Fourier spectral density of $X(n)$, say $f_X(\lambda)$, is given by $f_X(\lambda) = f_S(\lambda) + f_\varepsilon(\lambda)$. Now, let $X(0), \ldots, X(N-1)$ be a sample from the time series (7) and let $d_N(\lambda_{j(\ell),N})$, $\ell = 1, \ldots, M$, denote the finite Walsh-Fourier transforms of the sample where $\lambda_{j(\ell),N}$ are as in Corollary 2.1. Then by Corollary 2.1 and Skorokhod's Representation Theorem (Skorokhod, 1956), the smoothed periodogram has the representation

$$\tilde{I}_N(\lambda) = M^{-1} \sum_{\ell=1}^{M} d_N^2\left(\lambda_{j(\ell),N}\right) = M^{-1}[f_S(\lambda) + f_\varepsilon(\lambda)]\chi_M^2 \text{ a.s.}$$

as $N \to \infty$. If no signal is present, then $f_S(\cdot) \equiv 0$ and the smoothed periodogram behaves asymptotically like a constant times a χ^2 variate divided by its degrees of freedom. Otherwise, the binary signal dominates the behavior of $\tilde{I}_N(\lambda)$. We note here that models of the type given in (7) will extend easily beyond the two state process to time series which take on values in a discrete finite set.

How well the Walsh-Fourier transform describes continuous valued processes has been explored empirically by various authors (Ahmed and Rao, 1975; Beauchamp, 1975; Harmuth, 1972; Robinson, 1972, to mention a few). How well the Walsh-Fourier transform describes discrete-valued time series, in particular binary time series, is explored by Panchalingam (1985). We discuss some of the findings here.

First, suppose that $\{X(n), n = 0, \pm1, \pm2, \ldots,\}$ is a binary time series generated by "clipping" or "hard limiting" (Kedem, 1980). That is, let $\{Z(n), n = 0, \pm1, \pm2, \ldots,\}$ be an unobservable strictly stationary, continuous valued time series and put

$$X(n) = \begin{cases} 1 & \text{if } Z(n) \geq u \\ 0 & \text{if } Z(n) < u, \quad u \text{ fixed.} \end{cases} \tag{9}$$

This is a reasonable model for various situations; for example, a person will have an allergic reaction when the pollen level crosses a certain threshold. Various binary processes were simulated and analyzed in the sequency domain using the model (9) of Panchalingam (1985). We present some of the examples here. In what follows, $\varepsilon(n)$ is an i.i.d. $N(0,1)$ sequence. Consider the following cases:

$$\begin{aligned} &(i) \quad Z(n) = 0.9Z(n-1) + \varepsilon(n) \\ &(ii) \quad Z(n) = 0.25Z(n-1) - 0.9Z(n-2) + \varepsilon(n) \\ &(iii) \quad Z(n) + 0.9Z(n-1) = \varepsilon(n) + 0.25\varepsilon(n-1). \end{aligned} \tag{10}$$

Figure 2 shows the Walsh-Fourier periodogram smoothed as follows over 5 sequencies

$$\tilde{I}_N(m) = \frac{1}{5} \sum_{k=-2}^{2} d_N^2\{(m+k)/N\} \tag{11}$$

from a sample of size $N = 2^7$ from the clipped process 10(i) with $u = 0$ in (9). Similarly, Figures 3 and 4 show the smoothed periodogram (11) for the clipped processes 10(ii) and 10(iii), respectively, for samples of size $N = 2^7$. Each figure is plotted on a ln scale.

Second, we consider the discrete ARMA (DARMA) processes described by Jacobs and Lewis (1978). Let $\{I(n), n = 0, 1, \ldots\}$, $\{J(n), n = 0, 1. \ldots\}$ and $\{Y(n), n = 0, 1, \ldots\}$ be mutually independent i.i.d. Bernoulli sequences

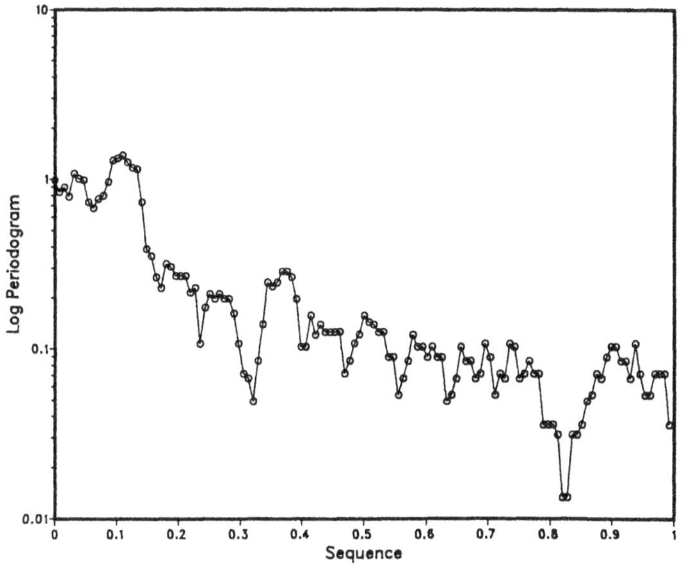

Figure 2. *Clipped AR (1)*

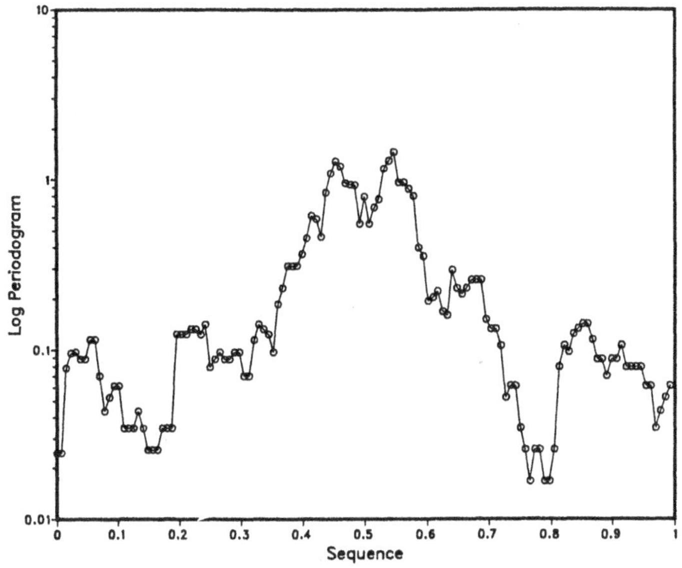

Figure 3. *Clipped AR (2)*

such that $P\{I(n) = 1\} = p$, $P\{J(n) = 1\} = q$ and $P\{Y(n) = 1\} = \theta$. The binary DAR(1) model is written (with $X(0) = Y(0)$) as follows:

$$X(n) = I(n)X(n-1) + (1 - I(n))Y(n), \qquad n = 1, 2, \ldots. \tag{12}$$

Note that $E[X(n)] = \theta$ and $\mathrm{Corr}(X(n), X(n+h)) = p^h, h = 0, 1, 2, \ldots$. Figure 5 shows the smoothed Walsh-Fourier periodogram (11) plotted on ln scale for $N = 2^7$ observations from (12) with $p = 0.75$ and $\theta = 0.50$.

Next we consider the DARMA (1,1) which is written as follows:

$$X(n) = I(n)X(n-1) + (1 - I(n))B(n), \tag{13}$$

where

$$B(n) = J(n)Y(n) + (1 - J(n))Y(n-1).$$

In this case $E[X(n)] = \theta$ and $\mathrm{Corr}(X(n), X(n+h)) = p^{h-1}\{p + (1-p)^2 q(1-q)\}, h = 1, 2, \ldots$. Figure 6 shows the smoothed Walsh-Fourier periodogram (11) plotted on ln scale for $N = 2^7$ observations from (13) with $p = 0.90$, $q = 0.25$, and $\theta = 0.50$.

As a real data example, we compare the state of the diastolic blood pressure (DBP), classified either "high" or "normal", of a mild hypertensive being treated by diet over two periods of time. The first period is 2 months of data (2 observations per day) after an initial 2 months on the diet. The second period is 2 months of data (2 observations per day) after 4 months on the diet. Figure 7 shows the periodogram for the data from the first period, and Figure 8 shows the periodogram for the second period. In each case, the periodogram has been smoothed by a simple average over 5 sequences as given in (11). From Figure 7 we see that there is virtually no signal, that is, there is extremely low power and the periodogram is relatively flat. However, in Figure 8 we note that there is power at the lower sequencies (that is, the data is predominantly long runs of "highs" and "normals"). We note the similarity between Figure 8 and the clipped AR(1) periodogram given in Figure 2. One could conjecture that there is a Markov signal and although the patient is on treatment, whether or not his DBP is "high" or "normal" is conditionally dependent on the state of the most recent past.

5. SUMMARY AND CONCLUSIONS

It has been known, primarily in the engineering disciplines, that Walsh spectral analysis is superior to Fourier spectral analysis for non-sinusoidal time series. This is primarily due to the empirical analyses performed by Ahmed and Rao (1975), Beauchamp (1975), Harmuth (1972), and Robinson

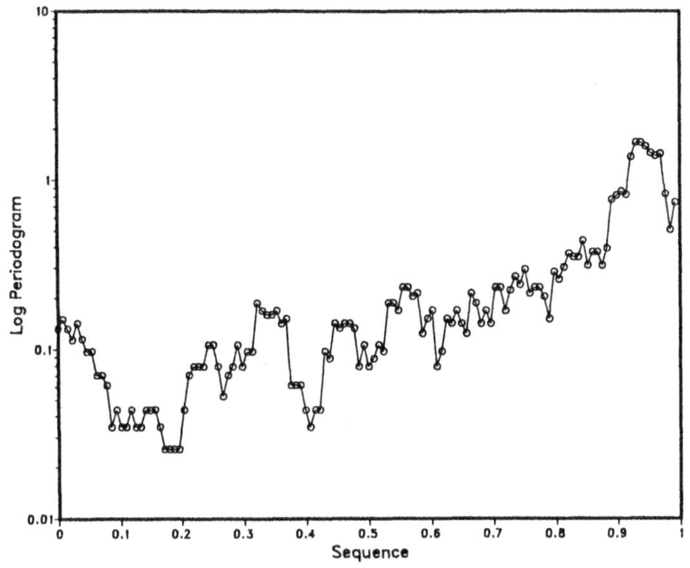

Figure 4. *Clipped ARMA (1,1)*

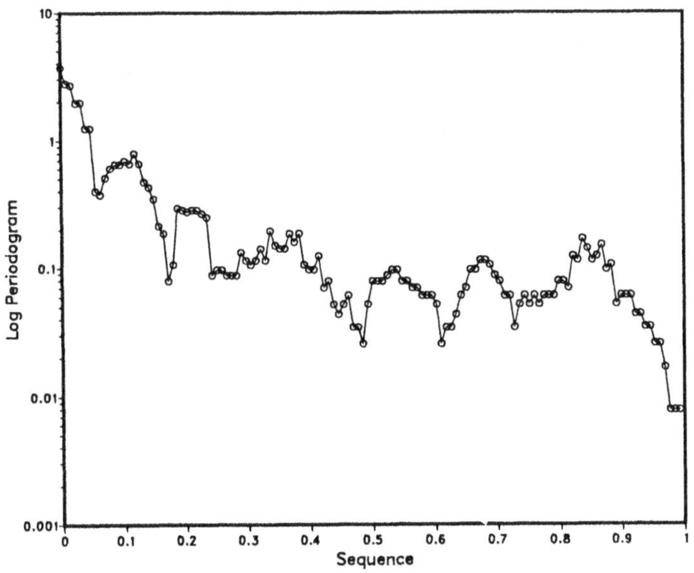

Figure 5. *DAR (1)*

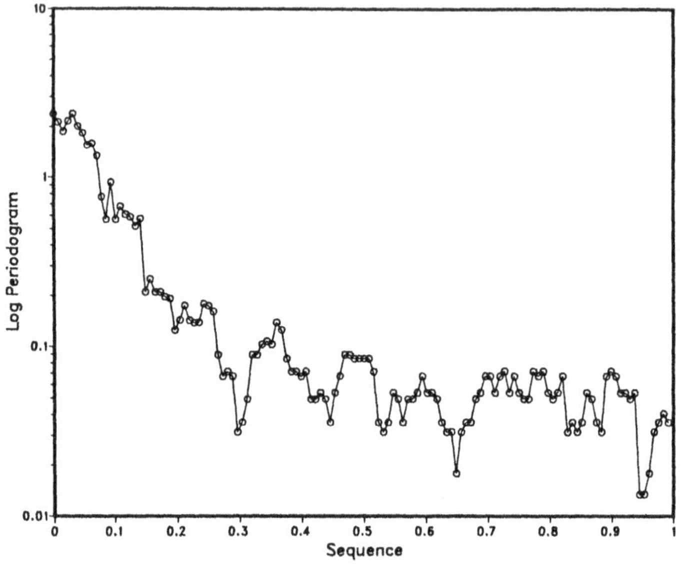

Figure 6. *DARMA (1,1)*

(1972). As we have demonstrated in Section 4, Walsh-Fourier analysis is also informative in the analysis of binary time series and may also be superior in analyzing discrete-valued and non-Gaussian continuous-valued processes. As previously mentioned, very little work has been done on the real time statistical theory of the Walsh-Fourier analysis of discrete-valued time series and hence there are many avenues open for development.

In particular, one may develop models and corresponding analyses for discrete-valued time series along the lines of the binary signal-plus-noise model given in Section 4. Also, we may consider analysis of power for discrete-valued time series where one observes $Z_q(n), q = 1, \ldots, Q, n = 0, 1, \ldots, N-1$, which are Q independent repeated observations on a discrete-valued signal-plus-noise process. One particular hypothesis of interest is whether or not the Q series have a common signal. Such analyses may be carried out in the sequency domain along the lines of the analysis of power for Gaussian time series described by Brillinger (1975, Section 7.9). In such cases a test of $\max_{0 \le \lambda < 1} f(\lambda)$ would be of interest.

We have also seen models in which the signal and the noise are correlated, such as the DARMA models given by Jacobs and Lewis (1978) for which a Walsh-Fourier analysis similar to the uncorrelated case could be developed. The feasibility and practicality of models and filters for discrete valued stationary time series where spectral analysis in the sequency domain

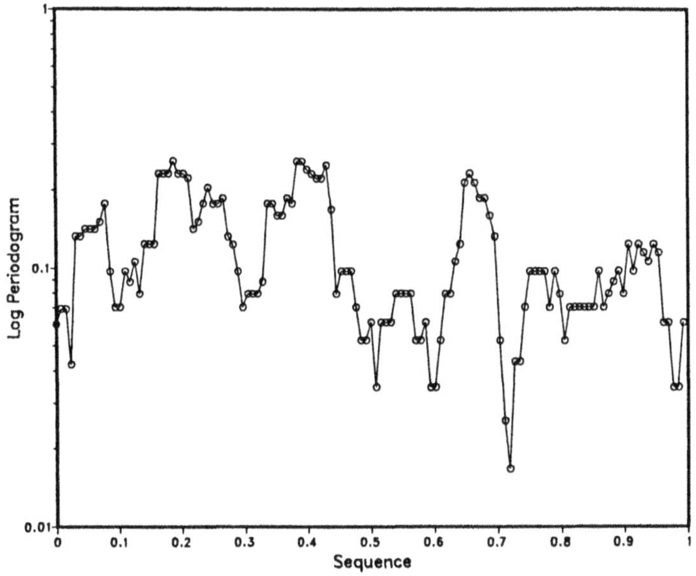

Figure 7. *DBP data, period one*

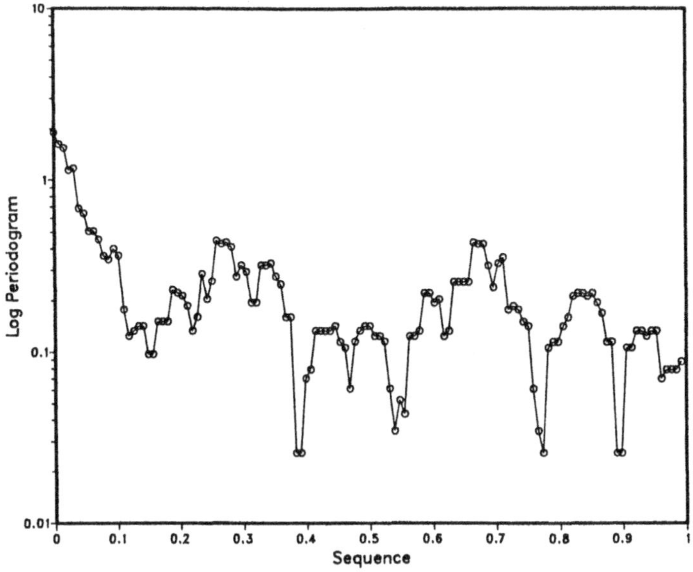

Figure 8. *DBP data, period two*

is of primary interest must be explored further. Moreover, appropriate statistical theory such as applicable central limit theorems and ergodic results must be developed. At present, most of the work in this area is empirical.

ACKNOWLEDGMENT

Part of the work of D. S. Stoffer was supported by the Air Force Office of Scientific Research under Contracts F49620-85-C-0008 and AFOSR-84-0113.

REFERENCES

Ahmed, N., and K. R. Rao (1975), *Orthogonal Transforms for Digital Signal Processing*. New York: Springer-Verlag.

Andrews, H. C. (1970), *Computer Techniques for Image Processing*. New York: Academic Press.

Beauchamp, K. G. (1975), *Walsh Functions and Their Applications*. London: Academic Press.

Brillinger, D. R. (1975), *Time Series: Data Analysis and Theory*. New York: Holt, Rinehart and Winston.

Brillinger, D. R., and J. W. Tukey (1982), "Spectrum estimation and system identification relying on a Fourier transform." Technical Report No. 5, Department of Statistics, Berkeley, California.

Fine, N. J. (1949), "On the Walsh functions." *Transactions of the American Mathematical Society* **65**, 372–414.

Fine, N. J. (1950), "The generalized Walsh functions." *Transactions of the American Mathematical Society* **69**, 66–77.

Fine, N. J. (1957), "Fourier-Stieltjes series of Walsh functions." *Transactions of the American Mathematical Society* **86**, 246–255.

Harmuth, H. (1972), *Transmission of Information by Orthogonal Functions*. Berlin: Springer-Verlag.

Jacobs, P. A., and P. A. W. Lewis (1978), "Discrete time series generated by mixtures. I: correlational and runs properties." *Journal of the Royal Statistical Society*, Series B **40**, 94–105.

Kedem, B. (1980), *Binary Time Series*. New York: Marcel Dekker.

Kohn, R. (1980a), "On the spectral decomposition of stationary time series using Walsh Functions, I." *Advances in Applied Probability* **12**, 183–199.

Kohn, R. (1980b), "On the spectral decomposition of stationary time series using Walsh Functions, II." *Advances in Applied Probability* **12**, 462–474.

Morettin, P. A. (1974a), "Limit theorems for stationary and dyadic-stationary processes." *Boletim de Sociedade Brasileira de Matematica* **5**, 97–104.

Morettin, P. A. (1974b), "Walsh-function analysis of a certain class of time series." *Stochastic Processes and their Applications* **2**, 183–193.

Morettin, P. A. (1981), "Walsh spectral analysis." *SIAM Review* **23**, 279–291.

Morettin, P. A. (1983), "A note on a central limit theorem for stationary processes." *Journal of Time Series Analysis* **4**, 49–52.

Ott, J., and R. A. Kronmal (1976), "Some classification procedures for multivariate binary data using orthogonal functions." *Journal of the American Statistical Association* **71**, 391–399.

Panchalingam, T. (1985), "Estimation of Walsh-Fourier spectral density for binary time series." Masters Thesis, University of Pittsburgh.

Robinson, G. S. (1972), "Discrete Walsh and Fourier power spectra." *Proceedings of the Symposium on the Applications of Walsh Functions*. Washington, DC. 298–309.

Skorokhod, A. V. (1956), "Limit theorems for stochastic processes." *Theory of Probability and its Applications* **1**, 261–290.

Stoffer, D. S. (1985), "Central limit theorems for finite Walsh-Fourier transforms of weakly stationary time series." *Journal of Time Series Analysis* **6**, 261–267.

Oliver G. Jensen [1] and Lalu Mansinha [2]

EXCITATION OF GEOPHYSICAL SYSTEMS WITH FRACTAL FLICKER NOISE

ABSTRACT

Geophysicists often model their measurements, derived from natural processes, as the linear superposition of a simple rational system function and a purely random excitation process. For many geophysical processes, the assumption of linearity for its deterministic component is sufficient but the assumption of a purely random excitation often and easily leads to a misidentification of the system function. Many geophysical systems are excited by stochastic processes which appear to be stationary even on geological time scales but which possess a preponderance of long period components. Self-similar, fractal stochastic processes form a class of possible geophysical excitations having "power spectrum" of the form $1/\mid f \mid^k$. Of this class, flicker-noise processes, for which $k = 1$ exist, on the boundary between the stationary and evolutionary subsets. No fractal stationary random excitation can provide for greater weighting of long period components.

The Chandler wobble of the earth's rotation axis can be essentially described as a single-pole linear system. The multitude of natural forces which contribute to its excitation combine as a stochastic process which is heavily weighted in long periods. Because of its basic importance in astronomy, navigation, time-keeping, etc., the wobble has been carefully measured since the turn of this century. Recent advances in geodetic and astronometric technology has provided a reliable, homogeneous data set which can be directly decomposed into a linear, deterministic wobble function and stochastic excitation. The use of the flicker-noise excitation model allows for the direct identification of the theoretically-simple, single-pole resonance. A subset of the pole position record, obtained from the Bureau International de l'Heure in Paris is analysed in terms of a modified autoregressive data model com-

[1] Geophysics Laboratory, McGill University, Montreal, Quebec H3A 2A7
[2] Department of Geophysics, The University of Western Ontario, London, Ontario N6A 5B7

I. B. MacNeill and G. J. Umphrey (eds.), Time Series and Econometric Modelling, 165–188.

prising an all-pole system function excited by a minimum-power, flicker-noise process.

1. INTRODUCTION

Linear data models are often used by geophysicists in the description and analysis of measurements which are regarded as being derived from the excitation of a deterministic system by some stochastic process. The excitation of the system represents some generally unknown but natural geological or geophysical variation. The system includes the instrumentation used in the observation and the current geophysical theory of the phenomena involved. Analysis and subsequent interpretation of the measurements allows the discovery of essential properties of the system and its excitation, that is of the characteristic geological condition and the manifest geophysical phenomena. Applying a linear data model, the geophysicist recognizes the excitation as a stochastic process corresponding to the model innovation; the system which is usually, but not generally, linear and rational contains certain undetermined parameters and corresponds to the operator function of the data model.

Continuing technological developments improve the quality of measurements while the geophysical theory becomes an ever more detailed and complete description of natural phenomena. As a consequence, we are forced to elaborate the statistical models of the excitations of the systems. In this paper, we propose to argue, by example, in support of a particular statistical excitation model which we believe to be appropriate for the description of measurements derived from a wide class of geophysical phenomena.

2. STOCHASTIC MODELS OF GEOPHYSICAL DATA

White-Gaussian or filtered white-Gaussian processes have long been used directly in (or implied by the standard methods of) spectral analysis and signal decomposition employed in traditional geophysical analysis. The autoregressive (AR), moving-average (MA) and autoregressive-integrated-moving-average (ARIMA) linear data models are commonly used in geophysical data modelling. These models are based upon the assumption that the signal or measurement derives from a linear operation on uncorrelated Gaussian noise. In geophysics, and particularly in the analysis of seismic reflection data, the AR data model has been found to be most useful. The predictive deconvolution (then called decomposition) method for the analysis of the seismic reflection records was first suggested by Wadsworth *et al.* (1953). Robinson's now classical "MIT GAG" report and Ph.D. thesis (Robinson 1954; republished 1967) formalized the method and properly recognized the roots of the technique in the work of Yule (1927). Common predictive deconvolution as practised in geophysical analysis is essentially a variation on the Box-Jenkins (1970) forecasting theory for autoregressive time series. Within the geophysical community, this AR-model-based deconvolution process has been much developed and elaborated to account for evolutionary (Clarke, 1968) and multi-channel systems (Burg, 1964; Davies and Mercado, 1968; Treitel, 1970). Much work has been devoted to the development of efficient and accurate algorithms for the determination of the appropriate AR model from vast seismic data sets (e.g. Burg, 1967, 1975; Wiggins and Robinson, 1965; Ulrych and Clayton, 1976; Barrodale and Erickson, 1980; Tyraskis and Jensen, 1985).

In the analysis and decomposition of seismic reflection data, the components of the AR model correspond to real physical elements. The models are "structural" (Akaike, 1985): the model innovation corresponds to the excitation of the linear geophysical system by a subjectively random geological condition. Because the geophysical system (comprising the instrumentation used in the seismic surveying technique, the source of seismic wave energy and the theoretically deterministic part of the seismic wave propagation phenomenon) is essentially resonant, a purely autoregressive model filter operator is most appropriate for its description. Little advantage has been found in using linear data models with a moving-average component.

Most recent geophysical interest in the seismic linear data modelling problem has involved the non-Gaussian and/or self-correlated properties of the structural innovation function. Wiggins (1978) introduced the so-called minimum-entropy deconvolution method which obtains that linear filter operator which, while consistent with the seismic data, maximizes the kurtosis of the structural innovation. Postic *et al.* (1980) have elaborated this method to allow for maximization of arbitrary fractional-order moments of the probability density function of the innovation. These methods are recognized to be clearly superior to the classical predictive deconvolution method which uses the minimum-variance (or least-squares) of the innovation as the solution criterion. Hosken (1980) showed that a stochastic model of the seismic reflectivity sequence, equivalent to the structural innovation in the modelling problem, as derived empirically from geological logs from several major petroleum-bearing sedimentary basins, is strongly leptokurtotic. This discovery justifies the current preference for maximum kurtosis, rather than minimum variance, as the criterion of choice in sophisticated seismic deconvolution. Vafidis (1984) and Jensen and Vafidis (1986) have extended these concepts to allow for extremal skewness and kurtosis as criteria in the solution of a more general class of inverse problems.

Hosken (1980) also showed that the acoustic impedance as a function of depth in a sedimentary geological sequence does not show a white spectral character and is therefore self-correlated. In particular, he showed that the velocity-depth function, which determines the seismic reflectivity sequence, has the nearly $1/|f|$ spectral characteristic of flicker noise, where f is the frequency. This accounts directly for the well-known fact that a seismic reflectivity sequence is deficient in low-frequency power density in comparison to an uncorrelated sequence. The prior assumption that the seismic reflectivity sequence is uncorrelated and Gaussian, that is, the assumption of a purely random structural innovation in the data model, is not justifiable on empirical grounds.

Non-white and/or non-Gaussian stochastic models are important in many areas of geophysical analysis apart from seismic reflection deconvolution. Indeed, Mandelbrot (1983) has discussed the ubiquitousness of "fractal" stochastic processes of the nearly $1/|f|$ type (i.e. flicker noise) in many natural and geophysical phenomena. Jensen and Mansinha (1984) have shown that the Earth's rotation pole-path is well modelled as the excitation of the Chandler resonance (Munk and MacDonald, 1960) by a fractal flicker-noise process. Also, Jensen (1982, unpublished) has shown that the basic geological excitation of airborne electromagnetic prospecting systems is best modelled as a flicker-noise process. One might reasonably expect that general "geophysical landscapes" of topography, reduced gravity anomaly, electrical resistivity variations, magnetic anomaly or susceptibility, for example, can be best described as fractal flicker noise.

Eventually, we must develop linear data modelling methods which simultaneously allow for self-correlated and non-Gaussian structural innovations. Gauss (1839) solved the first geophysical inverse problem, modelling the worldwide observations of the geomagnetic field as a finite order expansion in terms of associated Legendre functions. Employing the least-squares criterion and assuming spatially uncorrelated differences between his model and data set, he proved that the Earth's magnetic field is of internal origin. While Gauss made no claim that the model structural differences were necessarily uncorrelated and of minimum variance, he was forced to choose such criteria for computational convenience. Geophysicists are not today so technologically disadvantaged; given the power and precision of contemporary computing machines, we need not fall back to Gauss' choice of criteria. We can accommodate actual (geo)physical knowledge of the structural innovation in our data modelling procedures. We must now develop analytical methods which allow for non-Gaussian and self-correlated stochastic composition.

3. FRACTAL STOCHASTIC PROCESSES

In numerous papers and articles culminating in his book, *The Fractal Geometry of Nature,* Mandelbrot (1983) developed the concept of fractal (*fract*ional dimension*al*) curves and surfaces. He defines his new geometry as a set for which the Hausdorff-Besicovitch dimension strictly exceeds the topological dimension. A most important feature of the fractal geometry is its self-similarity: each piece of the geometry is similar to its whole except for scale. Fractal curves and surfaces may be either deterministic, in the sense that they may be generated by a regular rule, or stochastic, and having inherent randomness. A subset of the stochastic fractal geometry comprises the spectrally "scaling noises" which are characterized most simply by their $1/f^k$, $0 \leq k \leq 2$, power density spectrum. Mandelbrot states:

> "Many scaling noises have remarkable implications in their fields, and their ubiquitous nature is a remarkable generic fact."

One can demonstrate the fractal self-similarity of a time series of a frequency band-limited scaling noise as follows.

1. Low-pass filter the sequence to reduce its bandwidth from $-f_0 < f < f_0$ to $-f_0' < f < f_0'$,
2. rescale time: $t' = t\ f_0'/f_0$,
3. rescale amplitude: $a' = a(f_0/f_0')^{1/2}$.

The form of the sequence is thus preserved. Within this class of noises, flicker noise for which $k = 1$ is unique in that it represents that scaling noise which possesses the largest value of k (consequently the greatest weight of low frequencies) while remaining properly stationary. For $k > 1$, the variance from the mean (or initial state) must increase with time and, therefore strictly, no power spectrum exists. However, the energy density spectrum of a sample of such noise does describe the $1/f^k$ form over the range of frequencies that may be estimated given the length of sample and its sampling increment. Uncorrelated or white Gaussian noise is spectrally scaling with $k = 0$. A random walk in amplitude with Gaussian steps is spectrally scaling with $k = 2$. Figure 1 shows the form of white noise, flicker noise and "brown" noise (the random walk with Gaussian steps) obtained from pseudo-random generators of these noises which were developed by one of the authors (O. G. J.).

Geophysicists are, or should be, attracted to the flicker noise description of the randomness in the phenomena which they observe for several reasons. Geophysicists apply physics as their tool for the description of the conditions and phenomena of the Earth, planets, satellites and the solar-system environment. Properly covariant physics must apply everywhere in the universe and for all time. More locally and specifically, many of the geological

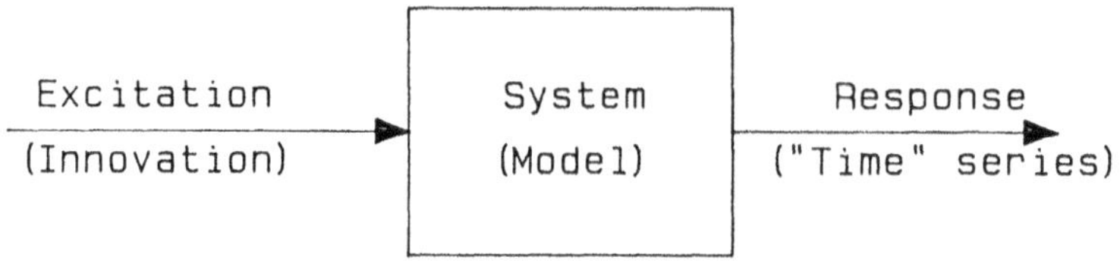

Figure 1. *A linear system model representing the formation of geophysical measurements. Note the time-series equivalent terms shown bracketed.*

phenomena geophysicists attempt to describe by our theories are slowly evolutionary. We often presume that, during any time of observation which is short compared with geological time scales, the geophysical manifestations are essentially stationary. We further presume that our geophysical theories can equally well apply to any appropriate geological subset; that is, they should apply just as well in Africa as they do in Canada. Then, recognizing the colloquial fact that most geophysical observations in time or space show a strong preponderance of low-frequency composition which cannot often be accounted for by basic geophysical theory, we are led to the choice of flicker noise as the preferred form of excitation or structural innovation in data modelling procedures.

4. THE DECONVOLUTION PROBLEM

The response of a linear geophysical system excited by a stochastic geophysical or geological variation (Figure 2) is determined as the linear superposition of the excitation and system functions:

$$\begin{aligned} r(t) &= \int_{-\infty}^{\infty} e(s)h(t-s)ds \\ &= e(t) \, * \, h(t), \end{aligned} \tag{1}$$

where $r(t)$ is the response, $e(t)$, the excitation and $h(t)$, the system function. The symbol $*$ is used to represent the convolution or superposition integral form above. The stochastic excitation is assumed to correspond to some appropriate statistical model while being extreme in some sense. In the most simple deconvolution theory, the excitation is assumed to be purely random (frequency bandwidth-limited white-Gaussian noise) with minimum variance. The full complexity of the geophysical system, function $h(t)$, is

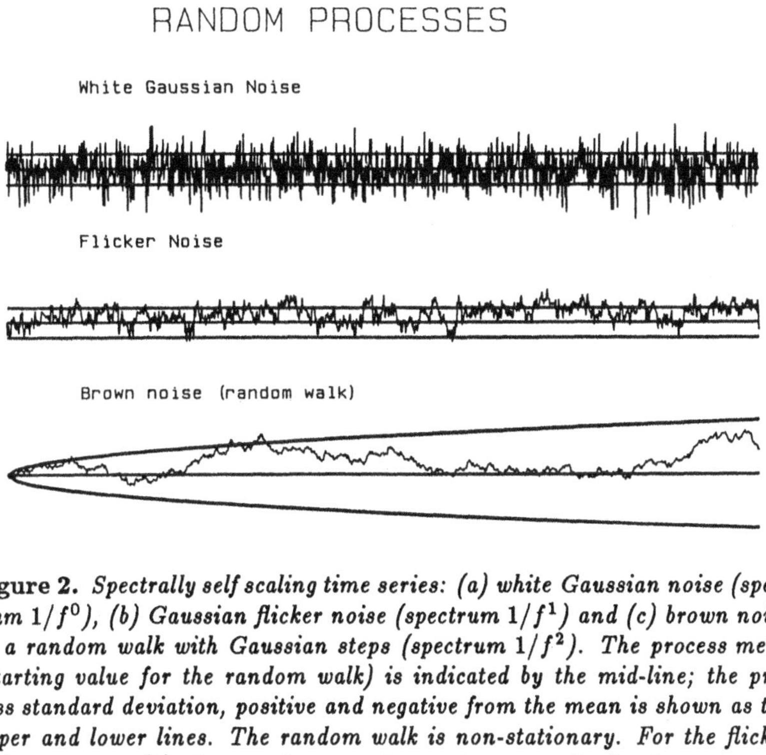

Figure 2. *Spectrally self scaling time series: (a) white Gaussian noise (spectrum $1/f^0$), (b) Gaussian flicker noise (spectrum $1/f^1$) and (c) brown noise or a random walk with Gaussian steps (spectrum $1/f^2$). The process mean (starting value for the random walk) is indicated by the mid-line; the process standard deviation, positive and negative from the mean is shown as the upper and lower lines. The random walk is non-stationary. For the flicker noise sequence (b), the sample mean does not necessarily closely correspond to the process mean.*

generally assumed to be unknown. We, however, desire its inverse, $h^{-1}(t)$, so that under convolution of this function with the observed and recorded response, $r(t)$, we may determine the excitation, $e(t)$ as follows:

$$e(t) = r(t) \, * \, h^{-1}(t)$$

since

$$h^{-1}(t) \, * \, h(t) = \delta(t),$$

where $\delta(t)$ is the Dirac impulse function. Nature and geophysical theory constrain the properties of $h(t)$. Often, we expect this function to be properly causal, i.e.,

$$h(t) = 0, \quad t < 0;$$

stable, i.e.,

$$\int_0^\infty h^2(t)dt \text{ finite,}$$

so that energy be conserved, and usually but not always, we may expect that the function is of minimum-delay characteristic. This latter condition, also called the minimum-phase condition, is apparently appropriate for all completely described and passive temporal geophysical systems (Ulrych and Lasserre, 1966). It is essentially a variation of Fermat's principle which holds that any physical signal follows the shortest or longest possible path. In this case, the power of the excitation is most quickly transferred through the system to provide its response. The causality and minimum-delay conditions do not necessarily hold for all geophysical deconvolution problems; in particular, because space, unlike time, does not evolve in a single direction, deconvolution problems involving geophysical space series (one or more spatial variables substituting for the time dependence in a time-series analogy) cannot presume these properties. Here, we shall restrict our attention to the simpler, proper time-series deconvolution problem for which we have $h(t)$ causal and of minimum delay. This, then, allows that $h^{-1}(t)$ is also causal and of minimum delay; the system is invertible.

We are required to find that causal, stable, minimum-delay inverse function, $h^{-1}(t)$, from our observations record such that the excitation corresponds to the appropriate prior-assumed statistical model. In the classical predictive deconvolution problem (Robinson, 1954), the excitation is assumed to be purely random and of minimum variance.

5. POLAR MOTION

The rotation axis of earth is not fixed to the earth but has periodic and secular motions within the earth. On the other hand, an observer in space would notice that the rotation axis is more or less stationary in space and it is the earth that has slow motions (in addition to the spin) about the rotation axis. The change in orientation of the earth in space also implies that, at any given point on the earth, all the stars would be displaced by identical amounts. Since the latitude of a place is measured by observing reference stars, the measured latitude at any given location will also reflect the apparent stellar displacement. The change in longitude appears as an error in the spin rate of the earth, which is often expressed as a change in the length-of-day (l.o.d.). The terms "variation of latitude", "polar motion" and "wobble" are used interchangeably to describe the same physical phenomenon (see Munk and MacDonald, 1960; Lambeck, 1980).

The wobble has secular (i.e., long period), seasonal, annual, and 14 monthly components. Other minor components are also present. The 14-month period was first detected by S. C. Chandler and is usually referred to as the "Chandler wobble". On the basis of the theory of rotating rigid bodies, Leonhard Euler predicted a 10-month free wobble for the earth in the eighteenth century. The lengthening of the period to 14 months is due to elastic yielding of the earth, as well as the presence of the ocean and the fluid outer core. Occasionally, the Chandler wobble is referred to as the "free Eulerian nutation".

The Chandler wobble arises whenever the axis of rotation is not coincident with the polar axis of figure which is the axis of maximum moment of inertia. The pole then executes a slowly decaying spiral around the axis of figure with a 14-month periodicity until the two axes are coincident. One can view this motion as that of a damped harmonic oscillator. Thus, regardless of the value of the "Q" of the oscillator, the rotation axis and the figure axis should have become coincident over geological time. At present, neither the damping mechanism nor the excitation source have been definitely identified. Speculations abound. Geophysical interest in the phenomenon is spurred by the hope that identification of the two mechanisms will provide insight into the structure and properties of the earth.

There is also a more utilitarian interest. Variation of the latitude and longitude affects the geographical reference frames and the measurements of time. Thus interest in the wobble has been high among geodesists and timekeepers. In 1899, the International Latitude Service (ILS) was established. Since 1900, the ILS has been observing the position of the rotation pole. Originally the data was obtained from five ILS stations near the 39°N parallel. In 1962, the ILS was reorganized into the International Polar Motion Service (IPMS). The IPMS continues the work of the ILS, but now includes data from other observatories.

In 1955, the Bureau International de l'Heure (BIH) in Paris was entrusted with the task of determining and predicting the path of the rotation pole to aid in timekeeping. The BIH uses time and pole position data from all available sources (see Mueller, 1969). In addition to data from optical instruments, the BIH has been using pole positions determined from space geodesic and radio-interferometric methods.

Although the physics of the wobble appears simple enough, the measurements present problems of extraordinary complexity. An earth "fixed" coordinate system is not immune to slow and undetectable drift. Possible causes are shifts in the local vertical and physical motion of the observatories due to tectonic processes. The mean pole has shifted from its position around 1900. In 1962, the International Union of Geodesy and Geophysics defined the mean pole position during the epoch 1900 to 1905 as the Conventional International Origin (CIO). The location of the CIO is fixed to the earth only as well as the five original ILS stations are fixed. While the IPMS-ILS record has remained essentially homogeneous since its beginning, it has not been able to accommodate the remarkable improvements in the accuracy of measurements which have been achieved through contemporary technology and applied in the compilation of the inhomogeneous BIH records. The advent of satellite geodetic and navigation systems and of radio-astronomical interferometry methods of measuring the pole path has resulted in a current standard error of measurement of less than 10 cm in each of the two orthogonal coordinates (along 0° and 90°E longitude, origin CIO) for the pole position compiled at 5-day intervals by the BIH. Unfortunately, this BIH record of "raw" values (no averaging or smoothing of the computed pole position) is not homogeneous for a long period because of the recent incorporation of contemporary technology. As in any time series problem, we would prefer to use the longest and cleanest record available for our analyses. Unfortunately, geophysical data sets of most recent epoch and, consequently, of short duration are most free of error. We are forced to select between the long noisy record and the short clean one. Through 1982, the maximum amplitude of the pole position offset from the CIO has been about 10.5 m (0°) and 17.5 m (90°E) and consequently, the standard errors of measurement correspond to about 1 part in 100–200. The BIH pole path record of raw offsets at 5-day sampling intervals which we use in this article remained approximately homogeneous from the beginning of 1978 through to the end of 1982.

The pole-path record as compiled by the BIH comprises two major oscillations with centres offset from the CIO: the Chandler wobble and the annual wobble. The annual wobble is a forced oscillation of the body of the earth caused by regular meteorological variations in the atmospheric and hydrological mass balance. Longer period climactic trends and cycles cause a low-level amplitude modulation of the annual wobble's period of 365.2422 (solar) days. It forms a slightly elliptical path component which beats with the damped Chandler resonance. The Chandler resonance, which has a period of about 420 days, is continuously excited to an average amplitude which is similar to that of the annual wobble. Conventionally, we reduce the annual component by direct subtraction of that elliptical path which best correlates with the sample-mean-reduced pole-path record; essentially, this obtains the best fit of the annual wobble in a least-squares sense. This mean- and annual-reduced pole-path record can then be represented by a linear systems model as the convolution of the damped Chandler resonance transient with the "excitation pole function". That is, in absence of any measurement error, the path as a function of time can be described by equation (1) in the form

$$z(t) = c(t) * p(t),$$

where $z(t)$ is the path of annual-reduced pole positions, $p(t)$ is a stochastic excitation function and $c(t)$ is the damped Chandler resonance transient. We use a right-handed coordinate system describing the pole path as a complex-valued time function with

$$z(t) = x(t) + iy(t),$$

where $x(t)$ is the displacement of the pole from the C.I.O. along the Greenwich meridian (0°) and $y(t)$ is the displacement along the 90°E meridian.

The Chandler resonance function, $c(t)$, must be causal and stable and should have the minimum-delay or phase characteristic. Therefore, there exists a causal, stable, minimum-phase inverse, (t), such that

$$\gamma(t) \ * \ c(t) = \delta(t)$$

and we can deconvolve $z(t)$ to obtain the excitation pole function (Smylie *et al.*, 1970). If now $\gamma(t)$ is normalized such that

$$b(t) = \delta(t) - \gamma(t)$$

with

$$b(0) = 0,$$

then we may describe a continuous analog to an autoregressive linear data model of the pole path:

$$z(t) = b(t) \ * \ z(t) + p(t).$$

Sampled without aliasing with interval Δt, this continuous analog reduces to the discrete, infinite-order autoregressive model approximation:

$$z_n = b_n \ * \ z_n + p_n \tag{2a}$$

or alternately

$$\gamma_n \ * \ z_n = p_n, \tag{2b}$$

where the symbol * represents the discrete convolution operation. b_m, $m = 1, 2, \ldots$ is the infinite order autoregressive, one-step forecasting operator which is related to the autoregression or deconvolution operator as

$$\gamma_m = \delta_m^0 - b_m; \quad m = 0, 1, \ldots, \infty,$$

where δ_m^0 is the Kronecker delta operator. In a notation more commonly used by statisticians in time series analysis, equations (2a,b) can be represented as

$$\phi(B) z_n = p_n \tag{2c}$$

where

$$\phi(B) = 1 - \sum_{i=1}^{\infty} b_i B^i$$

and where B is the one-step backshift operator (Anderson, 1976). For a finite-length sequence of observations

$$z_n; \quad n = 0, 1, \ldots N$$

and a finite-order autoregressive forecasting operator,

$$b_m; \quad m = 1, 2, \ldots M,$$

equations (2a–c) reduce to a system of linear equations

$$z_n = \sum_{m=1}^{M} b_m z_{n-m} + p_n \tag{3a}$$

which may be rewritten in vector-matrix form

$$\mathbf{z} = \mathbf{Z} \cdot \mathbf{b} + \mathbf{p}. \tag{3b}$$

Properly, the Chandler resonance can be described by an autoregressive forecasting operator of order $M = 1$ where

$$b_1 = e^{iu}\Delta t,$$
$$b_0 = 0,$$
$$b_m = 0; \quad m > 1$$

and

$$u = w_0 + i/\tau,$$

where further,

$$P = 2\pi/w_0$$

is the period of the Chandler resonance and τ its damping time constant.

Geophysical theory provides that the actual excitation of the wobble will be that which possesses the minimum power whatever its self-correlation characteristics may be. Classical least-squares inversion for the M-order autoregressive forecasting operator obtains that estimate, $\hat{\mathbf{b}}$ (with elements $\hat{b}_m$, $m = 1, 2, \ldots, M$) of the vector, $\mathbf{b}$, which is consistent with a minimum-variance or minimum-power excitation. That is,

$$\hat{\mathbf{b}} = (\mathbf{Z}^t \cdot \mathbf{P}_N^{-1} \cdot \mathbf{Z})^{-1} \cdot (\mathbf{Z}^t \cdot \mathbf{P}_N^{-1}) \cdot \mathbf{z}, \tag{4a}$$

where the symbol t indicates the complex-conjugate-transposition of the appropriate vector or matrix. The autocovariance matrix of the assumed zero-mean, stationary excitation having variance $\sigma_p^2 = E[p_i \cdot p_i^*]$ is

$$\mathbf{P} = \sigma_p^2 \mathbf{P}_N \tag{4b}$$

with elements

$$p_{ij} = E[p_i \cdot p_j^*].$$

P_N is the prior-assumed, diagonal-normalized autocovariance matrix of the excitation process which is the structural innovation in the autoregressive data model. The least-squares analysis also obtains estimates of the excitation scale-variance,

$$\hat{\sigma}_p^2 = \hat{\mathbf{p}}^t \cdot \mathbf{P}_N^{-1} \cdot \hat{\mathbf{p}}/(N - 2M + 1), \tag{4c}$$

where the estimated or deconvolved excitation vector,

$$\hat{\mathbf{p}} = \mathbf{z} - \mathbf{Z} \cdot \hat{\mathbf{b}}. \tag{4d}$$

If we were to make the a *priori* assumption that the excitation process, equivalently the structural innovation of the linear data model, is stationary and uncorrelated, that is $\mathbf{P}_N = \mathbf{I}$, the identity matrix, this solution would reduce to the so-called "exact least-squares solution" to the autoregressive data modelling problem which was introduced to geophysics by Ulrych and Clayton (1976). Our understanding of the Chandler wobble suggests that it would be much more appropriate to assume that the structural innovation has a flicker-noise form. This will allow for the greatest possible weighting of long periods in the excitation process while retaining true stationarity and a desired fractal character. In the relatively short records which have so-far been compiled, the long periods in the excitation appear as a slow trend. Practically, we are forced to assume that the process is bandlimited in a range of frequencies $f_\ell < | f | < f_u$ where the upper limit is, at least, no greater than the Nyquist frequency, $f_{NYQ} = 1/2\Delta t$, and the lower limit corresponds to a period sufficiently longer than the duration of the record under analysis. In the absence of any knowledge to the contrary, we further assume that the real and imaginary components of the structural innovation are not cross-correlated; then the Toeplitz autocovariance matrix, P_N, will be real-valued.

What we have essentially described in equations (4a–d), above, is a direct least-squares solution for a complex-valued time series modelled as an ARIMA$(M; 1/2, 0)$ process. The equivalent half-order differencing (Granger and Joyeux, 1980) arises directly through the substitution of the minimum-variance, correlated flicker noise in replacement of the commonly-assumed purely random innovation. In problems where the data-sets are either much longer or perhaps less valuable and unique than those being discussed here, the direct use of half-order differencing and standard autoregressive estimation might suffice.

Using the theory described so far, Jensen and Mansinha (1984) analysed the BIH pole-path record of raw 5-day means during the period 1967.27 to 1981.73 in an attempt to better determine the period and damping time constant of the Chandler wobble and to obtain, by deconvolution, the excitation pole path for its interpretation in terms of the known geophysical and geological events which had been recorded during this epoch. They compared inversions of the data set based upon the flicker-noise and white-noise assumptions. AR model order $M = 4$ was indicated for this data set by the Akaike (1969, 1974) final-prediction error criterion applied to the classical model with a purely random innovation. The same model order was used in the flicker-noise inversions. The flicker inversion determined one significant long-period component with a period corresponding to 425 days and a $Q = 32$, which is equivalent to a damping time constant of about 12 years. This is the Chandler resonance.

Much of the geophysical interest in the phenomenon of the Chandler wobble concerns its damping time constant since a long time constant (very high Q) would allow for its excitation to observed levels with low power. It is difficult to account for the excitation of the wobble by any known phenomena if its Q is less than, perhaps, 30 (Lambeck, 1980). Jensen and Mansinha (1984) introduced their data modelling theory in the hope of defining the high Q of the Chandler resonance.

The standard AR(4) model with a solution obtained using the Ulrych and Clayton (1976) direct least-squares method resolved the Chandler resonance ($P = 429$ days, $Q = 28$) but also determined a substantial, and geophysically unaccountable, resonance ($P = -353$ days, $Q = 0.43$) representing a wobble in the negative rotation sense. Jensen and Mansinha (1984) recognized that this false wobble resonance was the result of an attempt by the classical AR data model to account for the preponderance of long periods in the data set as a property of the autoregression operator even though they are properly due to the properties of the stochastic excitation phenomenon. The classical AR model had, by implication, ascribed false character to the geophysical system. Above, we showed that the Chandler resonance which should rest as the only significant period in these data following annual and mean reduction can be described by a purely autoregressive forecasting operator of order $M = 1$. Any other periods which extend the order of the required best operator are false providing the description of our data model is complete. Unfortunately, a major component of the BIH pole-path data has not been included in the theory so-far described here and by Jensen and Mansinha (1984). We have not accounted for additive measurement errors; we have described the model appropriate to a measurement-noise-free data set.

Measurement noise in the data is revealed by the classical and flicker-noise AR models, just described, as equivalent low-Q resonances at shorter periods. The classical AR(4) model found two resonances to fill in the spectrum of the data set: $P = 105$ days, $Q = 0.66$ and $P = -148$ days, $Q = 0.44$. The flicker-innovation AR(4) model found two similar resonances: $P = 85$ days, $Q = 1.2$ and $P = -135$ days, $Q = 1.2$. The fourth period not yet accounted for in the AR(4) flicker-innovation model was found to be insignificant with a root magnitude of only 0.06. It is clear that resonances of larger amplitude are assigned by the flicker-innovation solution to account for measurement error. This is because the flicker innovation is relatively deficient in short period spectral components as compared to a purely random innovation. Consequently, the AR model operator must further amplify these short periods in comparison to the white-innovation operator in order to fill in the spectrum of the data set. Jensen and Mansinha (1984) recognized that both modelling procedures were incomplete because they could not account for a separately-additive measurement error. They proposed augmentation of the model described by equations (3a,b) as follows.

6. THE ADDITIVE NOISE-AUGMENTED/AUTOREGRESSIVE DATA MODEL

Only the signal component of the pole-path data must, by geophysical theory, follow the form of a structural autoregressive data model. The measurement error in the data are properly and separately additive. That is, the signal may be described by the AR model as follows:

$$s_n = b_n * s_n + p_n,$$

where the signal, s_n, is the difference between the observation, z_n, and the measurement error, e_n:

$$s_n = z_n - e_n.$$

If, now, we reform the "apparent structural innovation" as

$$q_n = p_n + e_n - b_n * e_n, \tag{5a}$$

$$= p_n + \gamma_n * e_n, \tag{5b}$$

we may employ an AR model for the data comprising both signal and additive noise as follows:

$$z_n = b_n * z_n + q_n. \tag{6}$$

We expect that the measurement error is stationary and uncorrelated, with zero mean and known variance, σ_e^2. We believe that the excitation pole path is also stationary, with zero mean, variance, σ_p^2, and is selfcorrelated like flicker noise.

Requiring the autocorrelation function of the excitation pole path

$$\phi_{p_k} = E[p_n \cdot p_{n+k}^*],$$

to follow a flicker noise model, we determine that the autocorrelation function of the apparent innovation described by equations (5a,b) has the form

$$\begin{aligned} \phi_{q_k} &= \phi_{p_k} + \sigma_e^2 \sum_{m=0}^{M} \gamma_m \gamma_{m+k}^*, \quad k = 0, 1, \ldots, M, \\ &= \phi_{p_k}, \quad k > M. \end{aligned}$$

Note that this autocorrelation function, from which the variance-covariance matrix required in a least-squares solution of the problem must be formed, depends upon the regression operator which is the subject of the inversion. A straight-forward iterative procedure for obtaining this operator is now described.

Normalizing the complex-valued, Toeplitz variance-covariance matrix of the apparent innovation as follows,

$$\mathbf{Q} = \sigma_p^2 \mathbf{Q}_N, \tag{7a}$$

we solve for a temporary estimate of the regression vector,

$$\hat{\mathbf{b}} = (\mathbf{Z}^t \cdot \mathbf{Q}_N^{-1} \cdot \mathbf{Z})^{-1} \cdot (\mathbf{Z}^t \cdot \mathbf{Q}_N^{-1}) \cdot \mathbf{z}, \tag{7b}$$

which minimizes the variance of the excitation power estimate,

$$\hat{\sigma}_p^2 = \hat{\mathbf{q}}^t \cdot \mathbf{Q}_N^{-1} \cdot \hat{\mathbf{q}}/(N - 2M + 1), \tag{7c}$$

and which essentially minimizes the variance of the apparent innovation determined by deconvolution,

$$\hat{\mathbf{q}} = \mathbf{z} - \mathbf{Z} \cdot \hat{\mathbf{b}}. \tag{7d}$$

Knowing the error variance, we recompute the variance-covariance matrix and solve again for the estimated regression vector and excitation variance and deconvolve for a new apparent innovation vector. We iterate until the excitation variance converges.

For the problem solved here, we initialized the inversion with the solution to the measurement error-free model described above and terminated iterations when the continuously decreasing estimate of the excitation variance changed by less than 1 part in 10^4 between successive steps. Then, reinitializing the inversion with an estimate of the excitation variance 1 part in 100 lower than that just found, and using the regression vector just found, we continued iteration until the continuously increasing estimate of the excitation variance changed again by less than 1 part in 10^4 between successive steps. For a flicker noise excitation path, this procedure appears to be robust; the eventual solution is quite unaffected by starting conditions and is insensitive to slight truncations of the pole path record. On the other hand this inversion procedure is not stable, often diverging after four or five iterations, when both the innovation and additive measurement error are assumed to be uncorrelated.

7. POLE PATH SPECTRA

The mean- and annual-reduced BIH pole-path record of raw 5-day means (1978–82) comprising 366 points and used in these analyses is shown in Figure 3. A classical AR data model, computed using the Ulrych-Clayton (1976) algorithm modified for complex-valued data (equations 4a–d with $P_N = I$), showed a minimum final-prediction error (Akaike, 1969) at model-order $M = 5$. Since we have not yet extended the FPE or AIC criteria for model selection to allow for generally self-correlated innovations or for the additive noise-augmented AR model, we have used model order $M = 5$ throughout our several solutions following. Factoring the AR(5) operator, we may determine its complex-valued roots or zeros. Each root represents a damped harmonic component of the system which is stimulated by the excitation pole. We are especially interested in that component which corresponds to the Chandler wobble.

Figure 4 shows the conventional AR model spectrum of the BIH pole-path record obtained in the usual way by division of the squared Fourier transform of the computed regression operator, γ_m, $m = 0, 1, \ldots, M$, into a white spectrum scaled by the computed innovation variance. The very sharp resonance ($Q \simeq 69$) of the Chandler wobble at 0.86 cycles per year (period 425 days) is the most evident feature of the spectrum. Geophysical theory predicts no other visible resonances in the spectrum; the background level corresponds to the BIH-reported level of measurement error in the data set. Because the innovation assumed in this data is not sufficiently

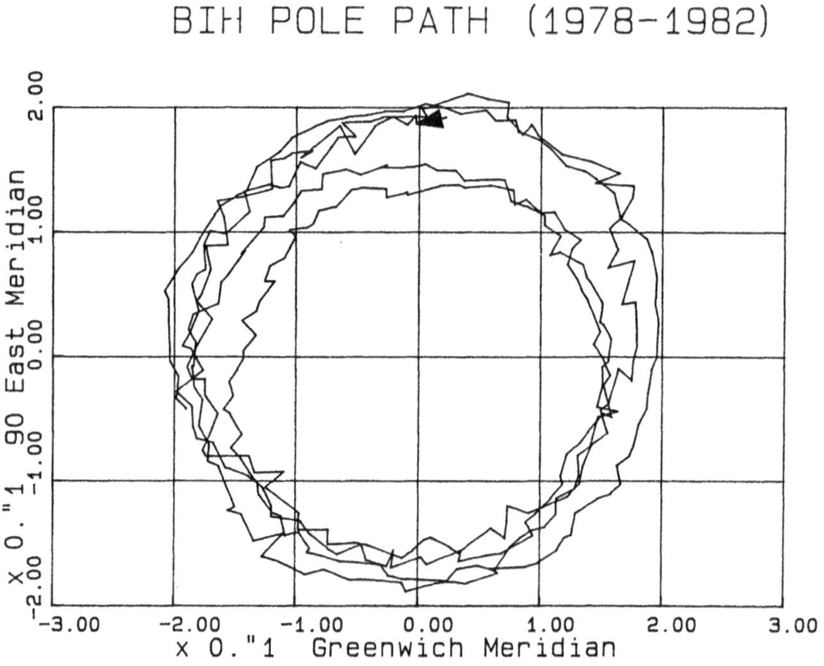

Figure 3. *The BIH pole path record (1978–82) following its mean and annual component reduction. The positive-phase sense of rotation is counter-clockwise. Straight lines join the data points which are separated by 5–day intervals.*

rich in low-frequency composition, the AR operator itself has been forced to account for the low-frequencies in the data set by exaggeration of the long-period Chandler resonance. We believe that this is the major reason for the otherwise attractively high Q found for the resonance. Moreover, this data model has not been found to be robust since the resonance frequency and Q obtained is quite sensitive to the removal of a few points from the beginning or end of the data set. For example, removing the last 10 points of 366 results in a reduced $Q \simeq 54$ and increased resonance period while removing the first 10 points results in an increased $Q \simeq 73$ and period. Finally, we know that the data model is incomplete and consequently, we have little confidence that the results reported here are geophysically meaningful.

Figure 5 shows the equivalent AR spectrum under the assumption that the innovation process has a flicker noise form. Here, both the zero-frequency

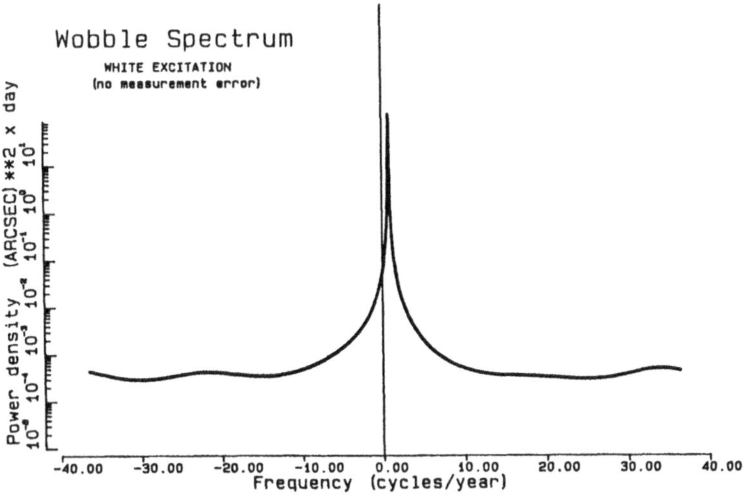

Figure 4. *The standard AR(5) wobble spectrum of the BIH data set of Figure 3 as computed by the Ulrych-Clayton "exact least squares method". The measurements are assumed to arise from the excitation of the autoregressive operator by a purely random, minimum-variance process. No additive measurement error is considered.*

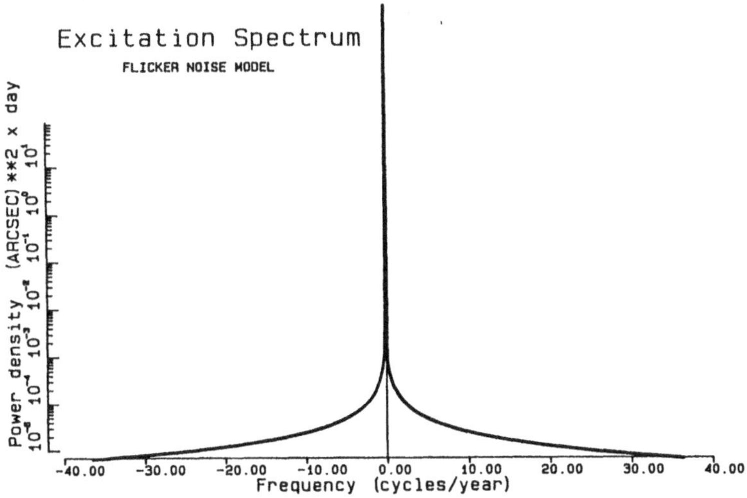

Figure 5. *The spectrum of the modelled flicker-noise excitation or innovation. Here, the excitation has been scaled to a variance of* 10×10^{-6} *arcsec*^{2}.

singularity of the assumed flicker excitation and the sharp ($Q \simeq 10$), symmetrical Chandler resonance at 0.87 cycles per year are resolved. This data model allows for a lower Q resonance because the regression operator obtained via equations (4a–d) does not have to account for a preponderance of very long period composition. The known level of measurement error in the data set, reconstructed by the scaling of the flicker innovation by the remaining four zeros of the calculated AR(5) operator, is somewhat overestimated. Figure 6 shows the spectrum of the prior-assumed flicker excitation which is the structural innovation for the latter data model. Results similar to those shown here were reported by Jensen and Mansinha (1984) in their flicker-noise AR modelling of a different data set.

The new data model, described above (equations 6, 5a,b), accounts for a self-correlated structural innovation to the autoregression operator and a separately additive, stationary and uncorrelated measurement error. This model closely corresponds to our current geophysical understanding of the BIH data set. Assuming a flicker innovation to allow for the expected preponderance of long period components in the excitation spectrum and a measurement error with the known variance, the spectrum shown in Figure 7 was calculated using the newly-elaborated algorithm (equations 7a–d). The spectrum derived from this composite model was determined as the sum of a white spectrum corresponding to the known level of measurement error and the spectrum obtained by division of the squared Fourier transform of the calculated regression operator into the variance-scaled $1/|f|$-spectrum. Essentially, only the zero-frequency singularity of the flicker noise innovation and the Chandler resonance exceed the spectral density of the white measurement error background. Our attempt to use this same data model under the assumption that the innovation was uncorrelated failed numerically after four or five iterations.

8. GEOPHYSICAL INTERPRETATION

The noise-augmented, flicker-noise excited AR model decomposition determines only one period in the BIH record which has a power density significantly in excess of that of the known level of measurement error. This component, with a period of 415 days and a $Q \simeq 21$, is obviously the Chandler resonance. It is interesting to note that this analysis allows for a shorter period of the Chandler resonance than is normally found using the standard procedures. Unfortunately, the Q appears to be too low. This latter failure is due, at least in part, to the unique character of this data set. Beginning about 1980, the record (Figure 3) shows an evidently rapid decrease in amplitude of the resonance. This could be due to a lack of excitation

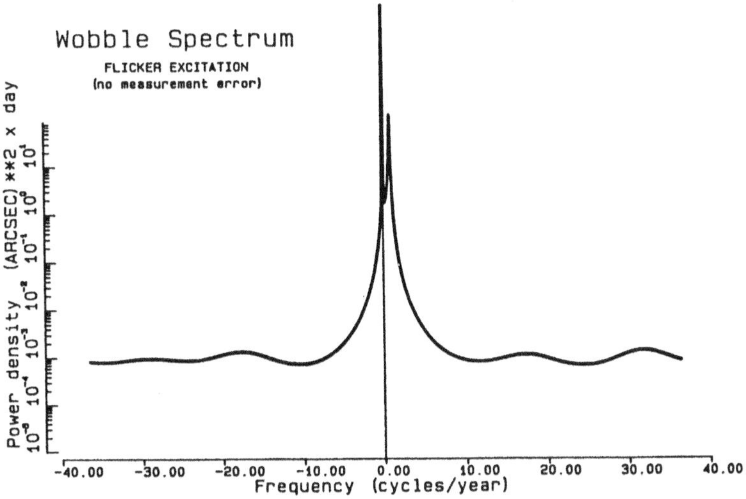

Figure 6. *The AR(5) wobble spectrum of the BIH data set of Figure 3 assuming a minimum-variance, flicker noise innovation. No additive measurement error is considered.*

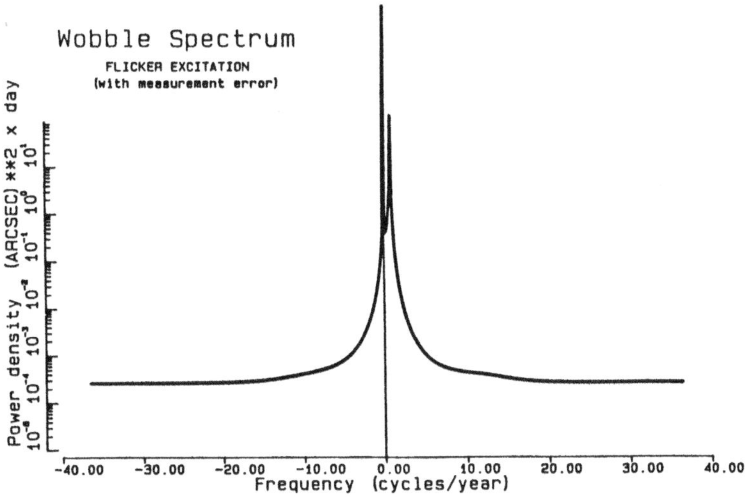

Figure 7. *The AR(5) wobble spectrum of the BIH data set of Figure 3 assuming a minimum-variance flicker noise innovation in the presence of additive white noise representing the measurement error with variance* 50×10^{-6} *arcsec*^{2} *which approximates the average level of standard error in the BIH data set.*

of the wobble during this epoch. If this is the case, we are observing the actual free decay of the resonance. Equally well, the wobble's collapse could have resulted from a large asynchronous excitation which partially cancelled the wobble by interference. Moreover, one cannot discount the possibility that this apparent collapse of the wobble is only an artifactual result of the method employed in reducing the mean and annual components of the record. To resolve which of these possibilities holds, we require a longer and continually homogeneous record. A longer record (to December 30, 1984) has already been published by the BIH in its Annual Reports for the year 1984. However, the continuing rapid improvement in the pole-position measurement technology has allowed the BIH to reduce the standard errors in measurement by a factor of 2 since the beginning of the 1978–82 data set used here. Our present method presumes stationarity and non-correlation of the additive measurement error. Adequate analysis of the BIH's recently-published extended record will require further elaboration of the method to allow for non-stationarity of the errors. Presently, the standard errors of measurement represent about 1 part in 200 or so referenced to the magnitude of the pole position. If these errors could be further reduced by an order of magnitude, the noise-augmented data model would not be required in order to determine the Chandler resonance period and quality unequivocally.

9. CONCLUSIONS

We have presented, by means of a single geophysical example, an argument for an elaborated structural data model which we believe to be appropriate for a wider class of problems in time series analysis. The geophysicist or natural scientist can almost always draw upon his understanding and theoretical description of nature in construction of an appropriate structural data model. All time series analysts are not so convenienced by their problems. Economic systems, for example, are evidently extremely complex, time variant and non-linear. No sufficient theory exists to describe almost any econometric time series and consequently one cannot hope to employ adequately elaborated structural data models in their analysis. Rather, and more appropriately, the time series analyst employs arbitrary models which he can only identify with respect to an optimum form and order subsequent to his analysis. We do not presume to criticize this conventional approach to time series analysis. As natural scientists whose major objectives are to understand and explain nature, we are grateful for the continuing developments in statistical methods which we may adapt and bring to bear in resolution of our problems. We are perhaps only attempting to warn ourselves in our eagerness to attack our data by some fashionable statistical method without first carefully considering whether or not it is appropriate to our problem.

ACKNOWLEDGMENT

This project was supported by the Natural Sciences and Engineering Research Council of Canada through separate operating grants to the authors.

REFERENCES

Akaike, H. (1969), "Fitting autoregression for prediction". *Annals of the Institute of Statistical Mathematics* **21**, 243–247.

Akaike, H. (1974), "A new look at the statistical model identification". *IEEE Transactions on Automatic Control* **AC-19**, 716–723.

Akaike, H. (1985), "Some reflections on the modelling of time series". Presented at the Symposia on Statistics and a Festschrift in Honor of Professor V. M. Joshi's 70th Birthday, London, Ontario, May, 1985.

Anderson, O. D. (1976), *Time Series Analysis and Forecasting, The Box-Jenkins Approach*. London: Butterworths.

Barrodale, I., and R. E. Erickson (1980), "Algorithms for least-squares linear prediction and maximum entropy spectral analysis—Part 1: Theory". *Geophysics* **45**, 420–432.

Box, G. E. P., and G. M. Jenkins (1970), *Time Series Analysis, Forecasting and Control*. San Francisco: Holden-Day.

Burg, J. P. (1964), "Three-dimensional filtering with an array of seismometers". *Geophysics* **29**, 693–713.

Burg, J. P. (1967), "Maximum entropy spectral analysis". Presented at the 37th Annual Meeting, Society of Exploration Geophysicists, Oklahoma City, OK, 1967. (Abstract: *Geophysics* **32**, Preprint: Texas Instruments, Dallas).

Burg, J. P. (1975), "Maximum entropy spectral analysis." Ph.D. thesis, Stanford University.

Clarke, G. K. C. (1968), "Time-varying deconvolution filters". *Geophysics* **33**, 936–944.

Davies, E. B., and E. J. Mercado (1968), "Multichannel deconvolution filtering of field recorded seismic data". *Geophysics* **33**, 711–722.

Gauss, C. F. (1839), *Allgemeine Theorie des Erdmagnetismus*, Leipzig. (Republished, 1877: Gauss, Werke, 5, Gottingen).

Granger, C. W. J., and R. Joyeux (1980), "An introduction to long-memory time series models and fractional differencing". *Journal of Time Series Analysis* **1**, 15–29.

Hosken, J. W. J. (1980), "A stochastic model of seismic reflections". Presented at the 50th Annual Meeting of the Society of Exploration Geophysicists, Houston. (Abstract G-69, *Geophysics* **46**, 419).

Jensen, O. G., and L. Mansinha (1984), "Deconvolution of the pole path for fractal flicker-noise residual". In *Proceedings of the International Association of Geodesy (IAG) Symposia 2*, p. 76–99. Columbus: Ohio State University.

Jensen, O. G., and A. Vafidis (1986), "Inversion of seismic records using extremal skewness and kurtosis". Manuscript in review.

Lambeck, K. (1980), *The Earth's Variable Rotation: Geophysical Causes and Consequences.* Cambridge: Cambridge University Press.

Mandelbrot, B. B. (1983), *The Fractal Geometry of Nature.* San Francisco: Freeman.

Mueller, I. I. (1969), *Spherical and Practical Astronomy.* New York: Frederick Unear.

Munk, W. H., and G. J. F. MacDonald (1960), *The Rotation of the Earth, a Geophysical Discussion.* Cambridge: Cambridge University Press.

Postic, A., J. Fourmann, and J. Claerbout (1980), "Parsimonious deconvolution". Presented at the 50th Annual Meeting of the Society of Exploration Geophysicists, Houston. (Abstract G-76, *Geophysics* **46**, p. 421).

Robinson, E. A. (1954), "Predictive decomposition of time series with application to seismic exploration". *Geophysics* **32**, 418–484. (Republication of MIT GAG Report No. 7, July 12, 1954; Ph.D. thesis, Massachusetts Institute of Technology, 1954).

Smylie, D. E., G. K. C. Clarke, and L. Mansinha (1970), "Deconvolution of the pole path". In *Earthquake Displacement Fields and Rotation of the Earth*, Astrophysics and Space Science Library Series. Dordrecht: Reidel.

Treitel, S. (1970), "Principles of digital multichannel filtering". *Geophysics* **35**, 785–811.

Tyraskis, P. A., and O. G. Jensen (1985), "Multichannel linear prediction and maximum-entropy spectral analysis using least squares modelling". *IEEE Transactions on Geoscience and Remote Sensing* **GE-23**, 101–109.

Ulrych, T. J., and R. W. Clayton (1976), "Time series modelling and maximum entropy". *Physics of the Earth and Planetary Interiors* **12**, 188–200.

Ulrych, T. J., and M. Lasserre (1966), "Minimum-phase". *Journal of the Canadian Society of Exploration Geophysicists* **2**, 22–32.

Vafidis, A. (1984), *Deconvolution of Seismic Data Using Extremal Skew and Kurtosis.* M.Sc. thesis, McGill University, Montreal.

Wadsworth, G. P., E. A. Robinson, J. G. Byran, and P. M. Hurley (1953), "Detection of reflections on seismic records by linear operators". *Geophysics* **18**, 539–586.

Wiggins, R. A., and E. A. Robinson (1965), "Recursive solution to the multichannel filtering problem". *Journal of Geophysical Research* **70**, 1885–1891.

Wiggins, R. A. (1978), "Minimum entropy deconvolution". *Geoexploration* **16**, 21–35.

Yule, G. U. (1927), "On a method of investigating periodicities in disturbed series, with special reference to Wolfer's sunspot numbers." *Philosophical Transactions of the Royal Society* **226**, 267–298.

Andrey Feuerverger [1]

ON SOME ECF PROCEDURES FOR TESTING INDEPENDENCE

ABSTRACT

This paper is concerned with the use of the empirical characteristic function (ecf) in nonparametric testing for independence. Properties of the ecf are briefly reviewed, and a new distributional convergence result (Theorem 2.3) included. Nonparametric testing for independence is discussed briefly, but with particular focus on asymptotic aspects. Some new procedures for testing independence based on the ecf are presented and developed, and a Monte Carlo study carried out. The asymptotic efficiency of the procedure is discussed and suggestions for further work and some open problems noted.

1. INTRODUCTION AND SUMMARY

In this paper we explore the use of certain characteristic function (cf) based quantities and their empirical estimates in the context of nonparametric testing for independence in multivariate i.i.d. data. The key underlying idea is that since independence can be characterised by factorizability of a cf into the product of its marginals, the empirical cf (ecf) should provide an effective tool for testing the hypothesis H_0 of independence.

If $X_j = (X_j^1, \ldots, X_j^p)'$, $j = 1, \ldots, n$ is an i.i.d. sample from a p-variate distribution whose cf is $c(t) = E[e^{it'X_1}]$ where $t = (t_1, \ldots, t_p)'$, then the ecf is defined as

$$c_n(t) = \frac{1}{n} \sum_{j=1}^{n} e^{it'X_j}.$$

The properties of this estimator are now fairly well understood and for convenience of the reader Section 2 below provides a brief review of certain key

[1] Department of Statistics, The University of Toronto, Toronto, Ontario M5S 1A1

I. B. MacNeill and G. J. Umphrey (eds.), Time Series and Econometric Modelling, 189–206.

results. (Theorem 2.3, however, is new.) In Section 3 we outline very briefly the current status of the nonparametric independence testing problem giving particular focus to asymptotic aspects. Our ecf based tests are proposed and developed in Section 4, and a Monte Carlo study is presented in Section 5. Section 5 also includes a discussion on efficiency of the tests and also itemizes some suggestions for further work.

2. THE ECF

Let $X_j = (X_j^1, \ldots, X_j^p)'$, $j = 1, \ldots, n$ be a sample of a p-variate distribution having cf $c(t)$ and define the ecf as $c_n(t) = \frac{1}{n}\sum_{j=1}^n e^{it'X_j}$, where $t = (t^1, \ldots, t^p)'$. The properties of $c_n(t)$ have been explored extensively by Feuerverger and Mureika (1977), Csorgo (1981a,b), Feuerverger and McDunnough (1981a,b), and references appearing in these papers, for example. It is easily seen that $c_n(t)$ is an average of n independent processes of the type $e^{it'X_1}$ and has mean $E[c_n(t)] = c(t)$. Defining the ecf process $Y_n(t) = \sqrt{n}(c_n(t) - c(t))$, the full covariance structure follows from the relation

$$\operatorname{cov}(Y_n(s), Y_n(t)) = EY_n(s)\overline{Y_n(t)} = c(s-t) - c(s)c(-t), \tag{2.1}$$

and from the fact that $Y_n(-t) = \overline{Y_n(t)}$. The following theorem records the three basic and successively stronger consistency results.

Theorem 2.1

(i) $c_n(t) \to c(t)$ a.s. for all t.

(ii) $\sup_{|t^1|\le T, \ldots, |t^p| \le T} \mid c_n(t) - c(t) \mid \to 0$ a.s. for T finite.

(iii) In (ii) we may replace T by $T_n = \exp(o(n))$ and this result is the best possible, in general.

Results (i) and (ii) are given by Feuerverger and Mureika (1977). Result (iii) for various suboptimal rates for $T_n \uparrow \infty$ is due to Feuerverger and Mureika (1977) and Csorgo (1981a,b). The sharp result is given by, for example, Csorgo and Totik (1983).

The asymptotic distributional characteristics are given by the following theorem in which four successively more general results are provided.

Theorem 2.2. Let $Y(t)$ be a zero mean complex valued Gaussian process with covariance structure identical to (2.1) and such that $Y(-t) = \overline{Y(t)}$. Then

(i) $Y_n(t_1), Y_n(t_2), \ldots, Y_n(t_k)$ converges in distribution to $Y(t_1), Y(t_2), \ldots,$ $Y(t_k)$ for all k and $t_1, t_2, \ldots, t_k \in R^p$.

(ii) $Y_n(t)$ converges weakly to $Y(t)$ over any compact set in R^p provided $E \parallel X_1 \parallel^{1+\delta} < \infty$ for some $\delta > 0$.

(iii) In (ii) the moment restrictions may be weakened to

$$E[\log^+ \parallel X_1 \parallel (\log^+ \log^+ \parallel X_1 \parallel)^{2+\delta}] < \infty,$$

for some $\delta > 0$.

(iv) $Y_n(t)$ converges weakly to $Y(t)$ over compact sets if and only if

$$\int_0^\epsilon \frac{\overline{\sigma}(h)}{h(\log \frac{1}{h})^{1/2}} dh < \infty,$$

where $\overline{\sigma}(h)$ is a non-decreasing rearrangement of the function $\sigma(t) = (1 - Re\ c(t))^{1/2}$. Specifically, if

$$m(y) = \lambda_d\{t : \parallel t \parallel < \frac{1}{2},\ \sigma(t) < y\},\ 0 \le y \le 1,$$

where λ_d is a d-dimensional Lebesgue measure, $\overline{\sigma}$ is its inverse:

$$\overline{\sigma}(h) = \sup\{y : m(y) < h\}.$$

Results (i) and (ii) are given by Feuerverger and Mureika (1977). Results (iii) and (iv) lie deeper. The logarithmic moment condition is given by Keller (1979) and is nearly sharp. The sharp result is from the work of Csorgo (1981a,b) and Marcus (1981). Note $\overline{\sigma}(h)$ is defined on $[0, m(1)]$ and has there the same distribution with respect to λ_1 as does $\sigma(t)$ on $\parallel t \parallel < \frac{1}{2}$ with respect to λ_d. Feuerverger and McDunnough (1981a) have observed that actual weak convergence of the ecf is generally not required in statistical contexts. Indeed, results of the following type may be established.

Theorem 2.3. The finite expression

$$A_0 + \int_{R^p} Y_n(t) dA_1(t) + \int_{R^{2p}} Y_n(t_1) Y_n(t_2) dA_2(t_1, t_2)$$
$$+ \cdots + \int_{R^{kp}} Y_n(t_1) \cdots Y_n(t_k) dA_k(t_1, \ldots, t_k), \qquad (2.2)$$

where each of the functions $A_l(t_1, \ldots, t_l)$ is of bounded variation on R^{lp}, $l = 1, 2, \ldots, k$, converges in distribution to the expression obtained from (2.2) by replacing the terms $Y_n(\cdot)$ by the Gaussian process $Y(\cdot)$.

Proof. Denote the expression in (2.2) by $Q(Y_n)$ and let Y_n^M, Y^M be Y_n and Y except based on truncating the X variable at $\pm M$. Then for fixed

$M < \infty$, $Q(Y_n^M) \xrightarrow{d} Q(Y^M)$ by Theorem 2.2. We may also readily establish that $Q(Y^M) \xrightarrow{d} Q(Y)$ as $M \to \infty$ and finally that $Q(Y_n^M) \xrightarrow{d} Q(Y_n)$ as $M \to \infty$ uniformly in n giving $Q(Y_n) \xrightarrow{d} Q(Y)$ as required.

Turning to statistical contexts, a significant motivation for the study of ecf procedures is the efficiency result for inference based on the ecf due to Feuerverger and McDunnough (1981a,b). This result essentially is as follows.

Theorem 2.4. Suppose $X_j = (X_j^1, \ldots, X_j^p)'$, $j = 1, 2, \ldots, n$ are i.i.d. with cf $c_\theta(t)$ where $t = (t^1, \cdots, t^p)'$, and $\theta = (\theta_1, \cdots, \theta_q)'$ takes on some true value θ_0 in $\Theta \subset R^q$. Let $t_1, t_2, \cdots, t_k$ be a fixed grid of k points in R^p such that $t_j \pm t_l \neq 0$ for $j \neq l$ and define the $2k$-dimensional column vectors Z_θ and Z_n as

$$Z_\theta = (c_\theta(-t_k), \ldots, c_\theta(-t_1),\ c_\theta(t_1), \ldots, c_\theta(t_k))'$$

and

$$Z_n = (c_n(-t_k), \ldots, c_n(-t_1),\ c_n(t_1), \ldots, c_n(t_k))'.$$

Define the matrix $\Sigma_\theta = cov(\sqrt{n} Z_n)$ whose entries are given by (2.1) and consider estimating θ by fitting Z_θ to Z_n using generalized non-linear least squares, i.e., minimizing

$$(Z_n - Z_\theta)'\ \Sigma^{-1}(Z_n - Z_\theta),$$

where Σ is a consistent estimator of Σ_{θ_0}. Then under mild conditions that ensure asymptotic optimality of the MLE the estimation procedure above is consistent, asymptotically normal, and asymptotically has covariance which may be made arbitrarily close to the Cramér-Rao bound by selecting the grid $\{t_j\}$ to be sufficiently fine and extended.

The procedure of Theorem 2.4 is termed *harmonic regression* and is only one of several efficient procedures given by Feuerverger and McDunnough (1981a,b), where a fuller discussion of regularity conditions may also be found.

3. NONPARAMETRIC TESTS FOR INDEPENDENCE

In this section we give a very brief overview of what has up to now been accomplished in the literature on the nonparametric independence testing problem. However, our focus is primarily on asymptotic aspects and some related open problems are noted.

In the context of bivariate normality, optimal inference concerning independence is based, of course, on the Pearson product-moment correlation coefficient with the resulting procedure being both UMP unbiased and UMP invariant (Lehmann, 1959, § 5.11 and problem 6.11). See also Anderson (1958, Chapter 4), for example. For departures from normality, it is somewhat remarkable that the robustness of the correlation test for bivariate independence is still not completely understood, although evidence of non-robustness is mounting. See, for example, Srivastava and Lee (1984) and related references.

Many rank tests of independence (or non-association) have been proposed, in general for motivations related specifically to robustness considerations. If (X_i, Y_i), $i = 1, \cdots, n$ is a bivariate sample, these tests are based on the data only through the collection of ranks (U_i, V_i), $i = 1, \cdots, n$ where U_i (V_i) is the rank of X_i (Y_i) amongst the $X's$ $(Y's)$ respectively. Well-known among such rank tests are Spearman's rank-correlation, Kendall's tau, Fisher-Yates' normal scores and the quadrant test, among others. See, for example, Hollander and Wolfe (1973, Chapter 8), Lehmann (1975, Chapter 7) and Kendall (1970). Table 1 gives the known asymptotic (Pitman) efficiencies of various rank tests of bivariate independence relative to the sample correlation coefficient test in the context of the bivariate normal family. The full asymptotic efficiency in this case of the normal scores test is of special interest. For bivariate families other than the normal, it is known that the asymptotic efficiencies relative to the correlation can in general be both greater and less than unity, depending upon the particular family. (See, for example, Stuart (1954) and Konijn (1956).)

Table 1. *Asymptotic efficiencies of nonparametric tests for association relative to correlation in the bivariate normal family.*

Test	A.R.E.
Fisher-Yates normal scores	1.00
Spearman Rank correlation	.91
Kendall Tau	.91
Hoeffding B_n	.78
Quadrant test	.41

The A.R.E. values shown are Pitman efficiencies except for B_n, which is the limiting Bahadur efficiency. (Wieand (1976) has given general conditions under which these efficiency measures coincide.)

The nonparametric tests of association so far discussed share one significant flaw, namely they are not consistent for many typical nonparametric classes of alternatives. (Indeed, examples can be readily constructed where the tests are not unbiased or even asymptotically unbiased.) It is the case, however, that these tests are generally consistent against alternatives involving dependence structure which is, loosely speaking, of "monotone character" (eg., Lehmann, 1966). And since it is this type of dependence which is characteristic of many typical applications, these tests have found widespread use. Nevertheless dependencies of less direct character do sometimes occur—as, for example, when Y has a non-monotone regression on X, and X is sampled randomly. In any case, the lack of consistency is, at least from a theoretical standpoint, unsatisfying.

There are, however, two exceptions in the literature of which we are aware. The first is the distribution-free (under H_0) rank test proposed by Hoeffding (1948) and by Blum *et al.* (1961) based upon

$$B_n = \int\int [F_n(x,y) - F_n(x)F_n(y)]^2 dF_n(x,y),$$

while the second is due to Rosenblatt (1975) and is based upon

$$T_n = \int\int [f_n(x,y) - f_n^X(x)f_n^Y(y)]^2 a(x,y)dxdy,$$

where the f_n are suitable kernel density estimates, and $a(x, y)$ is a weight function. Rosenblatt points out, however, (*ibid,* Section 3) that the asymptotic distributions of T_n are generally normal, while those of B_n are normal only for the alternatives, and under independence has the less familiar distribution of an infinite weighted sum of chi-squared variates. Nevertheless, Rosenblatt indicates that tests based on the sample distribution function are typically more powerful than those based on density estimates. Incidentally, observe that B_n is a rank test, while T_n is not, although rank versions of T_n are easily constructed by replacing the observations by rank functions prior to computing the densities. In any case, neither B_n nor T_n is asymptotically efficient for normal alternatives. In fact, B_n is shown in Table 3.1 as having an efficiency of 0.78.

From the discussion above, it emerges that no test of dependence or association appears to be known which is both consistent in general and asymptotically efficient in the normal case and that to obtain such a test is an open problem of some interest. Now a resolution of this problem is certainly possible along the lines of a two stage procedure in which the first stage consists of a consistent test sequence for bivariate normality with levels tending to zero at a suitable rate, and the second stage then uses either the normal scores test say, or the Hoeffding test—depending on the results at the first stage. However, a less artificial resolution of the question, not involving such multiplicity would certainly be of special interest.

4. AN ECF TEST FOR INDEPENDENCE

The results of Section 2 suggest the study of ecf based procedures for testing independence. To reestablish notation, recall our sample consists of n i.i.d. p-variate terms $(X_j^1, \ldots, X_j^p)'$, $j = 1, 2, \ldots, n$ from a distribution having cf $c(t)$ where $t = (t^1, \ldots, t^p)'$. At issue is the mutual independence of the terms in $(X_1^1, \ldots, X_1^p)'$. The ecf, as usual, is $c_n(t)$ and we introduce now the marginal quantities

$$c^l(t^l) = c(0, \ldots, 0, t^l, 0, \ldots, 0) = E\,[e^{it^l X_1^l}], \tag{4.1}$$

and

$$c_n^l(t^l) = c_n(0, \ldots, 0, t^l, 0, \ldots, 0) = \frac{1}{n}\sum_{j=1}^{n} e^{it^l X_j^l}, \tag{4.2}$$

for $l = 1, 2, \cdots, p$. Self-evident notations such as $c^{X_1}(t^1)$, $c^{X_1,X_2}(t^1, t^2)$, $c_n^{X_1}(t^1)$, $c_n^{X_1,X_2}(t^1, t^2)$ and so on may also be used. The underlying null hypothesis of independence may be expressed as

$$H_0 : c(t) = \prod_{l=1}^{p} c^l(t^l), \text{ all } t \in R^p.$$

Now the central quantity in ecf based tests for independence is

$$\Gamma(t) = c(t) - \prod_{l=1}^{p} c_l(t^l) \tag{4.3}$$

and its asymptotically normal empirical counterpart

$$\Gamma_n(t) = c_n(t) - \prod_{l=1}^{p} c_n^l(t^l). \tag{4.4}$$

We know from Theorem 2.1 that $\Gamma_n(t) \to \Gamma(t)$ a.s. for all $t \in R^p$ (and indeed uniformly over compact sets) and that the independence hypothesis may be expressed as $H_0 : \Gamma(t) = 0$, $t \in R^p$. Note, however, that it is not the case that $E\Gamma_n(t) = \Gamma(t)$. For example, in the two-dimensional case ($p = 2$) we have

$$E[c_n^1(t^1)c_n^2(t^2)] = E\left[\frac{1}{n}\sum_{j_1=1}^{n} e^{it^1X_{j_1}^1} \cdot \frac{1}{n}\sum_{j_2=1}^{n} e^{it^2X_{j_2}^2}\right]$$

$$= \frac{n}{n^2}\, c(t^1,t^2) + \frac{n(n-1)}{n^2}\, c^1(t^1)c^2(t^2), \tag{4.5}$$

so that

$$E[\Gamma_n(t^1,t^2)] = \frac{n-1}{n}\Gamma(t^1,t^2). \tag{4.6}$$

Thus, our null hypothesis can be expressed here as $H_0 : E\Gamma_n \equiv 0$, but this is the case only for $p = 2$.

To express the general result we define the following quantities:

$$v_s = \sum \prod_{a=1}^{s} c^{i_{a1}i_{a2},\ldots,i_{ap_a}}(t^{i_{a1}},t^{i_{a2}},\ldots,t^{i_{ap_a}}),\ 1 \le s \le p, \tag{4.7}$$

where the sum is over all partitions of $(1,2,\ldots,p)$ into s sets $(i_{11},i_{12},\ldots,i_{p_1}),(i_{21},i_{22},\ldots,i_{2p_2}),\ldots,(i_{s1},i_{s2},\ldots,i_{sp_s})$. Note that $v_1 = c(t)$. We have:

Theorem 4.1. For arbitrary $p \ge 1$,

$$E\Gamma_n(t) = c(t) - \sum_{s=1}^{p} \frac{n(n-1)\cdots(n-s+1)}{n^p}\, v_s, \tag{4.8}$$

$$= \Gamma(t) + O(\frac{1}{n}), \tag{4.9}$$

where the order term is uniform in t. Under H_0, $E\Gamma_n(t) \equiv 0$ all t, n and conversely.

Proof. (4.8) follows upon examination of the n^p terms arising in

$$E(c_n^1(t^1)\cdots c_n^p(t^p)) = E\left[\frac{1}{n}\sum_{j_1=1}^{n} e^{it^1X_{j_1}^1}\cdots\frac{1}{n}\sum_{j_p=1}^{n} e^{it^pX_{j_p}^p}\right], \tag{4.10}$$

and, noting that $v_p = c^1(t^1)\ldots c^p(t^p)$, (4.9) follows from (4.8). The last assertion now follows easily.

In the cases $p = 3$ and $p = 4$, for example, the theorem gives, in obvious notation

$$E[\Gamma_n(t^1,t^2,t^2)] = \frac{n^3-n}{n^3}c^{123} - \frac{n(n-1)}{n^3}[c^{12}c^3 + c^{13}c^2 + c^{23}c^1]$$

$$\frac{n(n-1)(n-2)}{n^3}\, c^1c^2c^3, \tag{4.11}$$

and

$$
\begin{aligned}
E\Gamma_n(t^1,t^2,t^3,t^4) &= \frac{n^4-n}{n^4}c^{1234} \\
&- \frac{n(n-1)}{n^4}[c^{123}c^4 + c^{124}c^3 + c^{134}c^2 + c^{234}c^1 + c^{12}c^{34} + c^{13}c^{24} \\
&+ c^{14}c^{23}] \\
&- \frac{n(n-1)(n-2)}{n^4}[c^{12}c^3c^4 + c^{13}c^2c^4 + c^{14}c^2c^3 + c^{23}c^1c^4 \\
&+ c^{24}c^1c^3c^{34}c^1c^2] - \frac{n(n-1)(n-2)(n-3)}{n^4}c^1c^2c^3c^4. \qquad (4.12)
\end{aligned}
$$

To construct our test statistics for H_0 based on $\Gamma_n(t)$ we require the covariance structure of $\Gamma_n(t)$ under H_0. The general result is as follows:

Theorem 4.2. Under the null hypothesis H_0 of mutual independence the covariance structure of

$$\Gamma_n(t) = c_n(t) - \prod_{l-1}^{p} c_n^l(t^l),$$

is determined by

$$
\begin{aligned}
&\operatorname{cov}(\Gamma_n(s),\Gamma_n(t)) \\
&= \left(\frac{1}{n}-\frac{1}{n^p}\right)\prod_{l=1}^{p} c^l(s^l-t^l) \\
&+ \left(\frac{n-1}{n} - \left(\frac{n-1}{n}\right)^p\right)\prod_{l=1}^{p} c^l(s^l)\overline{c^l(t^l)} - \sum_{q=1}^{p-1}\frac{(n-1)^{p-q}}{n^p}\,\omega_q. \qquad (4.13)
\end{aligned}
$$

Here

$$\omega_q = \sum_{\mu}\prod_{l\in\mu} c^l(s^l-t^l)\cdot\prod_{l\notin\mu} c^l(s^l)\overline{c^l(t^l)}, \qquad (4.14)$$

where μ ranges over the $\binom{p}{q}$ combinations of q indices selected from $(1,2,\ldots,p)$. The asymptotic structure is given by

$$
\begin{aligned}
&\lim_{n\to\infty} n\cdot \operatorname{cov}(\Gamma_n(s),\Gamma_n(t)) \\
&= \prod_{l=1}^{p} c^l(s^l-t^l) + (p-1)\prod_{l=1}^{p} c^l(s^l)\overline{c^l(t^l)}
\end{aligned}
$$

$$- \sum_{l=1}^{p} c^l(s^l - t^l) \prod_{\substack{m=1 \\ m \neq l}}^{p} c^m(s^m)\overline{c^m(t^m)}$$

$$= \left[\prod_{l=1}^{p} c^l(s^l - t^l) - \prod_{l=1}^{p} c^l(s^l)\overline{c^l(t^l)} \right]$$

$$- \sum_{l=1}^{p} \left[c^l(s^l - t^l) - c^l(s^l)\overline{c^l(t^l)} \right] \prod_{\substack{m=1 \\ m \neq l}}^{p} c^m(s^m)\overline{c^m(t^m)}. \quad (4.15)$$

Proof. Under H_0, $E\Gamma_n(t) \equiv 0$ so that

$$\text{cov}(\Gamma_n(s),\Gamma_n(t)) \equiv E(\Gamma_n(s)\overline{\Gamma_n(t)})$$
$$= E[c_n(s)\overline{c_n(t)}] - E\left[c_n(s) \cdot \overline{\prod_{l=1}^{p} c_n^l(t^l)} \right]$$
$$- E\left[\prod_{l=1}^{p} c_n^l(s^l) \cdot \overline{c_n(t)} \right] + E\left[\prod_{l=1}^{p} c_n^l(s^l) \cdot \overline{\prod_{l=1}^{p} c_n^l(t^l)} \right]. \quad (4.16)$$

We next replace the terms $c_n(\cdot)$ and $c_n^l(\cdot)$ by their defining summations and find that the four terms of (4.16) consist respectively of n^2, n^{p+1}, n^{p+1} and n^{2p} products of exponential terms. Upon further examination we find that each of the $n^2 + n^{p+1} + n^{p+1} + n^{2p}$ expectations now arising must be of the type

$$\prod_{l \in \mu} c^l(s^l - t^l) \cdot \prod_{l \notin \mu} c^l(s^l)\overline{c^l(t^l)}, \quad (4.17)$$

where μ is a subset of q elements of $(1, 2, \cdots, p)$, $q = 0, 1, 2, \cdots, p$. We shall refer to the q terms in the first product as being *linked* and the $p - q$ terms in the second product as being *unlinked*.

We shall now collect up, for each term (4.17), the coefficients that are contributed to it by the four expressions in (4.16). For the case where all terms are linked we have

$$\frac{n}{n^2} - \frac{n}{n^{p+1}} - \frac{n}{n^{p+1}} + \frac{n^p}{n^{2p}} = \frac{1}{n} - \frac{1}{n^p}. \quad (4.18)$$

For the case where all terms are unlinked we have

$$\frac{n(n-1)}{n^2} - \frac{n(n-1)^p}{n^{p+1}} - \frac{n(n-1)^p}{n^{p+1}} + \frac{n^p(n-1)^p}{n^{2p}}$$
$$= \frac{n-1}{n} - \left(\frac{n-1}{n} \right)^p. \quad (4.19)$$

In the remaining cases, the first term in (4.16) does not contribute. Each term involving q links, $0 < q < p$ will have the coefficient

$$0 - \frac{n(n-1)^{p-q}}{n^{p+1}} - \frac{n(n-1)^{p-q}}{n^{p+1}} + \frac{n^p(n-1)^{p-q}}{n^{2p}} = -\frac{(n-1)^{p-q}}{n^p}. \quad (4.20)$$

We have now established (4.13). The asymptotic expression (4.15) now follows directly upon taking the limits indicated; note that of the terms ω_q, $1 \leq q \leq p-1$, in (4.13) only ω_1 survives asymptotically.

We remark that for dimension $p = 2$ Theorem 4.2 gives

$$\begin{aligned}&\operatorname{cov}(\Gamma_n(s), \Gamma_n(t)) \\ &= \frac{n-1}{n^2}[c^1(s^1 - t^1) - c^1(s^1)\overline{c^1(t^1)}][c^2(s^2 - t^2) - c^2(s^2)\overline{c^2(t^2)}], \quad (4.21)\end{aligned}$$

but this attractive form is not preserved for higher dimensions. In any case, however, the value of $\operatorname{cov}(\Gamma_n(s), \Gamma_n(t))$ depends only upon the quantities $c^l(s^l - t^l)$, $c^l(s^l)$, $c^l(t^l)$, $l = 1, 2, \cdots, p$ and is thus readily determined. In the last of the two forms for the asymptotic expression (4.15) we may note that the quantities in square brackets are the covariances of the ecf itself in the p-variate case with the univariate cases. This is due to the fact that the asymptotic expression may be derived using a differential argument by expanding out

$$\Gamma_n(t) = (c(t) + [c_n(t) - c(t)]) - \prod_{l=1}^{p}(c^l(t^l) + [c_n^l(t^l) - c^l(t^l)]), \quad (4.22)$$

and dispensing with products of the squared bracket terms which are of higher than first order. We remark also that the general covariance structure, without the independence assumption, may be derived by the methods of Theorem 4.2. Those results, which are somewhat more complicated, are not quoted here. We observe as well that the process $\Gamma_n(t)$ will inherit certain weak convergence characteristics from the ecf which we do not pursue here. It is enough for us that for k fixed points $t_1, t_2, \ldots, t_k$, we have that $\Gamma_n(t_1), \Gamma_n(t_2), \ldots, \Gamma_n(t_k)$ is asymptotically Gaussian with mean and covariance structures as given by Theorems 4.1 and 4.2

We may now describe our ecf based tests for independence as follows. First we select k fixed p-vectors t_j, $j = 1, \ldots, k$ such that the $2k$ points $-t_k, \ldots, -t_1, t_1, \ldots, t_k$ are all distinct. Next we form the $2k$-dimensional vector $\hat{\xi} = (\Gamma_n(-t_k), \ldots, \Gamma_n(-t_1), \Gamma_n(t_1), \ldots \Gamma_n(t_k))'$. According to Theorem 4.1 we shall have $E\hat{\xi} = 0$ under H_0, and otherwise $E\hat{\xi} \neq 0$ for some selection of the t_j. Now the complex variance structure Σ_ξ for $\hat{\xi}$ is known under H_0 from Theorem 4.2 and indeed, we may use the asymptotic form (4.15). Further, we may estimate Σ_ξ as $\hat{\Sigma}_\xi$ by replacing all cf's by their respective ecf's. Our test will then consist of rejecting H_0 for large values of the test statistic $\hat{\xi}^* \hat{\Sigma}_\xi^{-1} \hat{\xi}$ (here * is conjugate transpose) which has, under H_0, asymptotically a chi-squared distribution with $2k$ degrees of freedom. In accordance with the usual multivariate theory, this test is asymptotically the uniformly most powerful invariant test based upon the statistics $\Gamma_n(\pm t_j)$.

We proceed now to demonstrate heuristically that the test procedure which has been proposed above inherits certain asymptotic optimalities by virtue of Theorem 2.4. Our sole concern for the moment is that Theorem 2.4 essentially assures that asymptotic efficiency can be arbitrarily closely attained by restricting attention to a sufficiently dense collection $\{c_n(t_j)\}$ of the p-variate ecf, while the test proposed here is based on Γ_n and not on c_n. We thus seek to establish that this has not compromised potential asymptotic optimalities.

Let us thus suppose that we wish to test H_0 using some collection $c_n(t_j)$, $j = 1, \cdots, k$ of values of the p-variate ecf, together with their asymptotic Gaussian distribution in a manner that will not sacrifice the asymptotic information in these statistics. Now the vector of expectations $E\, c_n(t_j)$ takes values in general in the complex space C^k while H_0 states that this vector lies in the appropriate "unit rank" subspace $M \subset C^k$ consistent with the factorizability of $c(t)$ under independence. Thus, asymptotically we may view this as a problem in generalized non-linear least squares where the error structure may be regarded as being Gaussian with known covariance. Strictly speaking, the spaces C^k and M should be further confined by permitting only points consistent with Bochner's characterization theorem, but not doing so does not affect the argument because asymptotically we find ourselves within permissible neighbourhoods. We would thus obtain in this way asymptotically UMP invariant tests among those based on the $c_n(t_j)$.

Now let us compare these to our Γ_n-based tests. Note that Γ_n involves not only the $c_n(t_j)$ but also the marginal quantities $c_n^{(l)}(t_j^{(l)})$, $l =, \ldots, p$. In terms of the ecf the restricted model specifying H_0 may be written as

$$\begin{aligned} Ec_n^{(l)}(t_j^{(l)}) &= a_j^l \qquad l = 1, \ldots, p;\; j = 1, \ldots, k, \\ Ec_n(t_j) &= \prod_{l=1}^{p} a_j^l, \quad j = 1, \ldots, k, \end{aligned} \tag{4.23}$$

while in the unrestricted case the expectations are essentially free. The likelihood ratio test of H_0 here is the asymptotically UMP invariant test based on the chosen points of the ecf. On the other hand the test which was proposed based on the Γ_n is seen to be the Wald's test (see Wald, 1943; Rao, 1965, Section 6e.3) and so is asymptotically locally as powerful as the likelihood ratio test. We caution, however, that while this appears satisfying, it does not translate automatically to optimality for the general *nonparametric* testing problem; further discussion appears in the section following.

Concerning the complex variable format of the $\hat{\xi}^*\hat{\Sigma}_\xi^{-1}\hat{\xi}$ test we remark that this is preferred to working with the real variables

$$U = (\mathrm{Re}\Gamma_n(t_1), \ldots, \mathrm{Re}R_n(t_k))' \text{ and } V = (\mathrm{Im}\Gamma_n(t_1), \ldots, \mathrm{Im}\Gamma_n(t_k))',$$

since the form of $\Sigma_{U,V}$ (which is readily obtained from Σ_ξ) is less natural than that of Σ_ξ. In any case the two tests will be identical:

$$\begin{pmatrix} U \\ V \end{pmatrix}' \hat{\Sigma}_{U,V}^{-1} \begin{pmatrix} U \\ V \end{pmatrix} = \hat{\xi}^*\hat{\Sigma}_\xi^{-1}\hat{\xi}. \tag{4.24}$$

To see this note that $\binom{U}{V} = A\xi$ for some nonsingular A and the two expressions in (4.24) are each identical to $\xi^* A^*(A^*)^{-1}\Sigma_\xi^{-1}A^{-1}A\hat{\xi}$.

Finally we remark that while the grid $\{t_j\}$ is being regarded here and throughout as fixed, finite as well as being fine and extended, in practice we may wish to use only a moderate number of data-dependent gridpoints. Both numerical experience (see Feuerverger and and McDunnough, 1981b) and heuristic arguments suggest no practical difficulties in this although a rigorous demonstration is not within our scope here.

5. A MONTE CARLO STUDY AND FURTHER REMARKS

The general method proposed in the previous section raises many questions—more than can be resolved here. In this section we undertake two specific investigations. The first is to carry out a numerical implementation of the procedure and a modest Monte Carlo simulation to demonstrate the effectiveness of the method in a simple case where the dependence is of a non-monotone character. The second is to examine the asymptotic relative efficiency of the method when the underlying family is bivariate normal. Some general remarks concerning applications and suggestions for further work are also made.

Numerical implementation of the procedure involving the test statistic $\hat{\xi}^*\Sigma^{-1}\hat{\xi}$ as described in the previous section was carried out in a FORTRAN program developed on the University of Toronto IBM 360/370. The inversion Σ^{-1} is based on IMSL subroutine LEQ2C. (The program may be obtained from the author by request.) Our Monte Carlo study was based on i.i.d. samples (X_i, Y_i), $i = 1, 2, \ldots, n$ from a uniform distribution on the unit circle in

R^2. The points at which the Γ_n statistic were computed were chosen rather arbitrarily as $t_1 = (1,1)$, $t_2 = (1,-1)$, $t_3 = (2,2)$, $t_4 = (2,-2)$ and the resulting test statistic was computed and compared to the upper 5 percent point of the chi-square distribution with 8 degrees of freedom. One hundred trials were conducted at each of the sample sizes $n = 100$, 200, 400. Table 2 shows the number of significant trials in each case. Our test evidently is effective and consistent in the present case. For comparison purposes, consider, for example, the Fisher-Yates test. By symmetry, the correlation of the normal scores is estimating zero. Consequently the resulting test will not even be consistent. Similar remarks apply to other rank tests designed against monotone dependence. The Hoeffding test, of course, will be consistent; however, we have not attempted any power comparisons here as the question of how best to select the points $t_1, \ldots, t_k$ in practice is not fully resolved. For a Monte Carlo study of related interest, see Koziol and Nemec (1979).

Table 2. *The proportion of trials (out of 100) significant at 5% for samples of size n from the uniform distribution on the unit circle.*

n	100	200	400
proportion significant	.42	.97	1.00

Turning now to the question of efficiency for Gaussian samples we obtain mixed results. On the positive side we have the following:

Theorem 5.1. In the case $k = 1$ where the test procedure $\hat{\xi}^* \Sigma_{\xi}^{-1} \hat{\xi}$ is based on a single point $t_1 = (s_1, s_2)$ and the underlying family is Gaussian, we have (under suitable interpretation) asymptotic efficiency (relative to the Pearson correlation test) arbitrarily close to unity as t_1 tends to $(0,0)$ without touching the axes.

To see this result, note that as $(s_1, s_2) \to (0,0)$ we have, on Taylor expanding,

$$\begin{aligned} \Gamma_n(s_1, s_2) &= \frac{1}{n} \sum_{j=1}^{n} e^{i(s_1 X_j + s_2 Y_j)} - \left(\frac{1}{n} \sum e^{i s_i X_j}\right) \left(\frac{1}{n} \sum e^{i s_2 Y_j}\right) \\ &= -s_1 s_2 \cdot \operatorname{cov}_{j=1}^{n}(X_j, Y_j) + o(s_1 s_2), \end{aligned} \tag{5.1}$$

so that the Γ_n statistic becomes equivalent to the covariance. In this case Γ_n and $\bar{\Gamma}_n$ become identical within $o(s_1, s_2)$ so that the 2×2 matrix $\sum_{\xi}$

tends to singularity. The appropriate limiting adjustment is to use generalised inversion, or equivalently (Feuerverger and Fraser, 1980) to reduce the procedure to a single real statistic and chi-square distribution with one degree of freedom.

On the other hand, if we use additional points ($k > 1$) of the ecf, then in the case of Gaussian samples some loss of efficiency generally occurs. To illustrate this point let (X_j, Y_j) $j = 1, \ldots, n$ be an i.i.d. sample from the bivariate normal distribution $N^2(\binom{0}{0}, \binom{1\ \rho}{\rho\ 1})$. The null hypothesis here is $H_0 : \rho = 0$. We shall first attempt to examine asymptotic relative efficiencies defined as the limiting ratio of approximate Bahadur slopes as $\rho \to 0$. (See Bahadur (1960) and Wieand (1976).) Here the optimal procedure is based of course on $T_1 = \frac{1}{n}\sum_{j=1}^{n} X_j Y_j$. Now as a "second point" we introduce, say $T_2(t) = \frac{1}{n}\sum_{j=1}^{n} \sin tX_j \sin tY_j$. The test procedure requires that we compute the covariance matrix of T_1 and $T_2(t)$ under H_0. This may be done using trigonometric identities, the formula for the cf of the Gaussian distribution, and substitutions such as $x = \lim_{s\to\infty}(\sin sx)/s$ and $x^2 = \lim_{s\to 0} 2(1 - \cos sx)/s^2$. The result under H_0 is

$$n \cdot \operatorname{cov}\begin{pmatrix} T_1 \\ T_2 \end{pmatrix} \equiv \Sigma_0 = \begin{bmatrix} 1 & t^2 e^{-t^2} \\ t^2 e^{-t^2} & \left[\frac{1}{2}\left(1 - e^{-2t^2}\right)\right]^2 \end{bmatrix}. \tag{5.2}$$

Our test statistic is, in fact, $Q = n(T_1, T_2)\Sigma_0^{-1}(T_1, T_2)'$. When $\rho \neq 0$ we have $Q = n \cdot q(\rho) + o_p(n)$, where $q(\rho) = (E_\rho T_1, E_\rho T_2)\Sigma_0^{-1}(E_\rho T_1, E_\rho T_2)'$, so that using Lemma (2.4) of Gregory (1980) we may compute the approximate Bahadur slope in this case:

$$c(\rho; T_1, T_2(t)) \overset{\text{a.s.}}{=} \lim_{n\to\infty} \left[-\frac{2}{n} \log P(Q \geq \text{ observed value })\right], \tag{5.3}$$

where P is the asymptotic distribution of Q under H_0. Thus

$$c(\rho; T_1, T_2(t)) = \lim_{n\to\infty} \left[-\frac{2}{n} \log P(\chi_2^2 \geq nq(\rho)\right] = q(\rho). \tag{5.4}$$

By a similar computation we find that the approximate Bahadur slope for the test based on T_1 alone (which is, in fact, the optimal test) to be

$$c(\rho; T_1) = \rho^2 \tag{5.5}$$

so that the ratio of approximate slopes is

$$e(\rho; t) = \frac{c(\rho; T_1, T_2(t))}{c(\rho; T_1)} = \frac{q(\rho)}{\rho^2}.$$

Now inverting Σ_0 we find

$$q(\rho) = \frac{\left[\frac{1}{2}(1 - e^{-2t^2})\right]^2 \rho^2 + (E_\rho T_2(t))^2 - 2t^2 e^{-t^2} \rho E_\rho T_2(t)}{\left[\frac{1}{2}(1 - e^{-2t^2})\right]^2 - (t^2 e^{-t^2})^2}, \tag{5.6}$$

where, after some computations

$$E_\rho T_2(t) = \frac{1}{2} e^{-t^2} [e^{\rho t^2} - e^{-\rho t^2}]. \tag{5.7}$$

The limiting $\rho \to 0$ Bahadur approximate efficiency may now be obtained and is

$$e_0(t) = \lim_{\rho \to 0} e(\rho; t) = 1, \tag{5.8}$$

independent of t. Unfortunately this is not the complete picture. Firstly, for $\rho \neq 0$ we have typically $e(\rho; t) > 1$ which is unsatisfactory. Secondly, the condition III* of Wieand (1976) can be verified in this case so that $e_0(t)$ can be interpreted as a Pitman efficiency, however, the Pitman concept appears not to apply in comparing chi-square tests with unequal degrees of freedom (see, for example, Gregory, 1980, section 3.) We thus have here a case where the limiting ratio of approximate Bahadur slopes has no known meaning and which further illustrates the rather tenuous relationship between the exact and approximate Bahadur concepts (Bahadur, 1967). The comparison between the tests is therefore not easily carried out theoretically but could certainly be studied by Monte Carlo methods. In any case, the optimal test (in the Gaussian case) is that which rejects for large T_1 whereas the "two points" test rejects for large $(T_1, T_2)\Sigma_0^{-1}(T_1, T_2)'$ which is a full rank fixed quadratic form in (T_1, T_2) and clearly somewhat less than fully optimal; the efficiency loss will typically be small, but clearly a rather complicated function.

In addition to those raised in the analysis above, some further open questions emerge that seem worthwhile. Firstly, if inference concerning general dependence is to be centered on the statistic

$$\Gamma_n(s, t) = c_n^{XY}(s, t) - c_n^X(s) c_n^Y(t), \tag{5.9}$$

then we remark that it may be worthwhile to consider first replacing the X_i and Y_i by their normal scores (or van der Waerden scores) to help standardize testing considerations. Note that $c_n^X(s)$ and $c_n^Y(t)$ will now be fixed so that the Γ_n covariance function must be adjusted, however, under H_0 we will now have that Γ_n is distribution free. We remark also that continuous type test statistics such as

$$\int\int |\Gamma_n(s, t)|^2 \, w(s, t) ds dt, \tag{5.10}$$

(or more general quadratic forms) yield fruitful procedures which are consistent and therefore will be competitive to and may be compared with the Hoeffding statistic. Finally, we remark that the multivariate independence testing problem has a time series analogue. In fact, the testing for independence of a stationary process may be used as a tool in fitting time series models, an application which we plan to pursue in a subsequent work.

ACKNOWLEDGMENTS

Portions of this work were carried out while the author was on leave at The Hebrew University, Jerusalem whose generous hospitality and facilities provided are gratefully acknowledged. It is a special pleasure to thank J. Aaronson, S. Haberman and D. Pfefferman for stimulating conversations. The ongoing support of a Natural Sciences and Engineering Research Council of Canada grant is gratefully acknowledged.

REFERENCES

Anderson, T. W. (1958), *An Introduction to Multivariate Statistical Analysis.* New York: Wiley and Sons.

Bahadur, R. R. (1960), "Stochastic comparison of tests". *Annals of Mathematical Statistics* **31**, 276–295.

Bahadur, R. R. (1967), "Rates of convergence of estimates and test statistics." *Annals of Mathematical Statistics* **38**, 303–324.

Blum, J. R., J. Kiefer, and M. Rosenblatt (1961), "Distribution free tests of independence based on the sample distribution function." *Annals of Mathematical Statistics* **32**, 485–498.

Csorgo, S. (1981a), "Limit behaviour of the empirical characteristic function." *Annals of Probability* **9**, 130–144.

Csorgo, S. (1981b), "Multivariate empirical characteristic functions." *Zeitschrift für Wahrscheinlichkeitstheorie und Verwandte Gebiete* **55**, 203–229.

Csorgo, S., and V. Totik (1983), "On how long an interval is the empirical characteristic function uniformly consistent?" *Acta Scientiarum Mathematicarum (Szeged)* **45**, 141–149.

Feuerverger, A., and D. A. S. Fraser (1980), "Categorical information and the singular linear model." *Canadian Journal of Statistics* **8**, 41–45.

Feuerverger, A., and P. McDunnough (1981a), "On the efficiency of empirical characteristic function procedures." *Journal of the Royal Statistical Society*, Series B **43**, 20–27.

Feuerverger, A., and P. McDunnough (1981b), "On some Fourier methods for inference." *Journal of the American Statistical Association* **76**, 379–387.

Feuerverger, A., and R. A. Mureika (1977), "The empirical characteristic function and its applications." *Annals of Statistics* **5**, 88–97.

Gregory, G. G. (1980), "On efficiency and optimality of quadratic tests." *Annals of Statistics* **8**, 116–131.

Hoeffding, W. (1948), "A nonparametric test of independence." *Annals of Mathematical Statistics* **19**, 546–557.

Hollander, M., and D. A. Wolfe (1973), *Nonparametric Statistical Methods*. New York: Wiley and Sons.

Keller, H. D. (1979), "Einige Untersuchungen zur empirischen charakterischen Funktion und deren Anwendungen." Dissertation. Universität Dortmund.

Kendall, M. G. (1970), *Rank Correlation Methods*, 4th edition. London: Griffin.

Konijn, H. S. (1956), "On the power of certain tests for independence in bivariate populations." *Annals of Mathematical Statistics* **27**, 300–323. Correction **29**, (1958) 935–936.

Koziol, J. A., and A. F. Nemec (1979), "On a Cramer-von Mises type statistic for testing bivariate independence." *Canadian Journal of Statistics* **7**, 43–52.

Lehmann, E. L. (1959), *Testing Statistical Hypotheses*. New York: Wiley and Sons.

Lehmann, E. L. (1966), "Some concepts of dependence." *Annals of Mathematical Statistics* **37**, 1137–1153.

Lehmann, E. L. (1975), *Nonparametrics: Statistical Methods Based on Ranks*. San Francisco: Holden-Day.

Marcus, M. B. (1981), "Weak convergence of the empirical characteristic function." *Annals of Probability* **9**, 194–201.

Rao, C. R. (1965), *Linear Statistical Inference and its Applications*. New York: Wiley and Sons.

Rosenblatt, M. (1975), "A quadratic measure of deviation of two-dimensional density estimates and a test of independence." *Annals of Statistics* **3**, 1–14.

Srivastava, M. S., and G. C. Lee (1984), "On the distribution of the correlation coefficient when sampling from a mixture of two bivariate normal densities: robustness and the effect of outliers." *Canadian Journal of Statistics* **12**, 119–133.

Stuart, A. (1954), "The asymptotic relative efficiencies of tests and the derivatives of their power functions." *Skandinavisk Aktuarietidskrift* **37**, 163–169.

Wald, A. (1943), "Tests of statistical hypotheses concerning several parameters when the number of observations is large." *Transactions of the American Mathematical Society* **54**, 426–483.

Wieand, H. S. (1976), "A condition under which the Pitman and Bahadur approaches to efficiency coincide." *Annals of Statistics* **4**, 1003–1011.

C. W. J. Granger [1]

ARE ECONOMIC VARIABLES REALLY INTEGRATED OF ORDER ONE?

1. INTRODUCTION

Many macro-variables have a fairly smooth appearance over long time periods. This smoothness can be translated into functions commonly considered by time-series analysts as:

(i) the series have a spectrum with a large peak at low frequencies, called the "typical spectral shape" (or TSS) by Granger (1966),

(ii) the correlogram declines very slowly as lag length increases, and

(iii) the differenced series will have a correlogram that is explainable using a stationary ARMA model.

These properties will be called the TSS properties. Some economic series may appear to need to be differenced more than once, but I will not discuss that case in any detail.

It was pointed out by Box and Jenkins (1970) that a first-order-integrated process (denoted $I(1)$) or more specifically the ARIMA $(p, 1, q)$ process has the TSS properties. Such a process produces a stationary ARMA series after differencing. It has become common to equate the two facts and conclude that many economic series are actually $I(1)$. In this paper I will raise the question posed in the title but will not reach a definitive answer, as the necessary empirical work has not yet been undertaken. Rather, I will note that there are some problems in matching economic reality with the $I(1)$ model and suggest some alternative models that circumvent these difficulties. I also point out that there is a wide class of models that produce series with the TSS properties, particularly once the possibility of a deterministic trend is taken into account.

A formal definition of an $I(1)$ series, denoted by x_t, is that it has the

[1] Department of Economics, University of California at San Diego, La Jolla, California 92093

I. B. MacNeill and G. J. Umphrey (eds.), Time Series and Econometric Modelling, 207–217.

property that

$$x_t - x_{t-1} = y_t, \tag{1.1}$$

where y_t is stationary. An example is the simple random walk, where y_t is just a zero mean, white-noise series, denoted ϵ_t, so that in this case

$$x_t - x_{t-1} = \epsilon_t. \tag{1.2}$$

To generate such a series, one merely needs the inputs y_t plus a starting x value.

A simple but interesting and instructive example is where

$$y_t = m + \epsilon_t,$$

where ϵ_t is white noise as above. Now (1.1) is a model generating a series known as a random walk with drift, in the case when $m \neq 0$. If the process starts at time $t = -N - 1$, with the starting value x_{-N-1}, then

$$x_t = (N+t)m + \sum_{j=-N}^{t} \epsilon_j + x_{-N-1}. \tag{1.3}$$

For convenience, it will be assumed that $x_{-N-1} = 0$. It is convenient to rewrite (1.3) as

$$x_t = h(t) + s_t,$$

where

$$h(t) = (N+t)m$$

is the deterministic trend in mean, and

$$s_t = \sum_{j=-N}^{t} \epsilon_j$$

is the zero-mean, stochastic component of x_t, having variance

$$V(t) = (N+t)\sigma^2,$$

where $\sigma^2 = \text{var}(\epsilon_t)$. For ease of exposition it will be assumed that the variance of ϵ_t is constant, although in practice this does not necessarily occur. I will generally take a white noise series merely to be an independent sequence. It is frequently claimed that a random walk has infinite variance. This will be achieved only if $m = 0$ and then letting $N \to \infty$. It should be noted that this limiting procedure cannot occur if $m \neq 0$, as then the level of the series would be at infinity. The importance in practical terms of there being an actual starting time is clearly seen, and replacing ϵ_t by any zero-mean stationary series makes no difference in this analysis. Thus, rather than assume an infinite variance for a random walk, it is more realistic to assume that both mean and variance of x_t are linear functions of time. The actual time at which the process starts is not particularly important; what matters is that it started a finite time ago.

It should also be noted that

$$E\left[s_t s_{t-k}\right] = \sigma^2(N + t - k)$$

so that

$$\begin{aligned}\rho_k &= \operatorname{corr}\left(x_t, x_{t-k}\right)\\ &= \frac{N+t-k}{\left[(N+t)(N+t-k)\right]^{1/2}}\\ &= \left[\frac{N+t-k}{N+t}\right]^{1/2},\end{aligned}$$

which is nearly one if $N + t$ is large compared to k. It is thus seen that this is an $I(1)$ series, starting up a finite time ago, will have the TSS properties, at least to a high degree of approximation.

The basic evidence that many economic macro-variables are $I(1)$ is provided by the many model identifications using the Box-Jenkins interpretation of the correlogram. Actual series often have trends in mean and autocorrelations that are large for many lags. As $I(1)$ series also have these properties, it is common practice to conclude that actual series ae $I(1)$, but as will be shown this is not necessarily the correct conclusion. In the next section, various other linear models having at least approximately the TSS properties are discussed. In the following section, the fact that many economic series are inherently positive will be faced and some non-linear models considered. The paper finishes with a general discussion and conclusions.

2. AN INTRODUCTION TO TRENDING SERIES

If $h_j, j = 0, 1, 2, \ldots$ is a sequence of positive constants, a series x_t may be generated by the application of a sequence of increasingly long filters applied to the input series e_t, giving

$$x_t = \sum_{j=0}^{t} h_j e_{t-j}. \tag{2.1}$$

Taking

$$e_t = m + \epsilon_t,$$

where ϵ_t is i.i.d., with zero mean and variance σ^2, one gets the decomposition

$$x_t = mh(t) + s_t,$$

where

$$h(t) = \sum_{j=0}^{t} h_j$$

and

$$s_t = \sum_{j=0}^{t} h_j \epsilon_{t-j}$$

is a zero-mean stochastic component with variance

$$V(t) = \sigma^2 \sum_{j=0}^{t} h_j^2.$$

The sequence h_j is said to form a divergent series if $h(t)$ continually increases without bound, in which case $mh(t)$ will provide a deterministic trend in mean provided $m \neq 0$. Further, the sequence h_j^2 may generate a trend in variance if $V(t)$ is an increasing, unbounded sequence. Clearly, the theory of divergent series may be used to analyze these functions. As an example, consider the sequence $h_j = 1/j$, then $h(t)$ tends to $\log t$ and $V(t)$ tends to a constant. Thus, in this case, the filters generate a trend in mean, but not in variance. For a second example, if $h_j = j^{\alpha-1}$ then $h(t)$ tends to $c_1 t^{\alpha}$ and $V(t)$ tends to $c_2 t^{2\alpha-1}$, where c_1, c_2 are constants.

If $h_1(t)$, $h_2(t)$ are a pair of trending functions, then $h_1(t)$ is said to dominate $h_2(t)$ if the ratio

$$\frac{h_2(t)}{h_1(t)} \to \text{zero as } t \to \infty.$$

Thus $h(t)$ is a trend if it dominates any constant. A number of features of the trend generating filters are considered by Granger (1985), which concentrates on the dominant components of any trend; it is shown that a wide variety of trends can be generated, the effects of trend-reducing filters are determined and the corresponding spectral shapes at low frequencies are discussed.

A particularly interesting class of filters arise from considering functions $g(B)$ of the lag operator B such that $g(1) = 0$. Let

$$\frac{1}{g(B)} = \sum_{j=0}^{\infty} g_j B^j$$

and now take $h_j = g_j$, $j = 0, 1, \ldots, t$. The resulting filters generate trends and the stochastic part s_t may be called generalized integrated. Examples include the fractional integrated processes with $g(B) = (1-B)^c$ where c may be a fraction, as considered by Granger and Joyeux (1980), Hoskins (1981) and others. If x_t is $I(d)$, d a fraction, then the trend in mean produced is dominated by ct^d, for some constant c.

Many of the series generated by the filters discussed in this section will have the TSS properties, at least approximately. The series will usually have trends in mean and variance, but the differenced series are

$$x_t - x_{t-1} = e_t + \sum_{j=0}^{t} (h_j - h_{j-1}) e_{t-j}$$

and if Δh_j is a convergent series, then this differenced series may appear to be stationary. It is also proved by Granger (1985), that provided $h_t/h_{t-k} \to 1$ as $t + \infty$ for k fixed and if $V(t)$ is a trend; it is shown that then $\text{corr}(s_t, s_{t-k}) \to 1$ as t increases. This property will hold for all the generalized integrated series with $d \geq 1/2$ as well as filters that generate explosive trends. This point can be put another way, as most series produced by a generalized integrated model will have a plot of x_t against x_{t-1} that lies tightly along the 45° line. It is an empirical fact that many macro-economic series have this property.

It is thus seen that many series can be generated, including some fractional integrated series, which have the TSS properties, at least approximately. It follows that it is not necessarily correct to identify a series as $I(1)$ if it appears to have the TSS properties.

3. THE REAL ECONOMY

Many economic series are intrinsically positive, such as production, imports, prices, wages and employment. Most of these also have a distinctive upward trend also and thus can be well modelled as either an $I(1)$ process or by one of the generalized integrated processes discussed in the previous section. This follows because for all trends of the form $h(t) = ct^\alpha$, then $h(t)/\sqrt{V(t)}$ will itself be an increasing trend and the central limit theorem will also apply, suggesting that for large enough t, a negative value becomes extremely unlikely.

However, some economic series are also necessarily positive but have no clear-cut trends. The best examples are interest rates, exchange rates over long periods, and some real prices, including stock market indices measured daily and weekly. Such positive series are often found to have the TSS properties but not to have any distinct trend. This combination of properties makes an $I(1)$ model, without drift, seem very inappropriate.

Any $I(1)$ process without trend, such as a simple random walk, will wander widely, will cross all levels rarely but occasionally and will certainly take negative values. For many of the economic variables just mentioned, these are not realistic properties. Not only are they confined to positive values, but there may be control mechanisms that prevent wandering into high or unusual values, such as profit taking in the stock market and actions by central banks and governments for exchange rates and interest rates. Thus, we need to consider different classes of models to represent the properties of such real data.

4. MODELLING APPROXIMATE INTEGRATED PROCESSES

It is convenient to start by considering the simple random walk without drift

$$x_t = x_{t-1} + \epsilon_t, \tag{3.1}$$

where ϵ_t is a white noise (mutually uncorrelated) series with mean zero. The problem with this model, if x_t is known to be positive, is that if x_{t-1} is small, a negative value of ϵ_t may take x_t into the unallowed negative region. Before considering more exotic models, it is worth first asking if this simple linear model can be changed to overcome the problem of generating negative values. There does exist a possibly relevant class of models of the form

$$x_t = \rho x_{t-1} + \delta_t e_t, \tag{3.2}$$

where δ_t, e_t are an independent pair of white noise processes with

$$\begin{aligned} \delta_t &= 0 \ \text{ with probability } \rho \\ &= 1 \ \text{ with probability } 1 - \rho \end{aligned}$$

and e_t is an i.i.d. sequence of positive random variables. Lawrance and Lewis (1980), for example, have considered the case where e_t are exponentially distributed and called the process exponential autoregressive (EAR). These models can provide series with a given marginal distribution (exponential) and a required ($AR(1)$) autocorrelation sequence but the joint distribution is often somewhat less attractive in that the series generated have a different appearance than actual economic data. This problem is clearer as ρ becomes larger as then the series consists of exponentially declining sections $\rho^k x_{t-k}$ with occasional random positive jumps. In particular, if one takes ρ near one the process will not look like an $I(1)$ sequence and the limit as ρ tends to one becomes quite inappropriate. Thus, these models do not look promising for the purpose of this paper.

An alternative approach is to change the distribution of ϵ_t depending on the value of x_{t-1}. For example, ϵ_t may be drawn from a unimodal, symmetric distribution (bounded below) for "usual" values of x_{t-1}, from a unimodal distribution with most weight below the mode if x_{t-1} is large and the reverse distribution shape when x_{t-1} is small. However, if x_{t-1} gets very small, that is near zero, the distribution would have to have all its weight above zero and thus the residual cannot have zero mean at this extreme. It is unclear if a sequence of residuals constructed in such a way would be white noise and they certainly could not have zero mean for all possible x_{t-1} values. The result would be a non-linear $AR(1)$ process with heteroscedasticity.

One natural extension of the usual linear model is to consider a linear model with time-varying parameters, such as

$$x_t = \alpha_t x_{t-1} + \epsilon_t. \tag{3.3}$$

A series generated in this way might look similar to an $I(1)$ process if the average value of α_t is near one. However, if α_t varies independently of x_{t-1} the range of x_t will still be extensive, and x_t can go negative. If one makes α_t a function of x_{t-1}, the resulting model is just a non-linear model of the form

$$x_t = f(x_{t-1}) + \epsilon_t. \tag{3.4}$$

These considerations suggest that the most promising model is of the form (3.4) when the function $f(\cdot)$ is chosen to approximate an $I(1)$ and the distribution of ϵ_t is chosen to ensure that $x_t > 0$.

A simple example is a pure random walk with reflecting barriers that hardly ever come into operation. A sample taken from a process generated in this way will certainly have TSS. One can soften the barriers and get the same result. For example, suppose that $f(x)$ is virtually equal to x for a wide range of values but with $f(x) < x$ for x sufficiently large. This will ensure that the process has TSS but has finite variance. An example is

$$x_{t+1} = x_t \cdot \exp(-x_t^2/m^2) + \epsilon_{t+1}. \tag{3.5}$$

Again, data generated by this equation, with variance (ϵ_t) small, will form a series that has TSS, as indicated by some experiments mentioned by Granger *et al.* (1984). Models such as (3.5) are then called "almost integrated" or AINT models.

The function $f(x)$ in (3.4) can be estimated using a non-parametric technique similar to spline curve fitting, as investigated by Engle *et al.* (1984). When used with some U.S. interest rate data, indication of the existence of a curve like that in (3.5) was found, but not completely convincingly. No similar evidence was found for exchange rate or balance of trade data, when the plots of x_t against x_{t-1} produced almost perfect 45° degree lines.

It is interesting to note that even bounded deterministic processes can have a property similar to TSS. It is noted by Aizawa and Kohyama (1984) that a process generated by

$$x_{t+1} = f(x_t),$$

where

$$\begin{aligned} f(x) &= x + 2^{B-1}(1-2\epsilon)x^B + \epsilon, && 0 \leq x \leq 1/2 \\ &= x - 2^{B-1}(1-2\epsilon)(1-x)^B - \epsilon, && 1/2 < x \leq 1, \end{aligned}$$

where $\epsilon > 0$ is a small, fixed perturbation and $B > 3/2$ gives a process x_t that is bounded by

$$0 < x_t < 1$$

and having a spectrum

$$f(w) \sim w^{-k},$$

where

$$k = 3 - B(B-1)$$

and with

$$w > w_c = O(\epsilon^{1/B}).$$

Thus, the spectrum will have the "typical shape" down to a low frequency but is bounded at yet lower frequencies. If ϵ is small, but positive, the resulting spectrum will be very difficult to differentiate from a fractional differenced process, $d = k/2$, and thus with $0 < d < 1$. As x_t generated in this fashion is bounded, it will necessarily have finite variance.

A class of generating mechanisms that is potentially relevant for producing positive series with TSS is to take

$$x_t = g(y_t),$$

where y_t is $I(d)$, say, and $g(y) \geq 0$ for all y. A simple example would be

$$x_t = y_t^2.$$

It is clear that if y_t contains a trend then so will x_t provided $g(y)$ is not bounded above. A complete theory for this class of models does not seem to be available, although Surgailis (1981) has results for the case $d < 1/2$, so that y_t has finite variance.

Considering just the case where x_t is y_t^2, for x_t to not have a trend, it will be necessary that y_t has no trend in mean and no trend in variance. For example, if y_t is a random walk without drift, starting t periods earlier, then y_t will have variance proportioned to t, so that x_t will have a mean proportional to t, and so will not be trend-free in mean. It thus seems that instantaneous transforms of generalized integrated series are unlikely to provide satisfactory models for trend free TSS series. They will often lack economic plausibility, as an accepted interpretation of the core or driving series y_t is usually difficult to provide.

5. CONCLUSIONS. WHERE DOES TSS COME FROM?

It is a well-established empirical fact that many macro-economic variables display the properties here denoted by TSS. The question considered in the majority of this paper is what models generate series with such a property, and this class is found to be wider than the $I(1)$ models considered by Box and Jenkins (1970). For trending series, there are many filters which generate trends and corresponding stochastic components, many of which have TSS. At this time, the analysis of trends is one of the least developed parts of time-series analysis and econometrics, although it is potentially a very important part of a complete analysis of economic data. For non-trending, positive series with TSS, the AINT, non-linear models appear to have more potential than many of the more obvious alternatives.

One very important question remains, why do we observe economic series having the TSS property? There appears to have been surprisingly little discussion of this question. For one group of variables, prices of speculative goods such as stocks, bonds, currencies, gold, silver and other commodities, the efficient market theory provides an immediate answer. Such speculative prices should follow a random walk (possibly with a small drift) because otherwise it would be possible to have an investment stragegy that guarantees a positive return. If such a "money-pump" strategy existed, it would naturally be used by all investors and it follows that it cannot exist, at least for time periods up to the decision horizon of most investors. Thus, these speculative prices should be $I(1)$ except possibly in the very long run, beyond speculation investment horizons, when a model such as AINT could apply. Of course, the vast majority of the empirical evidence supports this observation.

For most macro-economic series with TSS, speculation is not an important component. As I have noted elsewhere (Granger, 1984), most of these variables are aggregates over huge numbers of micro-decision makers, such as consumers, families or companies. For example, U.S. consumption of non-durables is the aggregate of such consumption of over 80 million families. If one considers a single family, it is unlikely that this consumption will be a pure $I(1)$ series as consumption must be positive, may not be smooth because of changes in employment status or in income, and will not range widely from current levels. Because of income or borrowing constraints a non-linear model, such as AINT, may again appear to be appropriate for univariate series modelling. Of course, a better model will be one linking the consumption and income series for each family. If all of the individual family consumption series are independent of each other, it has been noted before (Granger, 1980) that each can obey a simple model, such as a stationary $AR(1)$, but for the aggregate to be $I(d)$, $d > 0$, and thus to have the TSS property. In particular, if e_t is the consumption series for the jth family and it obeys the $AR(1)$ model

$$c_{jt} = m_j + \alpha_j c_{j,t-1} + e_{jt},$$

where e_{jt} is white noise, and the α_j are drawn from the beta distribution on $(0,1)$

$$f(x) = \begin{cases} \frac{2}{B(p,q)} x^{2p-1}(1-x^2)^{q-1} & 0 \leq x \leq 1 \\ 0 & \text{elsewhere, with } p > 0,\ q > 0, \end{cases}$$

then the aggregate consumption is shown to be $I(1 - q/2)$.

However, the assumption that the consumption series are independent is quite unacceptable and if a rather strong type of dependence is introduced, coming from a common factor, a quite different source of the TSS property becomes important. For example, suppose that

$$c_{jt} = c'_{jt} + \beta_j \tilde{c}_t,$$

where all pairs of series c'_{jt} are independent of each other and also of the common factor $\tilde{c}_t$. Then the aggregate consumption

$$Ac_t = \sum_j c_{jt}$$

will have two components,

$$\sum_j c'_{jt} \text{ with variance proportional to } N$$

and

$$\tilde{c}_t \sum_j \beta_j \quad \text{with variance proportional to } N^2,$$

provided $\sum \beta_j \neq 0$, where N is the number of families in the aggregate. Even if the common factor component of each family consumption contributes very little to the variance of an individual c_{jt} series it will completely dominate the aggregate series if N is very large. Thus, if the families use common factors in their consumption decisions, such as interest rates, speculative price, or aggregate income (the common factor of family incomes) or even inflation expectations and if any of these common factors have the TSS property this will be sufficient to make observable aggregate series have this property. I think that this possibility deserves empirical investigation.

REFERENCES

Aizawa, Y., and T. Kohyama (1984), "Asymptotically non-stationary chaos." *Progress of Theoretic Physics* **71,** 847–850.

Box, G. E. P., and G. Jenkins (1970), *Time Series Analysis, Forecasting and Control.* San Francisco: Holden-Day.

Engle, R. F., C. W. J. Granger, J. Rice, and A. Weiss (1986), "Semiparametric estimates of the relation between weather and electricity sales." *Journal of the American Statistical Association* **81,** 310–320.

Granger, C. W. J. (1966), "The typical spectral shape of an economic variable." *Econometrica* **34,** 150–161.

Granger, C. W. J. (1984), "Implications of aggregation with common factors." Discussion paper, Economics Department, UCSD.

Granger, C. W. J. (1985), "Models that generate trends." Discussion paper 85-5, Economics Department, UCSD.

Granger, C. W. J., F. Huynh, A. Escribano, and C. Mustafa (1984), "Computer investigation of some non-linear time series models." Working paper, Economics Department, UCSD.

Granger, C. W. J., and R. Joyeux (1980), "An introduction to long-memory time series models and fractional differencing." *Journal of Time Series Analysis* **1,** 15–29.

Hosking, J. R. M. (1981), "Fractional differencing." *Biometrika* **68,** 165–176.

Lawrance, A. J., and P. A. W. Lewis (1980), "The exponential autoregressive–moving average EARMA(p, q) process." *Journal of the Royal Statistical Society,* Series B **42,** 150–161.

Surgailis, D. (1981), "Convergence of sums of non-linear functions of moving averages for self-similar processes." *Soviet Mathematics—Doklady* **23,** 247–250.

Peter C. B. Phillips [1]

FRACTIONAL MATRIX CALCULUS AND THE DISTRIBUTION OF MULTIVARIATE TESTS

ABSTRACT

Fractional matrix operator methods are introduced as a new tool of distribution theory for use in multivariate analysis and econometrics. Earlier work by the author on this operational calculus is reviewed and to illustrate the use of these methods we give an exact distribution theory for a general class of tests in the multivariate linear model. This distribution theory unifies and generalizes previously known results, including those for the standard F statistic in linear regression, for Hotelling's T^2 test and for Hotelling's generalized T_0^2 test. We also provide a simple and novel derivation of conventional asymptotic theory as a specialization of exact theory. This approach is extended to generate general formulae for higher order asymptotic expansions. Thus, the results of the paper provide a meaningful unification of conventional asymptotics, higher order asymptotic expansions and exact finite sample distribution theory in this context.

1. INTRODUCTION

The purpose of this paper is to provide a short review of some new methods I have been working with recently in the field of econometric distribution theory. These methods have turned out to be surprisingly useful in furnishing solutions of a rather general nature to a wide range of problems that occur in finite sample econometrics. Since these problems are very similar to those that arise naturally in other areas of statistical theory, notably multivariate analysis, I hope that the methods I have been developing will be of some interest to mathematical statisticians who are working in these related fields.

[1] Cowles Foundation for Research in Economics, Yale University, Box 2125, Yale Station, New Haven, Connecticut 06520

I. B. MacNeill and G. J. Umphrey (eds.), Time Series and Econometric Modelling, 219–234.

The methods rely on the concept of matrix fractional differentiation and therefore belong to an operational calculus. At an abstract level the techniques may be interpreted within the framework of pseudo-differential operators on which there is a large mathematical literature (see, for example, Treves, 1980). At the algebraic and purely manipulative level it is hard to find any references in the literature beyond those which apply to scalar methods of fractional calculus. Even here most attention is concentrated on the Riemann-Liouville definition of a fractional integral (or derivative). Whereas in applications to statistical distribution theory, I have found that a form of Weyl calculus yields the simplest and most direct results. It is also the most amenable to matrix generalizations. For an introduction to scalar fractional operators of this type the reader is referred to the books by Ross (1974a), Spanier and Oldham (1974) and the review article by Lavoie *et al.* (1976).

The use of an operational calculus in problems of distribution theory has many natural advantages. In the first place, seemingly difficult problems may often be solved quite simply with rather elegant general solution formulae. The latter usually avoid the complications of series representations, including those that are expressed in terms of zonal or invariant polynomials which many researchers find daunting and difficult for numerical work. Second, the routine manipulation of operators frequently leads to simplifications which are not otherwise obvious. Both these advantages arise, of course, in other applications of operator methods. However, I have discovered that there are some advantages to operational methods which are peculiar to their use in statistical distribution theory.

Perhaps the most important of these is that the methods provide a simple means of unifying limiting distribution theory, asymptotic expansions and exact distribution theory. This is because the operator representation of the exact finite sample distribution often lends itself to the immediate derivation of the asymptotic distribution and associated expansions about the asymptotic distribution. Thus, all three forms of distribution theory may often be derived from the same general formulae. An example will be studied later in the paper.

A further special advantage of operational methods is that they help to resolve mathematical problems for which existing techniques of distribution theory are quite unsuited. One of the more prevalent of these in multivariate models, at least in the present stage of the development of the subject, arises from the presence of random matrices (usually sample covariance matrices) that are embedded in tensor formations. These tensor formations inhibit the use of conventional techniques such as change of variable methods. Prominent examples of such problems occur in econometrics with seemingly unrelated regression equations, and systems estimation methods like three stage least squares. In multivariate analysis many multivariate tests, such as the Wald test for testing coefficient restrictions in the multivariate linear model come into this category. Since this particular test includes so many commonly occurring statistics such as the F test, Hotelling's T^2 and the T_0^2 statistic we shall use it as the focus of our attention in this paper as a prototypical application of the operational method. For other examples and related work the reader may refer to some other papers by the author (1984a, 1984b, 1985, 1986).

2. FRACTIONAL OPERATORS

Historically, the concept of a fractional operator arose from the attempt by classical mathematicians, principally Leibnitz, Euler, Liouville and Riemann, to extend the meaning of the operation of differentiation (to an integral order) to encompass differentiation of an arbitrary order. These classical mathematicians addressed the following question: given the operator $D = d/dx$ and rules for working with D^n to the integer order n what, if any, meaning may be ascribed to D^α where α is fractional or possibly even complex? An interesting historical study of the evolution of ideas in this field is provided by Ross (1974b), who traces the origin of this search for an extended meaning of the differential operator to correspondence between Leibnitz and L'Hospital in 1695.

Using the integral representation of the gamma function a very simple intuitive approach to fractional (complex) operators may be developed. Thus, if $\mathrm{Re}(\alpha) > 0, \mathrm{Re}(z) > 0$ we have:

$$z^{-\alpha} = \Gamma(\alpha)^{-1} \int_0^\infty e^{zt} t^{\alpha-1} dt. \tag{1}$$

This formula, which is extensively used in applied mathematics, provides a simple mechanism for replacing an awkward power of a complex variable that occurs in a denominator by an integral involving an exponent which is much simpler to deal with. In a certain sense, this simple idea is the key to much of the subject and to its multivariate extensions that we shall examine below.

If we now consider replacing z in (1) by the operator $D = d/dx$ we note that whereas $D^{-\alpha}$ is difficult to interpret e^{Dt} is not. The operator e^{Dt} yields Taylor series representations for analytical functions and may be regarded as a simple shift operator. Thus

$$e^{Dt} f(x) = f(x - t) \tag{2}$$

for f analytic. This suggests that we may formally write:

$$D^{-\alpha} f(x) = \Gamma(\alpha)^{-1} \int_0^\infty f(x-t) t^{\alpha-a} dt. \tag{3}$$

Then if the right side of (3) is absolutely convergent it may be used as a definition for the fractional integral $D^{-\alpha} f(x)$. Quite general operators with complex powers such as D^μ may now be defined by writing

$$D^\mu f(x) = D^{-\alpha}\{D^m f(x)\},$$

where $\mu = m - \alpha$, m is a positive integer and $\mathrm{Re}(\alpha) > 0$. Operators of this type obey the law of indices and are commutative, although this is not true of general matrix extensions, of course. At an abstract level, these operators may be used to form algebraic systems such as operator algebras, which may in turn be used to justify routine manipulations of the operators as algebraic symbols.

After a change of variable on the right side (3) may be written as:

$$D^{-\alpha} f(x) = \Gamma(\alpha)^{-1} \int_{-\infty}^{x} f(x)(x-s)^{\alpha-1} ds, \tag{4}$$

which corresponds to one form of the Weyl fractional integral (see, for example, Miller, 1974).

It is easy to show with this definition that:

$$D^{-\alpha} e^{ax} = e^{ax} a^{-\alpha}. \tag{5}$$

This may be proved using (3) for $\mathrm{Re}(\alpha) > 0, \mathrm{Re}(a) > 0$. The result (5) then holds by analytic continuation for all complex $a \neq 0$ and for all complex α. Similar results extending the rules for differentiating elementary functions may be obtained in the same way. Another rule which is quite useful is:

$$D^\mu (1-x)^{-\beta} = \frac{\Gamma(\beta+\mu)}{\Gamma(\beta)} (1-x)^{-\beta-\mu}, \tag{6}$$
$$\mathrm{Re}(\beta) > 0, \qquad \mathrm{Re}(\beta+\mu) > 0.$$

(5) and (6) illustrate the great advantage that the Weyl operator (3) has over the Riemann-Liouville operator defined by

$$D_{x_0}^{-\alpha} f(x) = \Gamma(\alpha)^{-1} \int_{x_0}^{x} f(s)(x-s)^{\alpha-1} ds \tag{7}$$

for $\text{Re}(\alpha) > 0$. The finite limit of integration x_0 in (7) allows us to admit a wider class of functions into the definition (avoiding the conditions of convergence required by the improper integral involved in the Weyl definition (4)). However, when (7) is applied to elementary functions the results are usually much more complex than (5) and (6). For example, in the case of (5) we have

$$D_{x_0}^{-\alpha} e^{ax} = a^{-\alpha} e^{ax} \Gamma(\alpha)^{-1} \Gamma(\alpha, a(x - x_0)),$$

where $\Gamma(\alpha, z)$ is the incomplete gamma function. This complication turns out to be a significant drawback to the Riemann-Liouville opeator in multivariate extensions and in applications to distribution theory. I have, therefore, found it most useful in my own work to employ (3) and its various generalizations rather than (7).

Multivariate extensions follow from the matrix gamma integral:

$$(\det\ Z)^{-\alpha} = \Gamma_n(\alpha)^{-1} \int_{S>0} \text{etr}(-SZ)(\det\ S)^{\alpha-(n+1)/2} dS, \tag{8}$$

where Z is an $n \times n$ matrix with $\text{Re}(Z) > 0$ and $\text{Re}(\alpha) > (n-1)/2$, and $\text{etr}(\cdot)$ represents e to the power of the trace of the matrix. $\Gamma_n(\alpha)$ is the multivariate gamma function which may be evaluated as $\Gamma_n(\alpha) = \pi^{n(n-1)/4} \prod_{i=1}^{n} \Gamma(\alpha - (i-1)/2)$. The integral (8) is extensively used in multivariate analysis. Its significance was first brought into prominence in the remarkable paper by Herz (1955).

We may now proceed as in the scalar case by introducing the matrix operator $\partial Z = \partial/\partial Z$. Whereas $(\det\ \partial Z)^{-\alpha}$ is difficult to interpret $\text{etr}(-\partial Z S)$ is not. In fact, if $f(Z)$ is an analytic function of the matrix variate Z the operator $\text{etr}(-\partial Z S)$ yields the matrix Taylor series representation

$$\text{etr}(-\partial Z S) f(Z) = f(Z - S), \tag{9}$$

generalizing (2). We may therefore define

$$\begin{aligned} D_Z^{-\alpha} f(Z) &= \Gamma_n(\alpha)^{-1} \int_{S>0} f(Z - S)(\det\ S)^{\alpha-(n+1)/2} dS; \\ D_Z &= \det\ \partial Z \end{aligned} \tag{10}$$

provided the integral is absolutely convergent and $\text{Re}(\alpha) > (n-1)/2$. The general case of an arbitrary complex power of D_Z may be dealt with in the same way as the scalar case by setting

$$D_Z^{\mu} f(Z) = D_Z^{-\alpha} \{D_Z^{m} f(Z)\}$$

for $\mu = m - \alpha$ with m a positive integer and $\text{Re}(\alpha) > (n-1)/2$.

Elementary functions of matrix argument may be complex differentiated as before. Thus

$$D_Z^{-\alpha}\mathrm{etr}(AZ) = \mathrm{etr}(AZ)(\det\ A)^{-\alpha} \tag{11}$$

generalizes (5) and may be proved for $\mathrm{Re}(A) > 0$, $\mathrm{Re}(\alpha) > (n-1)/2$ using (10). The formula (11) holds by analytic continuation for all nonsingular A and for all complex α. In a similar way, we find

$$D_Z^{\mu}\det(I-Z)^{-\beta} = \frac{\Gamma_n(\beta+\mu)}{\Gamma_n(\beta)}\det(I-Z)^{-\beta-\mu},$$
$$\mathrm{Re}(\beta) > (n-1)/2, \quad \mathrm{Re}(\beta+\mu) > (n-1)/2, \tag{12}$$

generalizing (6).

It is also useful to work with more complicated operators than D_Z. For example, if R is a $q \times nm$ matrix of rank $q \leq nm$ and M is a positive definite $m \times m$ matrix, then we may define

$$\begin{aligned}&[\det(R(\partial Z \otimes M)R')]^{-\alpha} f(Z)\\ &= \Gamma_a(\alpha)^{-1}\int_{S>0}[\mathrm{etr}\{-R(\partial Z\otimes M)R'S\}\, f(Z)]\,(\det\ S)^{\alpha-(q+1)/2}dS\end{aligned} \tag{13}$$

if the integral converges absolutely. The exponent $R(\partial Z \otimes M)R'S$ in the integrand of (13) is linear in the operator ∂Z and we may write:

$$\mathrm{tr}\,[-R(\partial Z \otimes M)R'S] = \mathrm{tr}[-\partial Z Q(S)],$$

where the $n \times n$ matrix Q is linear in the elements of S. Thus, (13) has the form:

$$\Gamma_q(\alpha)^{-1}\int_{S>0} f(Z-Q(S))(\det\ S)^{\alpha-(q+1)/2}dS.$$

Extensions to more complex tensor formations of operators are possible in an analogous fashion. Some of these are given and applied in one of the author's paper (1985) on the subject. When $f(Z)$ is an elementary function like $\mathrm{etr}(ZA)$ one obtains extensions of rules such as (11):

$$\det(R(\partial Z\otimes M)R')^{-\alpha}\,\mathrm{etr}(AZ) = \mathrm{etr}(AZ)\det(R(A\otimes M)R')^{-\alpha}. \tag{14}$$

Once again (14) is proved for $\mathrm{Re}(Z) > 0$ and $\mathrm{Re}(\alpha) > (q-1)/2$ and then analytically continued for all nonsingular A and all complex α.

3. MULTIVARIATE TESTS

To illustrate the use of these operator methods in distribution theory we shall consider some commonly occurring multivariate tests. What we present here will in large part be a review of work already done by the author in (1984, 1986) and the reader is referred to these papers for full details and generalizations. However, we shall present some new results on asymptotic expansions and exact distribution functions.

We shall be concerned with the multivariate linear model

$$y_t = Ax_t + u_t; \qquad (t = 1, \ldots, T) \tag{15}$$

y_t is a vector of n dependent variables, A is an $n \times p$ matrix of parameters, x_t is a vector of nonrandom independent variables and the u_t are i.i.d. $N(0, \Omega)$ errors with Ω positive definite. Let us suppose that we are interested in a general linear hypothesis involving the elements of A, which we write in null and alternative form as:

$$H_0 : R \text{ vec } A = r, \qquad H_1 : R \text{ vec } A - r = b \neq 0, \tag{16}$$

where R is a $q \times np$ matrix of known constants of rank q, r is a known vector and vec(A) stacks the rows of A.

From least squares estimation of (15) we have:

$$A^* = Y'X(X'X)^{-1}, \qquad \Omega^* = Y'(I - P_X)Y/N \tag{17}$$

where $Y' = [y_1, \ldots, y_T]$, $X' = [x_1, \ldots, x_t]$, $P_X = X(X'X)^{-1}X'$ and $N = T - p$. We take X to be a matrix of full rank $p \leq T$ and define $M = (X'X)^{-1}$.

The Wald statistic for testing the hypothesis (16) is

$$\begin{aligned} W &= (R \text{ vec } A^* - r)' \{R(\Omega^* \otimes M)R'\}^{-1} (R \text{ vec } A^* - r) \\ &= N\ell' B\ell, \end{aligned} \tag{18}$$

where $\ell = R$ vec $A^* - r$, is $N(b, V)$ under H_1 with $V = R(\Omega \otimes M)R'$, and $B = \{R(C \otimes M)R'\}^{-1}$. $C = Y'(I = P_X)Y$ is central Wishart with covariance matrix Ω and N degrees of freedom.

We define $y = \ell' B\ell$ and write y in canonical form as

$$y = g'Gg, \tag{19}$$

where $g = V^{-1/2}\ell$ is $N(m, I_q)$, $m = V^{-1/2}b$ and $G^{-1} = V^{-1/2}\{R(C \otimes M)R'\}V^{-1/2}$. With this notation we see that y and W are simply positive definite quadratic forms in normal variates, conditional on C. The distribution problem becomes one of integrating up this conditional distribution over the distribution of C.

Important special cases of the statistic W are as follows.

(i) The regression F statistic

Set $n = 1$, $A = a$, $H_0 : Ra = r$, $\Omega^* = s^2$ and then

$$\begin{aligned} W &= (Ra^* - r)' \left[R(X'X)^{-1}R'\right]^{-1} (Ra^* - r)/s^2 \\ &\equiv c\, F_{q,N} \end{aligned} \tag{20}$$

where $F_{q,N}$ denotes a variate with an F distribution with q and N degrees of freedom. In (20) we use the symbol "$\equiv$" to signify equality in distribution and the letter "c" to represent a constant. These notations will be used throughout the paper.

(ii) Hotelling's T^2 statistic

Set $R = R_1 \otimes r_2'$, $H_0 : R_1 A r_2 = r$ and then

$$\begin{aligned} W &= (R_1A^*r_2 - r)'[R_1\Omega^*R_1']^{-1}(R_1A^*r_2 - r)/r_2'Mr_2 \\ &\equiv c\, x'S^{-1}x \equiv c\, F_{q,N-q+1}, \end{aligned} \tag{21}$$

with $x \equiv N(0, R_1\Omega R_1')$ and $S \equiv W_q(N, R_1\Omega R_1')$ under the null; x and S are of course independent.

(iii) The T_0^2 statistic

Set $R = R_1 \otimes R_2'$, $H_0 : R_1AR_2 = r$ with R_1 $q_1 \times n$ and R_2 $m \times q_2$. Then

$$\begin{aligned} W &= \text{vec}(R_1A^*R_2 - r)'[R_1\Omega^*R_1' \otimes R_2'MR_2]^{-1}\text{vec}(R_1A^*R_2 - r) \\ &= \text{tr}\left[(R_1A^*R_2 - r)'(R_1\Omega^*R_1')^{-1}(R_1A^*R_2 - r)(R_2'MR_2)^{-1}\right] \\ &= c\ \text{tr}\left(XX'S_2^{-1}\right) \\ &= c\ \text{tr}\left(S_1S_2^{-1}\right), \end{aligned} \tag{22}$$

with $X \equiv$ matrix $N(0, R_1\Omega R_2' \otimes I_{q_2})$, $S_2 \equiv W_{q_1}(N, R_1\Omega R_1')$ and $S_1 = W_{q_1}(q_2, R_1\Omega R_1')$ under the null. Because of invariance to the covariance matrix in (22) we may treat S_1 as $W_{q_1}(q_2, I_{q_1})$ and S_2 as $W_{q_1}(N, I_{q_2})$; S_1 and S_2 are independent.

Interestingly, the exact distribution of the statistic $\text{tr}(S_1S_2^{-1})$ has not been found in the statistical literature, in spite of apparently substantial efforts by many researchers (see Pillai (1976, 1977) and Muirhead (1982) for reviews). Many conjectures have been made about the form of the exact density of this statistic. The classic article by Constantine (1966) which gives a series representation that is valid over the unit interval $[0, 1]$ is still perhaps the most general treatment. We shall show below how the distribution may be found in the general case quite simply by operator algebra. A full treatment is available in the author's paper (1986).

4. THE NULL DISTRIBUTION

It is shown by Phillips (1986, equation (32)) that the null density of W in the general case (18) is given by:

$$\begin{aligned}\mathrm{pdf}(w) = \frac{w^{q/2-1}}{N^{q/2}\Gamma\left(\frac{q}{2}\right)} \Big[& \det\left(L(\partial X \otimes I)L'\right)^{1/2} {}_0F_0\left(-L(\partial X \otimes I)L', w/N\right) \\ & \cdot \det(I - X)^{-N/2}\Big]_{X=0}, \end{aligned} \tag{23}$$

where

$$L = \left[R(\Omega \otimes M)R'\right]^{-1/2} R\left(\Omega^{1/2} \otimes M^{1/2}\right). \tag{24}$$

The function ${}_0F_0(-L(\partial X \otimes I)L', w/N)$ in (23) is a linear operator which may be explicitly represented as:

$$\int_{V_{1,q}} \mathrm{etr}\left(-(w/N)L(\partial X \otimes I)L'hh'\right)(\underline{dh}),$$

where $(\underline{dh})$ denotes the normalized invariant measure on the sphere $V_{1,q} = \{h : h'h = 1\}$. An alternative representation in terms of an absolutely convergent operator power series is also available:

$$\sum_{j=0}^{\infty} \frac{(-1)^j (w/N)^j C_j\left(L(\partial X \otimes I)L'\right)}{j! C_j(I)},$$

where $C_j(\cdot)$ denotes the top order zonal polynomial of degree j, for which explicit formulae were given by James (1964).

The simplicity of (23) is unusually striking. Yet, as we shall see, all existing exact distribution theory for the null case is embodied in this formula. Moreover, (23) also delivers the appropriate asymptotic theory and asymptotic expansions with little effort. In the following specializations we shall use the notational reductions detailed for these special cases in Section 3.

(i) The regression F statistic

$$\begin{aligned}\mathrm{pdf}(w) &= cw^{q/2-1}\left[(\partial x)^{q/2} e^{-\partial x w/N}(1-x)^{-N/2}\right]_{x=0} \\ &= cw^{q/2-1}\left[e^{-\partial x w/N}(1-x)^{-N/2-q/2}\right]_{x=0} \\ &= cw^{q/2-1}(1+w/N)^{-(N+q)/2} \\ &\equiv cF_{q,N}.\end{aligned}$$

The reductions in the second and third lines above follow directly from the rules (5) and (6) given earlier for fractional differentiation.

(ii) Hotelling's T^2 statistic

Noting that $L = L_1 \otimes \ell_2'$ with $L_1 L_1' = I_{q_1}$, we find that the density of W is:

$$\begin{aligned}
\mathrm{pdf}(w) &= cw^{q/2-1}\Big[\det(L_1\partial X L_1')^{1/2}\ {}_0F_0(-L_1\partial X L_1', w/N) \\
&\quad \cdot \det(I-X)^{-N/1}\Big]_{X=0} \\
&= cw^{q/2-1}\Big[(\det\ \partial X_{11})^{1/2} a_0F_0(-\partial X_{11}, w/N) \\
&\quad \cdot \det(I-X_{11})^{-N/2}\Big]_{X=0} \\
&= cw^{q/2-1}\Big[\ {}_0F_0(-\partial X_{11}, w/N)\det(I-X_{11})^{-(N+1)/2}\Big]_{X_{11}=0} \\
&= cw^{q/2-1}\Big[\int_{V_{1,q}} \mathrm{etr} - (w/N)\partial X_{11} hh'(\underline{dh}) \\
&\quad \cdot \det(I-X_{11})^{-(N+1)/2}\Big]_{X_{11}=0} \\
&= cw^{q/2-1}\int_{V_{1,q}} \det(I+(w/N)hh')^{-(N+1)/2}(\underline{dh}) \\
&= cw^{q/2-1}(1+w/N)^{-(N+1)/2} \\
&\equiv cF_{q,N-q+1}.
\end{aligned}$$

In the second line of this argument X_{11} is a $q_1 \times q_2$ matrix of auxiliary variates obtained from the $q \times q$ matrix X by transforming $X \to PXP'$ where $P' = [L_1', K']$ is orthogonal. Note that under this transformation $\partial X \to P'\partial X P$ and $L_1 \partial X L_1' \to \partial X_{11}$, giving the stated result.

(iii) The T_0^2 statistic

$L = L_1 \otimes L_2$, $L_1 L_1' = I_{q_1}$, $q = q_1 q_2$ and the density of W is:

$$\begin{aligned}
\mathrm{pdf}(w) &= cw^{q,2-1}\Big[\det(L_1\partial X L_1')^{q_2/2}\ {}_0F_0(-L_1\partial X L_1' \otimes I_{q_2}, w/N) \\
&\quad \cdot \det(I-X)^{-N/2}\Big]_{X=0} \\
&= cw^{q/2-1}[\ {}_0F_0(-\partial X_{11} \otimes I_{q_2}, w/N) \\
&\quad \cdot \det(I-X_{11})^{-(N+q_2)/2}\Big]_{X_{11}=0} \\
&= cw^{q/2-1}\Big[\int_{V_{1,q}} \mathrm{etr}\{-(w/N)
\end{aligned}$$

$$\cdot\,(\partial X_{11}\otimes I)hh'\}\,(\underline{dh})\det(I-X_{11})^{-(N+q_2)/2}\Big]_{X_{11}=0} \tag{25}$$

$$= cw^{q/2-1}\int_{V_{1,q}}\det\left(I+(w/N)Q\right)^{-(N+q_2)/2}(\underline{dh}), \tag{26}$$

where $Q=\sum_{i=1}^{q_2}h_ih_i'$ and $h'=(h_1',\ldots,h_{q_2}')$. For $0\le w/N<1$ we may expand the determinantal expression in the integrand of (26) giving

$$\begin{aligned}\mathrm{pdf}(w) &= cw^{q/2-1}\sum_{k=0}^{\infty}\frac{(-w/N)^k}{k!}\sum_{\kappa}\left(\frac{N+q_2}{2}\right)_{\kappa}\int_{V_{1,q}}C_\kappa(Q)(\underline{dh})\\ &= cw^{q/2-1}\sum_{k=0}^{\infty}\frac{(-w/N)^k}{k!\left(\frac{q_1q_2}{2}\right)_k}\sum_{\kappa}\left(\frac{N+q_2}{2}\right)_{\kappa}\left(\frac{q_2}{2}\right)_{\kappa}C_\kappa(I_{q_1}), \quad (27)\end{aligned}$$

which is the series obtained by Constantine (1966) for the null distribution. The integration over $V_{1,q}$ leading to (27) may be obtained quite simply using operator methods. The reader is referred to Phillips (1984b) where full details are given.

An alternative everywhere convergent series is obtained by working from (25). Once again details are provided by Phillips (1984b). We state only the final result here:

$$\begin{aligned}\mathrm{pdf}(w) &= \frac{c\,w^{q/2-1}}{(N+q_2w)^{q_1(q_2+N)/2}}\sum_{k=0}^{\infty}\frac{\left(\frac{w}{N+q_2w}\right)^k}{k!}\\ &\cdot\sum_{\theta}\sum_{\kappa}\left\{\frac{b_\theta^\kappa\left(\frac{q_1q_2-1}{2}\right)_\kappa}{\left(\frac{q_1q_2}{2}\right)_\kappa}\right\}\left(\frac{q_2+N}{2}\right)_\theta C_\theta(I_{q_1}), \quad (28)\end{aligned}$$

where the summations are over all partitions θ, κ of k into $\le q_1$ parts, and the b_θ^κ are certain constants.

(iv) Asymptotic theory

We employ the simple asymptotic representation

$$\det(I-X)^{-N/2}\sim \mathrm{etr}(NX/2)$$

for $X\sim 0$ in (23) and deduce immediately that:

$$\mathrm{pdf}(w)\sim\frac{w^{q/2-1}e^{-q/2}}{2^{q/2}\Gamma\left(\frac{q}{2}\right)}\equiv\chi_q^2. \tag{29}$$

Thus, the asymptotic distribution appears as a special case of (23) in a single step.

(v) Higher order asymptotics

We transform $X \to Z = NX$ in (23) giving $\partial X = N\partial Z$ and:

$$\begin{aligned}\text{pdf}(w) = \Gamma(q/2)^{-1} w^{q/2-1} \Big[&\det\left(L(\partial Z \otimes I)L'\right)^{1/2} \\ &\cdot {}_0F_0\left(-L(\partial Z \otimes I)L', w\right) \det(I - Z/N)^{-N/2}\Big]_{Z=0}. \end{aligned} \tag{30}$$

We now expand the determinantal factor as $N \uparrow \infty$:

$$\begin{aligned}\det(I - Z/N)^{-N/2} &= \exp\left\{-\frac{N}{2}\ln\det\left(I - \frac{Z}{N}\right)\right\} \\ &\sim \exp\left\{\frac{N}{2}\sum_{k=1}^{\infty}\frac{1}{k}\text{tr}\left(\frac{Z}{N}\right)^k\right\} \\ &= \text{etr}\left(\frac{1}{2}Z\right)\exp\left\{\frac{1}{2}\sum_{j=1}^{\infty}\frac{\text{tr}(Z^{j+1})}{(j+1)N^j}\right\} \\ &= \text{etr}\left(\frac{1}{2}Z\right)\Bigg[1 + \sum_{k=1}^{\infty}\frac{1}{N^k}\sum_{\ell=1}^{k}\frac{(1/2)^\ell}{\ell!} \\ &\quad \cdot \sum\nolimits^*_{j_1,\ldots,j_\ell}\frac{\text{tr}\, Z^{j_1+1}\text{tr}\, Z^{j_2+1}\ldots\text{tr}\, Z^{j_\ell+1}}{(j_1+1)(j_2+1)\ldots(j_\ell+1)}. \end{aligned} \tag{31}$$

In the final expression (31) the summation $\sum^*$ is over all ℓ-tuples of positive integers $(j_1, \ldots, j_\ell)$ satisfying

$$\sum_{i=1}^{\ell} j_i = k, \quad j_i = 1, 2, \ldots, k; \quad (i \leq i \leq \ell).$$

We deduce from (30) and(31) the following general form for the asymptotic expansion of the density of W to an arbitrary order as $N \uparrow \infty$:

$$\begin{aligned}\text{pdf}(w) \sim\ & \frac{w^{q/2-1}e^{-w/2}}{2^{q/2}\Gamma(q/2)} \\ & + \sum_{k=1}^{\infty}\frac{1}{N^k}\sum_{\ell=1}^{k}\frac{(1/2)^\ell}{\ell!}\sum\nolimits^*_{j_1,\ldots,j_\ell}\frac{w^{q/2-1}}{\Gamma(q/2)(j_1+1)\ldots(j+1)} \\ & \cdot \Big[\det\left(L(\partial Z \otimes I)L'\right)^{1/2}\ {}_0F_0\left(-L(\partial Z \otimes I)L', w\right) \\ & \cdot \text{etr}\left(\frac{1}{2}Z\right)\text{tr}\left(Z^{j_2+1}\right)\text{tr}\left(Z^{j_2+1}\right)\ldots\text{tr}\left(Z^{j_\ell+1}\right)\Big]_{Z=0}. \end{aligned} \tag{32}$$

To $O(N^{-1})$ we have:

$$\begin{aligned}\mathrm{pdf}(w) = \chi_q^2 + \frac{w^{q/2-1}}{2N\ (q/2)} \Big[&\det\left(L(\partial Z \otimes I)L'\right)^{1/2} \\ &\cdot {}_0F_0\left(-L(\partial Z \otimes I)L', w\right) \mathrm{etr}\left(\frac{1}{2}Z\right) \mathrm{tr}\ Z^2\Big]_{Z=0} \\ &+ o(N^{-1}). \end{aligned} \tag{33}$$

The correction term of $O(N^{-1})$ in (33) may be evaluted using the rules of operator calculus given earlier. The final result may be shown to correspond to the expression obtained by more conventional methods by Phillips (1984c).

5. THE DISTRIBUTION FUNCTION

We may also derive the cdf of the null distribution of W. We shall use the incomplete gamma integral:

$$\int_0^Y e^{-y\varsigma} y^{\alpha-1} dy = \alpha^{-1} Y^{\alpha}\ {}_1F_1(\alpha,\ \alpha+1;\ -Y\varsigma),$$

where $\mathrm{Re}(\alpha)$, $\mathrm{Re}(\varsigma) > 0$ (Erdeyli, 1953, p. 266). We have:

$$\begin{aligned}\mathrm{cdf}(w) &= P(W \leq w) \\ &= \left[N^{q/2}\Gamma(q/2)\right]^{-1} \int_0^w y^{q/2-1} \Big[\det\left(L(\partial X \otimes I)L'\right)^{1/2} \\ &\qquad \cdot {}_0F_0\left(-L(\partial X \otimes I)L', y/Z\right) \det(I - X)^{-N/2}\Big]_{X=0} dy \\ &= \left[N^{q/2}\Gamma(q/2)\right]^{-1} \int_0^w \int_{V_{1,q}} y^{q/2-1} \Big[\det\left(L(\partial X \otimes I)L'\right)^{1/2} \\ &\qquad \cdot \exp\left\{(y/N)h'L(\partial X \otimes I)L'h\right\} (\underline{dh}) \det(I - X)^{-N/2}\Big]_{X=0} dy. \end{aligned}$$

Interchanging the orders of operation in the above expression, which is permissible in view of the continuity of the integrand and the compactness of the domains of integration, we obtain:

$$\begin{aligned}\mathrm{cdf}(w) = \frac{w^{q/2}}{N^{q/2}\Gamma(q/2+1)} \Big[&\det\left(L(\partial X \otimes I)L'\right)^{1/2} \\ &\cdot \int_{V_{1,q}} {}_1F_1\left(q/2, q/2+1; -(w/N)n'L(\partial X \otimes I)L'h\right) (\underline{dh})\end{aligned}$$

$$\cdot \det(I - X)^{-N/2}\Big]_{X=0}$$

$$= \frac{(w/N)^{q/2}}{\Gamma(q/2+1)} \Big[\det\left(L(\partial X \otimes I)L'\right)^{1/2}$$

$$\cdot\ {}_1F_1^{(q)}\left(q/2, q/2+1; -(w/N), L(\partial X \otimes I)L'\right)$$

$$\cdot \det(I - X)^{-N/2}\Big]_{X=0}.$$

In (34), ${}_1F_1^{(q)}$ is a confluent hypergeometric function with two matrix arguments (see James, 1964). In the present case one of the arguments is scalar and the function admits a series representation in terms of top order zonal polynomials.

6. THE NON-NULL DISTRIBUTION

Analysis of the non-null distribution of W proceeds along similar lines. The derivations are more complicated and the reader is referred to the author's paper (1986) for details. The final result for the density may be expressed as:

$$\text{pdf}(w) = \frac{w^{q/2-1}e^{-m'm/2}}{N^{q/2}\Gamma(q/2)} \Big[\det\left(L(\partial X \otimes I)L'\right)^{1/2}$$

$$\cdot \int_{V_{1,q}} \exp\{-(w/N)h'L(\partial X \otimes I)L'h\}$$

$$\cdot\ {}_0F_1\left(\frac{1}{2}, \frac{w}{2N}h'L(\partial X \otimes I)L'mm'h\right)(\underline{dh})$$

$$\cdot \det(I - X)^{-N/2}\Big]_{X=0}. \tag{35}$$

An alternative series representation of (35) is possible in terms of top order invariant polynomials (Davis, 1979) with two matrix argument operators. Specializations to the non-null distributions of the statistics in Section 4 and to the asymptotic theory of W under local alternatives are also given by Phillips (1986).

7. CONCLUSIONS

There seems to be considerable scope for applying the methods outlined here to other problems of distribution theory in multivariate analysis. The author (1984a) has used similar methods in studying the distribution of the Stein-rule estimator in linear regression. The latter results have recently been extended by Knight (1986) to nonnormal errors.

The technique of developing general formulae for asymptotic expansions from exact theory also seems to be very promising. This approach avoids much of the tiresome algebraic manipulation that is a feature of the traditional work on Edgeworth expansions. Moreover, the final formulae are simpler in form and may be used to obtain expansions to an arbitrary order, which is very difficult with the traditional approach.

Here and elsewhere in the application of these methods to problems of distribution theory it would be helpful to have a glossary of results on fractional and matrix fractional calculus. Until now I have been developing rules for working with these operators as the need for them arose. With a systematic set of formulae for the action of matrix fractional operators on elementary and commonly occurring special functions as well as rules for operation on products and compositions of functions of matrix argument it should be possible to make progress on many presently unsolved problems of multivariate distribution theory.

ACKNOWLEDGMENTS

My thanks, as always, go to Glena Ames for her skill and effort in typing the manuscript of this paper and to the NSF for research support under Grant No. SES 8218792.

REFERENCES

Constantine, A. G. (1966), "The distribution of Hotelling's generalized T_0^2". *Annals of Mathematical Statistics* **37**, 215–225.

Erdéyli, A. (1953), *Higher Transcendental Functions,* Vol. 1. New York: McGraw-Hill.

Herz, C. W. (1955), "Bessel functions of matrix argument". *Annals of Mathematics* **61**, 474–523.

James, A. T. (1964), "Distribution of matrix variates and latent roots derived from normal samples". *Annals of Mathematical Statistics* **35**, 475–501.

Knight, J. L. (1986), "The distribution of the Stein-rule estimator in a model with non-normal disturbances". *Econometric Theory* **2**, (to appear).

Lavoie, J. L. T. J. Osler, and R. Tremblay (1976), "Fractional derivatives and special functions". *SIAM Review* **18**, 240–268.

Miller, K. S. (1974), "The Weyl fractional calculus". In *Fractional Calculus and Its Applications,* ed. B. Ross. Berlin: Springer-Verlag.

Muirhead, R. (1982), *Aspects of Multivariate Statistical Theory.* New York: Wiley and Sons.

Phillips, P. C. B. (1984a), "The exact distribution of the Stein rule estimator". *Journal of Econometrics* **25**, 123–131.

Phillips, P. C. B. (1984b), "An everywhere convergent series representation of the distribution of Hotelling's generalized T_0^2". Cowles Foundation Discussion paper No. 723, Yale University, September.

Phillips, P. C. B. (1984c), "Finite sample econometrics using ERA's". *Journal of the Japan Statistical Society* **14**, 107–124.

Phillips, P. C. B. (1985), "The exact distribution of the SUR estimator". *Econometrica* **53**, 745–756.

Phillips, P. C. B. (1986), "The exact distribution of the Wald statistic". *Econometrica* (to appear).

Pillai, K. C. S. (1976), "Distributions of characteristic roots in multi-variable analysis: null distributions". *Canadian Journal of Statistics* **4**, 157–186.

Pillai, K. C. S. (1977), "Distributions of characteristic roots in multivariate analysis: non-null distributions". *Canadian Journal of Statistics* **5**, 1–62.

Ross, B. (1974a), *Fractional Calculus and Its Applications.* Berlin: Springer-Verlag.

Ross, B. (1974b), "A brief history and exposition of the fundamental theory of fractional calculus". In *Fractional Calculus and Its Applications,* ed. B. Ross. Berlin: Springer-Verlag.

Spanier, J., and K. B. Oldham (1974), *The Fractional Calculus.* New York: Academic Press.

Treves, S. (1980), *An Introduction to Pseudodifferential and Fourier Integral Operators.* New York: Plenum Press.

Victoria Zinde-Walsh and Aman Ullah [1]

ON ROBUSTNESS OF TESTS OF LINEAR RESTRICTIONS IN REGRESSION MODELS WITH ELLIPTICAL ERROR DISTRIBUTIONS

1. INTRODUCTION

Testing a set of linear restrictions in a regression model is usually performed with the help of the F-statistic, or the statistic based on the likelihood ratio (LR). More recently two other procedures, the Lagrangian Multiplier or Rao Score (RS) test due to Rao (1947) and Silvey (1959), and the Wald (W) test (1943), have become popular with econometricians; see, for example, Breusch and Pagan (1980) and Evans and Savin (1982).

A statistic can be called numerically robust over a class of error distributions if its values are independent of the specific error distribution from that class. If the statistic is such that no matter which error distribution from the class of distributions is considered the test criterion remains unchanged then the statistic is inferentially robust over that class.

If the statistics, F, LR, RS and W are constructed based on the assumption of the spherical normal error distribution (normal error with the covariance matrix $\sigma^2 I$), then F and LR are numerically robust against the class of all monotonically decreasing continuous spherical distributions, but RS and W are not. However, all these statistics are inferentially robust over this class, thus the test conclusions reached under the assumption of normality will not be overturned if the error distribution is spherical. These results are derived by Ullah and Zinde-Walsh (1984, 1985).

In this paper we consider the issues of numerical and inferential robustness of F, LR, RS and W tests, based on the assumption of spherical normality, against the general class of elliptical error distributions (errors with the nonscalar covariance matrix Σ). We provide the necessary and sufficient conditions for numerical robustness for the class of covariance matrices of-

[1] Department of Economics, The University of Western Ontario, London, Ontario N6A 5C2 (both authors)

I. B. MacNeill and G. J. Umphrey (eds.), Time Series and Econometric Modelling, 235–251.

ten used in econometrics, for example, autoregressive (AR), moving average (MA) and heteroskedastic. Our investigation shows that for these covariance matrices the numerical robustness of test statistics under consideration is rare. Our results are more general than those given by Ghosh and Sinha (1980) and Sinha and Mukhopadhyay (1981), who considered only intraclass covariance structure. Also, while Khatri (1981) gave conditions for numerical robustness in terms of pairs of data and covariance matrices, robustness over classes of covariance matrices considered here has not been examined in his paper.

Our investigation also showed the limitations of exact inferential rubustness. We, therefore, looked into the robustness of tests by developing bounds for critical values /hich will ensure that the conclusions based on the usual tests are not affected against a particular class of distributions. Bounds for critical values of test statistics for t and F-tests for first-order AR, MA and $ARMA$ processes have been tabulated (for normal errors) by Vinod (1976), Vinod and Ullah (1981) and Kiviet (1980). Their calculations require knowledge of all the eigenvalues of the matrices which characterize these processes and are quite complex. The situation becomes more complicated for higher order $ARMA$ processes. Our method offers bounds which are cruder for the specific processes considered by Vinod and Ullah and Kiviet, but they have the advantage of computational simplicity and generality; that is, they provide critical values that guarantee robustness of the test conclusions, for any Σ matrix, over wide classes of error distributions, and would utilize only the highest and lowest eigenvalues of the covariance matrix.

The plan of the paper is as follows. Section 2 develops the notation and definitions. Section 3 deals with the problem of numerical robustness and some applications. In Section 4 we examine the question of robustness of test conclusions and provide our bounds on the critical values of statistics. Finally, the proofs of the lemmas and theorems are presented in Section 5.

2. DEFINITIONS AND NOTATION

We consider the general linear regression model

$$y = X\beta + u, \tag{2.1}$$

where y is an $n \times 1$ vector of observations, X is an $n \times p$ known matrix of rank $p < n$, β is a $p \times 1$ unknown parameter vector and u is an $n \times 1$ disturbance vector whose probability density function is

$$\phi(u) = \sigma^{-n} \mid \Sigma \mid^{-\frac{1}{2}} f\left(\frac{u'\Sigma^{-1}u}{\sigma^2}\right) \tag{2.2}$$

with a monotonically decreasing f, and positive definite Σ. If $\Sigma = I$, the distribution given by (2.2) reduces to a spherically symmetric distribution.

Our problem is to test a set of r linear restrictions $H_0 : R\beta = 0$ against $H_1 : R\beta \neq 0$, where R is an $r \times p$ known matrix of rank r. Under this hypothesis, β can be partitioned as

$$\begin{bmatrix} \beta_0 \\ L\beta_0 \end{bmatrix},$$

where rank $L = r$; if $X = [X_1 : X_2]$ with X_1 containing $p - r$ columns of X and X_2 the remaining r, then for $X_0 = X_1 + X_2 L$ the model under H_0 can be rewritten as

$$y = X_0\beta_0 + u. \tag{2.3}$$

We denote by F, LR, RS, and W the values of the statistics calculated according to the usual formulae under the assumption of multivariate normality of u. F can be written as

$$F = \left[\frac{\left(y - X_0\hat{\beta}_0\right)'\left(y - X_0\hat{\beta}_0\right)}{\left(y - X\hat{\beta}\right)'\left(y - X\hat{\beta}\right)} - 1 \right] \frac{q}{r}, \quad q = n - p, \tag{2.4}$$

where $\hat{\beta}_0$ and $\hat{\beta}$ are the respective least squares estimators of β_0 and β, and LR, RS and W can be expressed through F, respectively, as

$$LR = n \log\left(1 + \frac{r}{q}F\right), \quad RS = n\frac{r}{q}F \Big/ \left(1 + \frac{r}{q}F\right), \quad W = n\frac{r}{q}F. \tag{2.5}$$

We introduce the following projection matrices:

$$P = X(X'X)^{-1}X'; \; A = I - P; \; P_0 = X_0\left(X_0'X_0\right)^{-1}X_0'; \; A_0 = I - P_0, \tag{2.6}$$

where I is the identity matrix. The following properties can be easily verified:

$$\text{rank } P > \text{ rank } P_0; \;\; \text{rank } A < \text{ rank } A_0; \; PP_0 = P_0;$$
$$AA_0 = A; \; PX = X; \; PX_0 = X_0.$$

Using (2.6) and noting that

$$\left(1 + \frac{r}{q}F\right) = \frac{u'A_0u}{u'Au},$$

we can rewrite (2.4) and (2.5) as

$$F = \frac{u'(A_0 - A)u}{u'Au}\frac{q}{r}; \qquad LR = n \log \frac{u'A_0 u}{u'Au};$$
$$RS = n\frac{u'(A_0 - A)u}{u'A_0 u}; \qquad W = n\frac{u'(A_0 - A)u}{u'Au}. \tag{2.8}$$

If, in fact, u has the spherical normal distribution, all the statistics have known distributions. If the error distribution is spherical, that is, the likelihood function is given by (2.2) with $\Sigma = I$, we denote the values of the statistics derived from this likelihood by $F_\phi (= F)$, LR_ϕ, RS_ϕ, and W_ϕ.

We denote the statistics calculated under the assumption of elliptical normality by F_Σ, LR_Σ, RS_Σ and W_Σ.

It is known that

$$F_\Sigma = \left[\frac{\left(y - X_0\hat{\beta}_{0\Sigma}\right)' \Sigma^{-1} \left(y - X_0\hat{\beta}_{0\Sigma}\right)}{\left(y - X\hat{\beta}_\Sigma\right)' \Sigma^{-1} \left(y - X\hat{\beta}_\Sigma\right)} - 1 \right] \frac{q}{r} \tag{2.9}$$

is the familiar F-statistic for testing H_0, with

$$\hat{\beta}_{0\Sigma} = \left(X_0'\Sigma^{-1}X_0\right)^{-1} X_0'\Sigma^{-1}y; \;\; \hat{\beta}_\Sigma = \left(X'\Sigma^{-1}X\right)^{-1} X'\Sigma^{-1}y. \tag{2.10}$$

Further, as in (2.5) we have

$$LR_\Sigma = n \log\left(1 + \frac{r}{q}F_\Sigma\right); \quad RS_\Sigma = n\frac{r}{q}F_\Sigma / \left(1 + \frac{r}{q}F_\Sigma\right);$$
$$W_\Sigma = n\frac{r}{q}F_\Sigma. \tag{2.11}$$

For a general elliptical distribution in (2.2), denote the appropriate statistics by $F_{\phi,\Sigma}$, $LR_{\phi,\Sigma}$, $RS_{\phi,\Sigma}$, $W_{\phi,\Sigma}$.

Ullah and Zinde-Walsh (1984, 1985) have analyzed the numerical robustness of LR, RS and W tests against spherically symmetric distributions. In particular, they have shown that

$$LR = LR_\phi, \;\; RS = \psi_\phi^{-1} RS_\phi, \text{ and } W = \rho_\phi^{-1} W_\phi, \tag{2.12}$$

where ψ_ϕ and ρ_ϕ are constants which depend on the spherical distribution $\phi(u)$. Thus LR is numerically robust against non-normality but RS and W are not.

The elliptical distribution (2.2) can be transformed into a spherical distribution by the substitution $u = \Sigma^{\frac{1}{2}} v$. Thus, for this case, from (2.12) we easily obtain

$$LR_\Sigma = LR_{\phi,\Sigma};\ RS_\Sigma = \psi_\phi^{-1} RS_{\phi,\Sigma};\ W_\Sigma = \rho_\phi^{-1} W_{\phi,\Sigma}. \tag{2.13}$$

Here LR_Σ is numerically robust against non-normality, but RS_Σ and W_Σ are not.

3. MAIN RESULTS ON NUMERICAL ROBUSTNESS

It was mentioned in Section 2 that Ullah and Zinde-Walsh (1984, 1985) analyzed the numerical robustness of F, LR, RS and W against spherically symmetric distributions. Robustness of F_Σ, LR_Σ, RS_Σ and W_Σ against non-normality in elliptically symmetric distributions follows from that analysis. Here we look into robustness of F, LR, RS and W (under spherical normality) against elliptical normal distributions by comparing the values of these statistics with the values F_Σ, LR_Σ, RS_Σ and W_Σ as in (2.9)–(2.11).

Conclusions about robustness against general elliptical distributions will follow in view of the relationships given in (2.13). We also note that we derive the results for parametric classes of Σ matrices often used in economic literature.

For deriving the conditions under which F_Σ (or LR_Σ) is numerically robust over some class of Σ given the data matrix X, we consider

$$\ell_\Sigma = 1 + \frac{r}{q} F_\Sigma \tag{3.1}$$

and examine the conditions under which $\ell_\Sigma = \ell_0 = 1 + \frac{r}{q} F$.

Consider a class Ω_ρ of matrices Σ with $\Sigma^{-1} = I - H_\rho$, where H_ρ is some symmetric matrix over some parameter space $B \in R^k$, $\rho = (\rho_1, \ldots, \rho_k) \in B$ with $H_0 = 0$ for $\rho_1 = \cdots = \rho_k = 0$ and $\mid y' H_\rho y \mid < y'y$ for all possible y, $\rho \in B$.

We now state the following lemmas which are used in the proof of Theorem 1.

Lemma 1. *For all $\Sigma \in \Omega_\rho$, ℓ_Σ can be represented as follows:*

$$\ell_\Sigma = \frac{y' A_0 y + y' A_0 T_\rho^0 A_0 y}{y' A y + y' A T_\rho A y}, \tag{3.2}$$

where

$$T_\rho = f(H,\ A);\ T_\rho^0 = f(H,\ A_0) \tag{3.3}$$

with the explicit form of the function f given by (5.5) $= A + AT_\rho A$. Here A *and* A_0 are *defined by* (2.6).

Lemma 2. *Suppose that for some symmetric matrix* T,

$$(y'A_0TA_0y)(y'Ay) = (y'ATAy)(y'A_0y). \tag{3.4}$$

Then $A_0TA_0 = \theta A_0$ and $TA = \theta A$, *where* θ *is some constant.*

Theorem 1. *Suppose that* H_ρ *is a polynomial or a convergent series in the parameters* $\rho_1, \ldots, \rho_k$ *with symmetric matrices as coefficients. If* $T(r_1, \ldots, r_k)$ *is the coefficient of* $\rho^{r_1}\rho^{r_2}\ldots\rho^{r_k}$ in $\sum_{r=1}^{m} H_\rho^r$, $m = \sum r_i$, and *if* $\ell_\Sigma = \ell_0$, *it follows that*

$$A_0T(r_1, \ldots, r_k)A_0 = \theta(r_1, \ldots, r_k)A_0 \tag{3.5}$$

for some constant $\theta(r_1, \ldots, r_k)$.

For proofs of the lemmas and Theorem 1, see Section 5.

Remark 1. *Suppose that* $A_0H_\rho^kA_0 = \theta_{k,\rho}A_0$, $k = 1, 2, \ldots$, *where* $\theta_{k,\rho}$ *is a scalar function of* $\rho_1, \ldots, \rho_k$. *Then* $\ell_\Sigma \equiv \ell_0$.

To prove the above statement one only needs to note that

$$AH_\rho^kA = AA_0H_\rho^kA_0A = A\theta_{k,\rho}A_0A = \theta_{k,\rho}A$$

and to substitute into (3.3) and (3.2).

Theorem 1 and Remark 1 give the necessary and sufficient conditions for the constancy of ℓ_Σ and, therefore, for the numerical robustness of F, RS, LR and W statistics against elliptically normal errors that can be described by a variance-covariance matrix $\Sigma \in \Omega_\rho$, $\Omega_\rho = \{\Sigma \mid \Sigma^{-1} = I - H_\rho$ with H_ρ being a polynomial or convergent series $\}$.

The stringency of these conditions makes numerical robustness an exception rather than the rule. No process with non-trivial H_ρ gives rise to robust statistics for all possible X and X_0; therefore, numerical robustness has applications mainly for experimental design. Also, of course, one can always check if the observation matrix X just happened to lead to statistics numerically robust against a particular process in the errors, but, if so, it would be strictly a matter of luck. We show that our results generalize those on experimental design with intraclass covariance structure by Ghosh and Sinha (1980) and examine the possibilities for numerical robustness over heteroskedastic and ARMA processes.

3.1 Implications for Intraclass Covariance Structures

The result of Ghosh and Sinha (1980, Theorem 3.1) follows as a special case of our theorem. Indeed, they considered $\Sigma = (1-\rho) + \rho 1_n \times 1_n'$, $-\frac{1}{n-1} < \rho < 1$, where 1_n is a column vector of ones, and hence $1_n \times 1_n' = nQ$, where Q is a projector of rank 1 into the subspace spanned by 1_n. Here

$$\Sigma^{-1} = \frac{1}{1-\rho}\left[I - \frac{\rho n}{(1+(n-1)\rho)}Q\right].$$

Direct application of Theorem 1 to Σ^{-1} implies that $A_0QA_0 = \theta A_0$. Since rank $A_0 >$ rank $A \geq 1$, it follows that $\theta = 0$, $A_0 1 \times 1'A_0 = 0$, and $A_0 1 = 0$; therefore, 1_n is the eigenvector of both P_0 and P as stated in the result of Ghosh and Sinha. It is also easy to verify that Theorem 3.2 of the same paper follows from our results.

3.2 Implications for Heteroskedastic Errors

Theorem 1 also provides a characterization of the class of heteroskedastic Σ for a given A and restriction R over which $\ell_\Sigma = \ell_0$: it is required that $\Sigma = \gamma_0 I + \Lambda$, where the diagonal matrix Λ is such that $\Lambda A_0 = 0$; this implies that the A matrix has block-diagonal structure with a block of zeros.

3.3 Implications of Theorem 1 for Autoregressive (AR) Error Structures

The matrix Σ^{-1} is known for autoregressive processes of order k, $AR(k)$. If we set all but one of the parameters of $AR(k)$, namely, $\rho_1, \rho_2, \ldots, \rho_k$, equal to zero, i.e., $\rho_k \neq 0$, $\rho_i = 0$ for $i \neq k$, then Σ^{-1} reduces to the matrix $I + \rho C_{1k} + \rho^2 C_{2k}$. Here C_{1k} is the matrix with elements $(C_{1k})_{ij}$ equal to -1 if $|i-j| = k$ and 0 otherwise, and C_{2k} is a diagonal matrix with elements $(C_{2k})_{ij}$ equal to 1 if $k < i = j < n-k$ and 0 otherwise. We shall denote this process by $AR(k,0)$. A necessary condition for constancy of ℓ_Σ for $AR(k,0)$ is that $A_0 C_{1k} A_0 = \theta A_0$, where A_0 is a projector of rank no less than 2. This implies that C_{1k} should have at least two identical eigenvalues, which is true only if $k \geq \frac{n}{2}+1$. If $k \geq \frac{n}{2}+1$ then C_{1k} has a kernel of dimension $n - 2(n-k) = 2k - n \geq 2$. In this case ℓ_Σ is constant for all A_0 that project into the intersection of the kernel of the matrix C_{1k} and either the image or the kernel of C_{2k}. Then, of course, $\theta = 0$ and $A_0 C_{2k} A_0 = \gamma A_0$ with $\gamma = 0$ or 1. It is not hard to check that for these k and A_0 this suffices for constancy of ℓ_Σ.

3.4 Implications for Moving Average Error Structures

The class of $MA(k,0)$ error structures, where all but the parameter of order k are zeros, is represented by $\Sigma = \sigma^2[(1+w^2)I + wC_{1k}]$, where C_{1k}

is the same as for the $AR(k)$ structure. We denote $\sigma^2(1+w^2)$ by γ and $w/(1+w^2)$ by ρ. Theorem 1 can be directly applied to this class of Σ after noting that $\Sigma^{-1} = (1+\gamma^2)(I+\rho C_{1k}+\rho^2 C_{1k}^2+\cdots)$. For this Σ, ℓ_Σ is robust only if $A_0 C_{1k} A_0 = \theta A_0$.

Here again if $k \geq \frac{n}{2}+1$, matrices A_0, which yield robust test statistics, exist. Such an A_0 would project onto the kernel of C_{1k}.

Thus, we conclude that there are some data structures that produce statistics that are robust over AR and MA error processes of sufficiently high orders (which do not include lower order components). We also notice that the higher the order of the error process the larger the class of data matrices that give robust statistics. This is hardly surprising, since in the limiting case, processes of order higher than the dimensions of the data will not affect the statistics at all.

We also note that, in general, the larger the number of equal eigenvalues, including zeros, of H (or the larger the dimension of any projector in the canonical representation of the symmetric matrix H) the more possibilities for numerically robust statistics.

Note that if $\ell_\Sigma = \ell_0$, then $F = F_{\phi,\Sigma}$ and $LR = LR_{\phi,\Sigma}$, but unless the distribution is elliptical normal $RS \neq RS_{\phi,\Sigma}$ and $W \neq W_{\phi,\Sigma}$.

4. INFERENTIAL ROBUSTNESS AND BOUNDS ON CRITICAL VALUES

If two test statistics are such that one is a monotonic function of the other, then any probabilistic statement about one implies a similar statement about the other. Thus, if one falls beyond a critical value for some level of the test, so does the other. Therefore, as was stated by Ullah and Zinde-Walsh (1985) (and can be seen immediately from (2.12)) RS and W are inferentially robust over the class of all spherical monotonic error distributions.

Here we examine the inferential robustness of the test statistics F, LR, RS and W, calculated under the assumption of spherical normality, for general elliptic distributions. To emphasize this we denote the statistics by $F(\Sigma)$, $LR(\Sigma)$, $RS(\Sigma)$ and $W(\Sigma)$. Since the test statistics are inferentially robust against spherical distributions it will not make any difference to our conclusions whether the statistics bear the subscript ϕ or not.

Consider the variate $S(\Sigma) = F(\Sigma)\frac{r}{q}$, where

$$S(\Sigma) = \frac{u'A_1 u}{u'A_2 u}, \tag{4.1}$$

with $A_1 = A_0 - A$, $A_2 = A$ as defined in (2.6). The critical values for $S(\Sigma)$ depend on the matrix Σ. Indeed, if one considers the transformation

$u = \Sigma^{\frac{1}{2}} v$, then

$$S(\Sigma) = \frac{v'\Sigma^{\frac{1}{2}} A_1 \Sigma^{\frac{1}{2}} v}{v'\Sigma^{\frac{1}{2}} A_2 \Sigma^{\frac{1}{2}} v}, \tag{4.2}$$

where v is spherically symmetric. Denote $S(I)$ by S.

We observe that as long as $S(\Sigma)$ is inferentially robust over a class Ω ot Σ matrices all the statistics $F(\Sigma)$, $LR(\Sigma)$, $RS(\Sigma)$ and $W(\Sigma)$ are inferentially robust over Ω as well. We assume that $I \in \Omega$.

Denote by $O(n)$ the group of orthogonal $n \times n$ matrices in the Euclidean space R^n. For any $T \in O(n)$ the distribution of S and of $S(\Sigma)$ in (4.2) is invariant with respect to the transformation $T : R^n \to R^n$.

Lemma 3. *For a positive definite matrix $\Sigma^{\frac{1}{2}}$ and any two mutually orthogonal idempotent matrices A_1, A_2, there exists $T \in O(n)$ such that $\Lambda_i = T'\Sigma^{\frac{1}{2}} A_i \Sigma^{\frac{1}{2}} T$ is a diagonal matrix for $i = 1, 2$.*

Proof. See Section 5.

This lemma allows us to rewrite $S(\Sigma)$ by substituting $w = Tv$ as

$$S(\Sigma) = \frac{w'\Lambda_1 w}{w'\Lambda_2 w}, \tag{4.3}$$

where we can write

$$\begin{aligned} \Lambda_1 &= \text{ diag } (\mu_1, \ldots, \mu_k, 0, \ldots, 0), \quad \mu_1 \le \cdots \le \mu_k, \quad k = p - m, \\ \Lambda_2 &= \text{ diag } (0, \ldots, 0, \mu_{p+1}, \ldots, \mu_n), \quad \mu_{p+1} \le \cdots \le \mu_n, \end{aligned} \tag{4.4}$$

where diag(...) denotes a diagonal matrix with given diagonal elements.

A similar transformation for S yields

$$S = \frac{w'Q_1 w}{w'Q_2 w} \tag{4.5}$$

with $Q_1 =$ diag $(1, \ldots, 1, 0, \ldots, 0)$, where the first k elements equal 1, and $Q_2 =$ diag $(0, \ldots, 0, 1, \ldots 1)$, where the last $n - p$ elements equal 1. Note that the transformation of S may be performed with a matrix from $O(n)$ different from T, but the distributions of $S(\Sigma)$ and S are not affected by an orthogonal transformation of the spherical variable.

Clearly the following inequality holds:

$$\frac{\mu_1}{\mu_n} S \le S(\Sigma) \le \frac{\mu_k}{\mu_{p+1}} S. \tag{4.6}$$

It follows from (4.3) that all the values for $S(\Sigma)$ within the bounds given by (4.6) are realized for some w. Therefore, a sufficient condition for inferential robustness is that $\mu_1/\mu_n = \mu_k/\mu_{p+1}$.

However, this type of condition is hardly less restrictive than those demanded for numerical robustness.

We thus seek bounds on the critical values of the statistics $F(\Sigma)$, $LR(\Sigma)$, $RS(\Sigma)$, $W(\Sigma)$ which will assure the test conclusions over some class Ω as long as the respective values calculated according to (2.4) and (2.5) are outside these bounds.

Since A_1, A_2 are projectors with eigenvalues equal to 0 or 1, the eigenvalues of $A\Sigma^{\frac{1}{2}}$ are bounded by the eigenvalues of $\Sigma^{\frac{1}{2}}$. Denote by $\lambda_{\max}$ the highest and $\lambda_{\min}$ the lowest eigenvalue of Σ. Also denote by δ the ratio $\lambda_{\max}/\lambda_{\min}$. Clearly

$$\lambda_{\min} w' Q_i W \leq w' \Lambda_i w \leq \lambda_{\max} w' Q_i w. \tag{4.7}$$

Therefore

$$\delta^{-1} S \leq S(\Sigma) \leq \delta S. \tag{4.8}$$

This inequality holds irrespective of A_1, A_2 and the particular Σ, and only reflects one characteristic of Σ—the ratio of the highest to lowest eigenvalues. The bigger δ is the more Σ is distinguished from I for which $\delta = 1$.

If for any two statistics S_1 and S_2, the inequality $S_1 \leq S_2$ holds everywhere, then their cumulative distributions $G_i(x) = \text{Prob}(S_i < x), i = 1, 2$, are related as follows:

$$G_1(x) \geq G_2(x),$$

and, therefore, for some level of the test the critical values satisfy

$$S_1^{cr} \leq S_2^{cr}.$$

From this observation and (4.8) we obtain the following theorem.

Theorem 2. *The critical values $F_{cr}(\Sigma)$ are located within the following intervals dependent on δ—the ratio of highest to lowest eigenvalues of Σ:*

$$\delta^{-1} F_{cr} \leq F_{cr}(\Sigma) \leq \delta F_{cr}. \tag{4.9}$$

Corollary. The following inequalities hold:

$$\begin{aligned} & LR_{cr} + n\log\left(\delta^{-1} - \left(1 - \delta^{-1}\right)/\ell_{cr}\right) \leq LR(\Sigma) \\ & \quad \leq LR_{cr} + n\log\left(\delta - (\delta - 1)/\ell_{cr}\right), \\ & \ell_{cr} = \exp\left(LR_{cr}/n\right); \\ & \frac{\delta^{-1} RS_{cr} n}{n - \left(1 - \delta^{-1}\right) RS_{cr}} \leq RS_{cr}(\Sigma) \leq \frac{\delta_n RS_{cr}}{n + (\delta - 1) RS_{cr}}; \\ & \delta^{-1} W_{cr} \leq W_{cr}(\Sigma) \leq \delta W_{cr}. \end{aligned} \tag{4.10}$$

The inequalities given in (4.10) are derived easily from (4.8) and (2.5).

4.1 Discussion of the Results

The relationship given by (4.9) has the following immediate interpretations for the F-test. Firstly, if a class Ω of Σ matrices is such that the biggest value of δ is limited by some δ^*, then the test conclusions are the same for any Σ as long as either $F/F_{cr} > \delta^*$ or $F_{cr}/F > \delta^*$, where F and F_{cr} are, respectively, the value of the test statistic according to (2.4) and the critical value for the hypothesis test under the spherical normal. Secondly, if $F > F_{cr}$ ($F < F_{cr}$), then the test conclusions are robust over the class Ω of Σ matrices with δ, the ratio of maximum to minimum eigenvalues, such that $\delta(\Sigma) < F/F_{cr}$ ($\delta(\Sigma) < F_{cr}/F$).

Since the relationship for W in (4.10) is similar to (4.9) for F, the same conclusions apply. A simple examination of (4.10) shows that the bounds on the critical values for $RS(\Sigma)$ are inside the interval $[\delta^{-1}RS_{cr},\ \delta RS_{cr}]$, thus, the conclusions made above hold for RS as well.

The following example demonstrates how our bounds compare to those obtained by Vinod (1976) and Vinod and Ullah (1981) for the t statistic under an $AR(1)$ process. Suppose that $\rho = .5$. Then the eigenvalues, of the variance-covariance matrix are contained between the asymptotic ($n \to \infty$) maximum and minimum eigenvalues, $1+\rho^2+2\rho = 2.25$ and $1+\rho^2-2\rho = .25$. Thus, the bounds on the critical value of the t statistic can be calculated based on the square root of the ratio $\sqrt{2.25/.25} = 3$. The critical values given by Vinod and Ullah are tabulated according to the number of restrictions p and sample size n. If $n = 50$, $p = 5$, for instance, their Table 4.1 gives 1.14 and 3.93 as the lower and upper bounds, respectively, at the 5% level, whereas our calculation, which involves only dividing and multiplying the standard critical value by 3, gives .671 and 6.042 as the lower and upper bounds, respectively.

However, there are three ways in which our results are an improvement. Firstly, they relate to any Σ matrices, not just those generated by an $AR(1)$ or $MA(1)$ process. Secondly, they require the calculation of the maximum and minimum eigenvalues of Σ only, whereas Vinod and Ullah utilized all the eigenvalues in a much more complicated calculation. Thirdly, our bounds are independent of A and A_0 matrices.

Note that if the bounds on the positive eigenvalues of $A\Sigma A$ and $(A_0 - A)\Sigma(A_0 - A)$ can be established they will provide more accurate intervals for critical values as can be seen from (4.6) and the fact that

$$\frac{\mu_1}{\mu_n} > \delta^{-1} \text{ and } \frac{\mu_k}{\mu_p} \leq \delta.$$

Recall that for the multivariate normal error distribution, Evans and

Savin (1982) have derived for $| W/n | < 1$ the relationship

$$W - LR \simeq LR - RS \simeq W^2/(2n), \tag{4.11}$$

which generalizes the known inequality:

$$W \geq LR \geq RS. \tag{4.12}$$

Ullah and Zinde-Walsh (1984) have shown that a more complex relationship exists between the statistics W_ϕ, LR_ϕ, RS_ϕ when the distribution is spherical but non-normal. Here, once again, straightforward inequalities relating the bounds on the statistics can be derived.

For any of the statistics F, LR, RS or W, denote the upper and lower bounds given by (4.9) and (4.10) by upper or lower bars. Next define F_U and F_L as follows:

$$F_U = (\overline{F} - \underline{F})/\overline{F} \text{ and } F_L = (\overline{F} - \underline{F})/\underline{F}. \tag{4.13}$$

In a similar notation, define LR_U and LR_L, RS_U and RS_L, and W_U and W_L. These ratios show the length of the interval between the bounds in relation to its upper and lower point, respectively. Thus, they measure the "tightness" of the bounds on the critical values of the statistics, and the following theorem establishes a ranking of the statistics with respect to this characteristic.

Theorem 3. *The following relationships hold:*

$$F_U = W_U \geq LR_U \geq RS_U \text{ and } F_L = W_L \geq LR_L \geq RS_L. \tag{4.14}$$

Proof. See Section 5.

This theorem demonstrates the relative robustness of the bounds on critical values for the different statistics. The bounds are tightest for RS and are worst when the F-test or the W-test is used.

5. PROOFS OF THE LEMMAS AND THEOREMS

Proof of Lemma 1. From (2.9) and (3.1) we can write

$$\ell_\Sigma = \frac{\left(y - X_0\hat{\beta}_{0\Sigma}\right)' \Sigma^{-1} \left(y - X_0\hat{\beta}_{0\Sigma}\right)}{\left(y - X\hat{\beta}_\Sigma\right)' \Sigma^{-1} \left(y - X\hat{\beta}_\Sigma\right)}. \tag{5.1}$$

We transform the denominator of (5.1) by substituting

$$\hat{\beta}_\Sigma = (X'\Sigma^{-1}X)^{-1}X'\Sigma^{-1}y$$

and obtain

$$y'\Sigma^{-1}\left[I - X\left(X'\Sigma^{-1}X\right)^{-1}X\Sigma^{-1}\right]y. \tag{5.2}$$

Consider

$$\Sigma^{-1} - \Sigma^{-1}X(X'\Sigma^{-1}X)^{-1}X'\Sigma^{-1}, \tag{5.3}$$

where $\Sigma^{-1} = (I - H)$ with $\mid y'Hy \mid < y'y$.

We can expand part of (5.3) in a geometric series as follows:

$$\begin{aligned} &X\left[X'(I-H)X\right]^{-1}X' \\ &= X(X'X)^{-\frac{1}{2}}\left[I - (X'X)^{-\frac{1}{2}}X'HX(X'X)^{-\frac{1}{2}}\right]^{-1}(X'X)^{-\frac{1}{2}}X' \\ &= P + PHP + PHPHP + \cdots + P(HP)^k + \cdots, \end{aligned}$$

where P is defined by (2.6). If one substitutes into (5.3) one obtains

$$\begin{aligned} &I - H - P - PHP - \cdots - P(HP)^k - \cdots \\ &\quad + HP + HPHP + \cdots \\ &\quad + (HP)^k + \cdots - HPH - HPHPH - \cdots - H(PH)^k - \cdots \\ &\quad + PH + PHPH + \cdots \\ &\quad + (PH)^k + \cdots = A - AHA - AHPHA - \cdots - AH(PH)^kA - \cdots \end{aligned}$$

Indeed the last term is obtained in the following way:

$$-P(HP)^kHP + (HP)^kHP - H(PH)^k + P(HP)^kH = -AH(PH)^kA$$

where we have substituted $P = I - A$. We therefore show that (5.3) can be represented as follows:

$$A - AHA - AHPHA - \cdots - AH(PH)^kA - \cdots \tag{5.4}$$

Further, we can replace P by $I - A$ everywhere in (5.4) to obtain

$$\begin{aligned} &A - AHA - AH^2A + AHAHA + \cdots \\ &\quad + (-1)^{k+1}AH^{i_1}AH^{i_2}A\ldots H^{i_{k+1}}A + \cdots \end{aligned} \tag{5.5}$$

This formula can be easily verified by substitution of $P = I - A$ into (5.4). The numerator of ℓ_Σ can be transformed in an analogous manner. This concludes the proof of Lemma 1.

Proof of Lemma 2. For a given vector y we define mutually orthogonal unitary vectors y_0, y_1, and y_2, such that

$$y_i'y_i = 1,\ i = 0,\ 1,\ 2;\ y_i'y_j = 0 \text{ for all } 0 \leq i < j \leq 2;$$

$$Ay = \alpha y_1;\ (A_0 - A)\, y = \beta y_0; \text{ and } P_0 y = \gamma y_2. \tag{5.6}$$

If one substitutes into (3.4) one obtains:

$$\alpha^2 \left(\alpha^2 y_1'Ty_1 + \alpha\beta y_1'Ty_0 + \alpha\beta y_0'Ty_1 + \beta^2 y_0'Ty_0\right) = \left(\alpha^2 + \beta^2\right)\left(\alpha^2 y_1'Ty_1\right). \tag{5.7}$$

We equate the coefficients of all the monomials in α and β in (5.7) to obtain:

$$y_1'Ty_1 = y_0'Ty_0 \tag{5.8}$$

and

$$y_1'Ty_0 + y_0'Ty_1 = 0. \tag{5.9}$$

Since T is symmetric, (5.9) implies that

$$y_1'Ty_0 = y_0'Ty_1 = 0. \tag{5.10}$$

Conditions (5.8) and (5.10) hold for any y. We can denote $y_1'Ty_1$ by θ, where θ is a constant. For any y,

$$y_1'Ty_1 = y'ATAy/\alpha^2 = \theta,$$

where $\alpha^2 = y'Ay$, and we have $y'ATAy = \theta y'Ay$.

Similarly, using (5.10) in addition to (5.8), we can show that

$$y'A_0TA_0y = \theta y'A_0y.$$

Proof of Theorem 1. Consider the expression (3.2) for ℓ_Σ. We can write it as follows:

$$\ell_\Sigma = \ell_0 \frac{1 + y'A_0T_\rho^0A_0y/y'A_0y}{1 + y'AT_\rho Ay/y'Ay}.$$

If $\ell_\Sigma = \ell_0$ it follows that

$$\left(y'A_0T_\rho^0A_0y\right) y'Ay = \left(y'AT_\rho Ay\right) y'A_0y. \tag{5.12}$$

The expressions on each side of (5.12) are series in the parameters $\rho_1, \ldots, \rho_k$ of H_ρ. If the two series of the right and left sides in (5.12) are to be identical, all the coefficients of monomials have to coincide. Consider the coefficient of ρ_1 in T_ρ, it is some symmetric matrix T_1 (it is also the coefficient of ρ_1 in H_ρ).

We get for T_1

$$(y'A_0T_1A_0y)\, y'Ay = (y'AT_1Ay)\, y'A_0y;$$

thus, by Lemma 2,

$$A_0T_1A_0 = \theta_1 A_0 \text{ (and } AT_1A = \theta_1 A).$$

Similar equalities hold for all coefficients of $\rho_2, \ldots, \rho_k$ in T_ρ.

Any coefficient of a monomial $\rho_1^{r_1} \ldots \rho^{r_k}$ in T_ρ^0 can be represented as the sum of such a coefficient in

$$\sum_{r=1}^{r_1+r_2+\cdots+r_k} H_\rho^r,$$

denoted by $T(r_1, \ldots, r_k)$, and products of coefficients of lower power monomials with A_0 in between. We can now use induction to show that

$$A_0T\,(r_1, \ldots, r_k)\, A_0 = \theta\,(r_1, \ldots, r_k)\, A_0. \tag{5.13}$$

If (5.13) holds for all coefficients of monomials of lower power, we can replace such A_0TA_0 by the appropriate θA_0 and will arrive at (3.4).

This concludes the proof of Theorem 1.

Proof of Lemma 3. Let $T_i \in O(n)$ be a matrix that diagonalizes $\Sigma^{\frac{1}{2}}A_i\Sigma^{\frac{1}{2}}$; that is, $T_i'\Sigma^{\frac{1}{2}}A_i\Sigma^{\frac{1}{2}}T_i = \Lambda_i$, where Λ_i is diagonal. Denote by Q_i the orthogonal projector onto the space of non-zero eigenvectors of $\Sigma^{\frac{1}{2}}A_i\Sigma^{\frac{1}{2}}$.

We show that Q_1 and Q_2 are mutually orthogonal; thus each T_i can be represented by the same matrix $M_1+M_2+M_3$ with M_i mutually orthogonal, with the columns of $M_1(M_2)$ formed by the orthonormal system of non-zero eigenvectors of $\Sigma^{\frac{1}{2}}A_1\Sigma^{\frac{1}{2}}(\Sigma^{\frac{1}{2}}A_2\Sigma^{\frac{1}{2}})$. Suppose for some vector ξ, $\Sigma^{\frac{1}{2}}A_i\Sigma^{\frac{1}{2}}\xi = \lambda\xi$ with $\lambda \neq 0$. Then $Q_i\xi = \xi$. We have $A_i\Sigma^{\frac{1}{2}}\xi = \lambda\Sigma^{-\frac{1}{2}}\xi = A_i^2\Sigma^{\frac{1}{2}}\xi = \lambda A_i\Sigma^{-\frac{1}{2}}\xi$. Therefore $A_i\Sigma^{-\frac{1}{2}}\xi = \Sigma^{-\frac{1}{2}}\xi$. Clearly then for any η such that $\eta = Q_i\eta$ we have $A_i\Sigma^{-\frac{1}{2}}\eta = \Sigma^{-\frac{1}{2}}Q_i\eta$. For a vector η for which $Q_1\eta = Q_2\eta = \eta$ one would have

$$\begin{aligned}\eta &= \Sigma^{\frac{1}{2}}A_2\Sigma^{-\frac{1}{2}}\eta = \Sigma^{\frac{1}{2}}A_1\Sigma^{-\frac{1}{2}}\eta = \left(\Sigma^{\frac{1}{2}}A_2\Sigma^{-\frac{1}{2}}\right)\left(\Sigma^{\frac{1}{2}}A_1\Sigma^{-\frac{1}{2}}\right)\eta \\ &= \Sigma^{\frac{1}{2}}A_2A_1\Sigma^{-\frac{1}{2}}\eta = 0,\end{aligned}$$

since $A_2A_1 = 0$. Therefore, Q_1 and Q_2 are orthogonal projectors.
This concludes the proof.

Proof of Theorem 3. We represent all the bounds as functions of $S = S_{cr}$, by combining (2.11), (4.9) and (4.10):

$$\begin{aligned} &\overline{F} = (q/r)\delta S; \overline{W} = n\delta S; \overline{LR} = n\log(1+\delta S); \overline{RS} = n\delta S/(1+\delta S);\\ &\underline{F} = (q/r)\delta^{-1}S; W = n\delta^{-1}S; \underline{LR} = n\log\left(1+\delta^{-1}S\right);\\ &\underline{RS} = n\delta^{-1}S/\left(1+\delta^{-1}S\right). \end{aligned} \tag{5.14}$$

Next we derive directly that

$$F_U = W_U = 1-\delta^{-2}; F_L = W_L = \delta^2 - 1; \tag{5.15}$$

$$W_U - LR_U = \ln\left(1+\delta^{-1}S\right) / \ln(1+\delta S) - \delta^2 \tag{5.16}$$

$$W_L - LR_L = \delta^2 - \ln(1+\delta S)/\ln(1+\delta^{-1}S); \tag{5.17}$$

$$LR_U - RS_U = \delta^{-2}(1+\delta S)/(1+\delta^{-1}S) - \ln\left(1+\delta^{-1}S\right)/\ln(1+\delta S); \tag{5.18}$$

$$LR_L - RS_L = \ln(1+\delta S)/\ln\left(1+\delta^{-1}S\right) - \delta^2\left(1+\delta^{-1}S\right)/(1+\delta S). \tag{5.19}$$

It immediately follows from (5.15) that whatever conclusions will be proved to hold here with respect to W will apply to F as well.

Examine (5.16). The expression $\ln(1+\delta^{-1}S) - \delta^{-2}\ln(1+\delta S)$ is always non-negative since it equals zero for $S = 0$ and its derivative with respect to S is

$$\frac{\delta^{-1}\left(\delta S - \delta^{-1}S\right)}{\left(1+\delta^{-1}S\right)(1+\delta S)}$$

and is thus positive. This proves that (5.16) is positive for positive S. Similarly, we show that $\delta^2\ln(1+\delta^{-1}S) - \ln(1+\delta S)$ is positive for $S > 0$; thus, (5.17) is positive.

Next, consider the expression

$$\delta^{-2}(1+\delta S)\ln(1+\delta S) - (1+\delta^{-1}S)\ln(1+\delta^{-1}S)$$

related to (5.18). It is zero for $S = 0$; its derivative is equal to

$$\delta^{-1}\ln\frac{1+\delta S}{1+\delta^{-1}S}$$

and is positive. Therefore, (5.18) is positive. Similarly, since

$$(1+\delta S)\ln(1+\delta S) - \delta^2(1+\delta^{-1}S)\ln(a+\delta^{-1}S)$$

is positive (identical proof), it follows that (5.19) is positive.

This concludes the proof of Theorem 3.

ACKNOWLEDGMENTS

The authors gratefully acknowledge research support from SSHRC and NSERC, respectively. They are also thankful to B. K. Sinha, R. Carter and the members of the U.W.O. Econometrics Workshop for useful comments and discussions on the subject matter of this paper.

REFERENCES

Breusch, T. S., and A. R. Pagan (1980), "The Lagrange multiplier test and its applications to model specification in econometrics." *Review of Economic Studies* **47**, 239–253.

Evans, G. B. A., and N. E. Savin (1982), "Conflict among the criteria revisited: the W, LR, and LM tests." *Econometrica* **50**, 737–748.

Ghosh, M., and B. K. Sinha (1980), "On the robustness of least squares procedures in regression models." *Journal of Multivariate Analysis* **10**, 332–342.

Khatri, C. G. (1981), "Study of F-tests under dependent models." *Sankhya*, Series A **43**, 107–110.

Kiviet, J. F. (1980), "Effects of ARMA errors on tests for regression coefficients: comments on Vinod's article; improved and additional results." *Journal of the American Statistical Association* **75**, 353–358.

Rao, C. R. (1947), "Large sample tests of statistical hypotheses concerning several parameters with applications to problems of estimation." *Proceedings of the Cambridge Philosophical Society* **44**, 50–57.

Silvey, S. D. (1959), "The Lagrangian multiplier test." *Annals of Mathematical Statistics* **30**, 389–407.

Sinha, B. K., and B. B. Mukhopadhyay (1981), "A note on a result of Ghosh and Sinha." *Calcutta Statistical Association Bulletin* **30**, 169–177.

Ullah, A., and V. Zinde-Walsh (1984), "On the robustness of LM, LR, and W tests in regression models." *Econometrica* **52**, 1055–1066.

Ullah, A., and V. Zinde-Walsh (1985), "Estimation and testing in regression models with spherically symmetric errors." *Economics Letters* **17**, 127–132.

Vinod, H. D. (1976), "Effects of ARMA errors on the significance tests for regression coefficients." *Journal of the American Statistical Association* **71**, 929–933.

Vinod, H. D., and A. Ullah (1981), *Recent Advances in Regression Methods*. New York: Marcel Dekker.

Wald, A. (1943), "Tests of statistical hypotheses concerning several parameters when the number of observations is large." *Transactions of the American Mathematical Society* **54**, 426–482.

Radhey S. Singh, [1] A. Ullah [2] and R. A. L. Carter [2]

NONPARAMETRIC INFERENCE IN ECONOMETRICS: NEW APPLICATIONS

ABSTRACT

In this paper, nonparametric estimators of a multivariate density, its conditional mean (regression function) and its conditional variances (heteroskedasticity) are presented. Among other results, we establish central limit theorems for the estimators and build up confidence intervals based on these estimators. These techniques are applied to obtain new results in two areas of econometrics: Monte Carlo investigations of the exact distributions of test statistics, and the treatment of heteroskedasticity in linear regression.

1. INTRODUCTION

In econometrics and in many other scientific disciplines (such as medical sciences, sociology and psychology) one often has to deal with several variables simultaneously, each in some sense dependent on the others. A common inference problem in such sciences, especially in econometrics, is to see how a particular variable on the average is dependent on others, so that prediction (estimation) of the value (or average values) of the variable in question can be made at any specified values of the other variables. A second common inference problem in such sciences, though somewhat related to the first one, is to see how the chosen variable varies (over various spots, items, individuals as the case may be) when other variables are held fixed at certain specified values of interest. The first problem, known as the regression problem, and the second problem, known as the heteroskedasticity problem

[1] Department of Mathematics and Statistics, University of Guelph, Guelph, Ontario N1G 2W1

[2] Department of Economics, The University of Western Ontario, London, Ontario N6A 5C2

I. B. MacNeill and G. J. Umphrey (eds.), Time Series and Econometric Modelling, 253–278.

in regression, are invariably handled in various sciences by postulating a certain fixed model (functional form) for the regression and by assuming fixed conditional variance (homoskedasticity) of the variable in question. However, it is now well known that the set of all suitable functional forms of the regression (or of the distributions of disturbances) is quite often large, and any postulations regarding the form of the regression and the value of the conditional variance (the variance of the disturbances in the regression) are questionable, and their violations have varying effects on the econometric inferences and policy implications.

The only way of avoiding the misspecification of the functional form of the regression model or of the conditional variance is, in fact, to assume no specific parametric functional form of the regression or of the conditional variance; and to estimate the conditional mean and the conditional variance completely nonparametrically. This in turn can be achieved by estimating nonparametrically the joint probability density function (p.d.f.) of all the variables involved. For example, we can estimate the conditional mean and variance of a variable x_1 given $p-1$ other related variables $(x_2, \ldots, x_p)$, if we can estimate the joint p.d.f. of $(x_1, \ldots, x_p)$ and then the conditional p.d.f. of x_1 given $(x_2, \ldots, x_p)$.

Nadaraya (1964), Watson (1964), Rosenblatt (1969), Noda (1976) and Collomb (1979, 1981) are among the first to consider estimation of a regression function nonparametrically using Rosenblatt (1956), Parzen (1962) and Cacoullos (1966)-type kernel estimates of a density function.

In this paper we present nonparametric estimates of a multivariate density, and of the conditional mean and the conditional variance of a variable given the others. These techniques are then applied to produce new results in two areas of econometrics. The integration of density estimation with the Monte Carlo technique of doing finite sample econometrics is explored. Also, the nonparametric estimate of the variance is used to analyze the problem of heteroskedasticity in linear regression.

The plan of the paper is as follows. In Section 2 we present the method of constructing nonparametric estimates of a multivariate p.d.f. and the conditional mean and the conditional variance of a variable given the others. In Section 3 we state various results with regard to the consistency, the variance and the distribution of each of these estimators. The confidence intervals for the joint density, the conditional mean and the conditional variance are also presented. In Section 4 we give proofs of the main results stated in Section 3. Finally, in Section 5 we illustrate the performance of our estimators through applications to certain econometric problems.

2. ESTIMATORS OF THE JOINT P.D.F., THE CONDITIONAL MEAN AND THE CONDITIONAL VARIANCE

Suppose we have n independent observations $w_t = (x_{t1}, \ldots, x_{tp})$, $t = 1, \ldots, n$ on p random variables $x_1, \ldots, x_p$ of our interest. We wish to draw an inference about the conditional mean of a variable, say x_1, given the remaining variables $x_2, \ldots, x_p$. As mentioned in Section 1, we achieve this by first considering the nonparametric estimation of the joint p.d.f., say f, of $\mathbf{x} = (x_1, \ldots, x_p)$, and then considering the estimation of the conditional density, say g, of x_1 given $\mathbf{x}' = (x_2, \ldots, x_p)$.

Throughout the remainder of this paper, we denote $\int w^j f(w_1, \mathbf{w}')\, dw_1$ by $\ell_j(\mathbf{w}')$, where $\mathbf{w}' = (w_2, \ldots, w_p)$ is a point in the $(p-1)$-dimensional Euclidean space R^{p-1}. We shall, however, use only ℓ_0, ℓ_1, ℓ_2 and ℓ_4. Notice that ℓ_0 is the marginal p.d.f. of $\mathbf{x}' = (x_2, \ldots, x_p)$; $g(w_1 \mid \mathbf{w}') = f(\mathbf{w})/\ell_0(\mathbf{w}')$ is the conditional p.d.f. of x_1 at w_1 given $\mathbf{x}' = \mathbf{w}'$;

$$M(\mathbf{w}') = \ell_1(\mathbf{w}'/\ell_0(\mathbf{w}'))$$

is the conditional mean of x_1 given $\mathbf{x}' = \mathbf{w}'$; and

$$V(\mathbf{w}') = (\ell_2(\mathbf{w}')/\ell_0(\mathbf{w}')) - M^2(\mathbf{w}')$$

is the conditional variance of x given $\mathbf{x}' = \mathbf{w}'$.

As by Singh (1981), for an integer $s > 1$ and for $i = 1, \ldots, p$, let $\mathcal{K}_i^s$ be the class of all Borel-measurable real valued bounded functions, on the real line, symmetric about zero, such that for a $K_i \in \mathcal{K}_i^s$,

$$\int y^j K_i(y) dy = \begin{cases} 1 & \text{if } j = 0 \\ 0 & \text{if } j = 1, \ldots, s-1, \end{cases} \tag{2.1}$$

$\int \mid y^s K_i(y) \mid dy < \infty$ and $\mid yK_i(y) \mid \to 0$ as $\mid y \mid \to \infty$. For example, for $s = 2$, take $K_i(y) = \frac{1}{2} I(-1 < y < 1)$ or $(2\pi)^{-\frac{1}{2}} \exp(-y^2/2)$ for all i; and for $s = 3$ or 4 take $K_i(y) = (2\pi)^{-\frac{1}{2}} [2 \exp(-y^2/2) - 2^{-\frac{1}{2}} \exp(-y^2/4)]\, I(-\infty < y < \infty)$ or $(2\pi)^{-\frac{1}{2}} (\frac{1}{2})(3 - y^2) \exp(-y^2/2)\, I(-\infty < y < \infty)$ for all i. Other examples of function K_i are given by Singh (1979, 1981). Define K on R^p by

$$\begin{aligned} K(y_1, \ldots, y_p) &= K_1(y_1) K_2(y_2) \cdots K_p(y_p) \\ &= \prod_{i=1}^{p} K_i(y_i). \end{aligned}$$

Remark 2.1. We have chosen the above kernel only for the sake of simplicity. All the results of this paper remain valid if K in the various results is replaced by a more general K, namely a Borel-measurable, real valued, bounded function on R^p symmetric about the origin such that:

$$\int y_1^{j_1},\ldots,y_p^{j_p}K(y_1,\ldots,y_p)=1 \text{ or } 0 \text{ according to whether}$$

$$j_1=\cdots=j_p=0 \text{ or } 0<j_1+\cdots+j_p<s-1;$$

$$\int \mid y_1^{j_1}\ldots y_p^{j_p}K(y_1,\ldots,y_p)\mid <\infty \text{ if } j_1+\cdots+j_p=s$$

and $\|\mathbf{y}\| \mid K(\mathbf{y})\mid\to 0$ as $\|\mathbf{y}\|\to\infty$ where $\|\mathbf{y}\|$ is the usual Euclidean norm on R^p.

For $i=1,\ldots,p$, let $0<h_i=h_i(n)$ be functions of the sample size n such that $h_i\to\infty$ as $n\to\infty$. (A suitable choice of h_i's will be given later.)

We estimate the joint p.d.f. of $\mathbf{x}=(x_1,\ldots,x_p)$ at $\mathbf{w}=(w_1,\ldots,w_p)$ by

$$\begin{aligned}\hat{f}(\mathbf{w})&=n^{-1}\sum_{t=1}^{n}\left(\prod_{i=1}^{p}h_i^{-1}\right)K\left(\frac{x_{t1}-w_1}{h_1},\ldots,\frac{x_{tp}-w_p}{h_p}\right)\\&=n^{-1}\sum_{t=1}^{n}\left\{\prod_{i=1}^{p}\left(h_i^{-1}K_i\left(\frac{x_{ti}-w_i}{h_i}\right)\right)\right\}.\end{aligned}\tag{2.2}$$

Singh (1981) and Singh and Ullah (1984) have considered the estimator (2.2) with $h_1=\cdots=h_p=h$. For other estimators, we define, for $j=0,1$,

$$\begin{aligned}\hat{\ell}_j(\mathbf{w}')&=n^{-1}\sum_{t=1}^{n}\left(x_{t1}^{j}\right)\left\{\prod_{i=2}^{p}\left(h_i^{-1}K_i\left(\frac{x_{ti}-w_i}{h_i}\right)\right)\right\}\\&=n^{-1}\sum_{t=1}^{n}x_{t1}^{j}\left(\prod_{i=2}^{p}h_i^{-1}\right)K'\left(\frac{\mathbf{x}_t'-\mathbf{w}'}{h'}\right),\end{aligned}$$

where $\mathbf{x}_t'=(x_{t2},\ldots,x_{tp}),\mathbf{w}'=(w_2,\ldots,w_p)$, and

$$K'\left(\frac{\mathbf{x}_t'-\mathbf{w}'}{\mathbf{h}'}\right)=\prod_{i=2}^{p}\left(K_i\left(\frac{x_{ti}-w_i}{h_i}\right)\right).$$

A nonparametric estimate of the marginal p.d.f. of $\mathbf{x}'=(x_2,\ldots,x_p)$ evaluated at $\mathbf{w}'$ is therefore

$$\int \hat{f}(w_1,\ldots,w_p)\,dw_1=\hat{\ell}_0(\mathbf{w}').$$

The estimate of the conditional density of x_1 at w_1 given $\mathbf{x}' = \mathbf{w}'$ is

$$\hat{g}\,(w_1 \mid \mathbf{w}') = \frac{\hat{f}(\mathbf{w}')}{\hat{\ell}_0(\mathbf{w}')}.$$

With the estimate of the conditional p.d.f. of x_1 given $\mathbf{x}'$ in view, a nonparametric estimate of the conditional mean $M(\mathbf{w}')$ of x_1 given $\mathbf{x}' = \mathbf{w}'$ is therefore

$$\hat{M}(\mathbf{w}') = \int w_1 \hat{g}\,(w_1 \mid \mathbf{w}')\, dw_1 = \frac{\hat{\ell}_1(\mathbf{w}')}{\hat{\ell}_0(\mathbf{w}')}. \tag{2.3}$$

Finally, our estimator of the conditional variance $V(\mathbf{w}')$ of x_1 given $\mathbf{x}' = \mathbf{w}'$ is

$$\hat{V}(\mathbf{w}') = \frac{\hat{\ell}_2(\mathbf{w}')}{\hat{\ell}_0(\mathbf{w}')} - \hat{M}^2(\mathbf{w}'). \tag{2.4}$$

We will show in the next section that the statistics $\hat{f}$, $\hat{M}$ and $\hat{V}$ are consistent estimators for f, M and V respectively.

3. PROPERTIES OF ESTIMATORS AND CONFIDENCE INTERVALS

Under certain regularity conditions on f we show in this section that the statistics $\hat{f}(\mathbf{w}), \hat{M}(\mathbf{w}')$ and $\hat{V}(\mathbf{w}')$ are consistent estimators for $f(w), M(w') = E(x_1 \mid \mathbf{x}' = \mathbf{w}')$ and $V(\mathbf{w}') =$ var $(x_1 \mid \mathbf{x}' = \mathbf{w}')$ respectively. We further obtain the variances and the estimates of the variances of $\hat{f}$, $\hat{M}$ and $\hat{V}$. We also prove the asymptotic normality of these estimators. Finally, using the estimates of the variances of $\hat{f}$, $\hat{M}$ and $\hat{V}$ and their distributional properties we obtain $100(1-\alpha)\%$ confidence intervals for f, M and V. Proof of the results will be presented in Section 4.

Theorem 3.1. *Let all the sth order partial derivatives of f be continuous at $\mathbf{w}$. Then taking*

$$h_i \propto n^{-1(2s+p)}, \tag{3.1}$$

we have

$$\hat{f}(\mathbf{w}) = f(\mathbf{w}) + O_p(n^{-s/(2s+p)}); \tag{3.2}$$

and with

$$a_n = n \prod_{i=1}^{p} h_i \tag{3.3}$$

we nave

$$a_n \operatorname{var}\big(\hat{f}(\mathbf{w})\big) = \Lambda_0(\mathbf{w}) + o(1), \tag{3.4}$$

where

$$\Lambda_0(\mathbf{w}) = f(\mathbf{w}) \int K^2; \int K^2 = \prod_{i=1}^{p} \left(\int K_i^2(y)dy \right) \tag{3.5}$$

and

$$a_n^{\frac{1}{2}} \left(f(\mathbf{w}) - E(\hat{f}(\mathbf{w})) \right) \to N\left(0, \Lambda_0(\mathbf{w})\right), \tag{3.6}$$

provided $a_n \to \infty$.

Remark 3.1. Since, as we will see in the proof of Theorem 3.1,

$$E[\hat{f}(\mathbf{w})] = f(\mathbf{w}) + O\left(\max\left\{h_1^s, \ldots, H_p^s\right\}\right), \tag{3.7}$$

if we take h_i's so that

$$a_n^{\frac{1}{2}} \left(\max\left\{h_1^s, \ldots, h_p^s\right\}\right) = o(1), \tag{3.8}$$

for example take h_i's proportional to $n^{-1/(2s+p-\epsilon)}$ for any $\epsilon > 0$, then

$$a_n^{\frac{1}{2}} \left(\hat{f}(\mathbf{w}) - f(\mathbf{w}) \right) \to N\left(0, \Lambda_0(\mathbf{w})\right). \tag{3.9}$$

Theorem 3.2. *Let the sth order partial derivatives of ℓ_0 and ℓ_1 be continuous at $\mathbf{w}'$. Further, let ℓ_2 be continuous at $\mathbf{w}'$. Then taking*

$$h_i \propto n^{-1/(2s+p-1)}, \tag{3.1$'$}$$

we have

$$\hat{M}(\mathbf{w}') = M(\mathbf{w}') + O_p(n^{-s/(2s+p-1)}); \tag{3.10}$$

and with

$$a_n' = n \prod_{i=2}^{p} h_i \tag{3.3$'$}$$

we have

$$a_n' \operatorname{var}\left(\hat{M}(\mathbf{w})\right) = \Lambda_1(\mathbf{w}') + o(1), \tag{3.11}$$

where

$$\begin{aligned} \Lambda_1(\mathbf{w}') &= \frac{(\ell_2(\mathbf{w}')/\ell_0(\mathbf{w}')) - M^2(\mathbf{w}')}{\ell_0(\mathbf{w}')} \cdot \int (K')^2 \\ &= \left[\operatorname{var}\left(x_1 \mid \mathbf{x}' = \mathbf{w}'\right)/\ell_0(\mathbf{w}')\right] \int (K')^2, \end{aligned} \tag{3.12}$$

with

$$\int (K')^2 = \prod_{i=1}^{p} \left(\int K_i^2(y)dy \right);$$

and

$$(a_n')^{\frac{1}{2}} \{\hat{M}(\mathbf{w}') - E[\hat{M}(\mathbf{w}')]\} \to N(0,\ \Lambda_1(\mathbf{w}')), \tag{3.13}$$

provided $a_n' \to \infty$.

Remark 3.2. From the proof of Theorem 3.2, it follows that

$$E[\hat{M}(w')] = M(w') + O\left(\max\left\{h_2^s, \ldots, h_p^s\right\}\right).$$

Therefore, if h_i's are chosen in such a way so that

$$(a_n')^{\frac{1}{2}} \left(\max\left\{h_2^s, \ldots, h_p^s\right\}\right) = o(1),$$

that is, take h_i's proportional to $n^{-1/(2s+p-1-\epsilon)}$ for any $\epsilon > 0$, then

$$a_n'\left(\hat{M}(\mathbf{w}') - M(\mathbf{w}')\right) \to N\left(0, \Lambda_1(\mathbf{w}')\right). \tag{3.14}$$

Theorem 3.3. *Let the* sth *order partial derivatives of* ℓ_0, ℓ_1 *and* ℓ_2 *be continuous at* $\mathbf{w}'$. *Further, let* ℓ_4 *be continuous at* $\mathbf{w}'$. *Then taking* h_i's *as in* (3.1)′, *we have*

$$\hat{V}(\mathbf{w}') = V(\mathbf{w}) + O_p(n^{-s/(2s+p-1)}) \tag{3.15}$$

and

$$a_n' \ \mathrm{var}\left(\hat{V}(\mathbf{w}')\right) = \Lambda_2(\mathbf{w}') + o(1), \tag{3.16}$$

where

$$\begin{aligned} \Lambda_2(\mathbf{w}') &= \frac{(\ell_4(\mathbf{w}')/\ell_0(\mathbf{w}')) - (\ell_2(\mathbf{w}')/\ell_0(\mathbf{w}'))^2}{\ell_0(\mathbf{w}')} \int (K')^2 \\ &= \left[\ \mathrm{var}\left(x_1^2 \mid x' = w'\right)/\ell_0(\mathbf{w}')\right] \int (K')^2 \end{aligned} \tag{3.17}$$

and

$$(a_n')^{\frac{1}{2}} \{\hat{V}(\mathbf{w}') - E[\hat{V}(\mathbf{w}')]\} \to N(0,\ \Lambda_2(\mathbf{w}')). \tag{3.18}$$

Remark 3.3. It will be seen in the proof of Theorem 3.3 that $E\left(\hat{V}(\mathbf{w}')\right) = V(\mathbf{w}') + O\left(\max\left\{h_2^s, \ldots, h_p^s\right\}\right)$. Thus if the h_i satisfy the hypothesis of Remark 3.2, then

$$(a_n')^{\frac{1}{2}} \left(\hat{V}(\mathbf{w}') - V(\mathbf{w}')\right) \to N\left(0, \Lambda_2(\mathbf{w}')\right). \tag{3.19}$$

The computable confidence intervals of f, M and V, respectively, can easily be written from (3.9), (3.14) and (3.19) after replacing f and ℓ_j, $j = 0, 1, 2$ and 4 by their consistent estimates $\hat{f}$ and $\hat{\ell}_j$.

4. PROOFS OF THE THEOREMS IN SECTION 3

In this section we give proofs of our theorems in Section 3.

Proof of Theorem 3.1. Since $(x_{t1}, \ldots, x_{tp})$, $t = 1, \ldots, n$, are i.i.d. with joint p.d.f. f, then if we take the expectation of $f(\hat{\mathbf{w}})$ in (2.2) with respect to the joint distribution of $\mathbf{x} = (x_1, \ldots, x_p)$ and use the transformation theorem, we obtain

$$E[\hat{f}(\mathbf{w})] = \int K(\mathbf{y}) f(\mathbf{w} + \mathbf{hy}) d\mathbf{y}, \tag{4.1}$$

where $(\mathbf{w}+\mathbf{hy}) = (w_1 + h_1 y_1, \ldots, w_p + h_p y_p)$. Now, if we replace $f(\mathbf{w}+\mathbf{hy})$ by its Taylor-series expansion at $\mathbf{w}$ with Lagrange's form of the remainder at the sth stage, apply the properties of K_i, and then use the continuity of the sth order partial derivatives of f at $\mathbf{w}$, we obtain by the arguments in the proof of Theorem 1 of Singh (1981) that

$$E[\hat{f}(\mathbf{w})] = f(\mathbf{w}) + O\left(\max\left\{h_1^s, \ldots, h_p^s\right\}\right). \tag{4.2}$$

Further, since $(x_{t1}, \ldots, x_{tp})$ for $t = 1, 2, \ldots, n$ are i.i.d. with joint p.d.f. f,

$$\begin{aligned} \operatorname{var}\left(\hat{f}(\mathbf{w})\right) &= n^{-1} \operatorname{var}\left(\prod_{i=1}^{p} h_i^{-1} K\left(\frac{x_{1i} - w_i}{h_i}\right)\right) \\ &= n^{-1}\left[\left(\prod_{i=1}^{p} h_i^{-1}\right) \int K^2(\mathbf{y}) f(\mathbf{w} + \mathbf{hy}) - \left(E[\hat{f}(\mathbf{w})]\right)^2\right] \\ &= n^{-1}\left(\prod_{i=1}^{p} h_i^{-1}\right)\left[f(\mathbf{w}) \int K^2 + o(1)\right], \end{aligned} \tag{4.3}$$

where $\int K^2 = \prod_{i=1}^{p}\left(\int K_i^2(y_i) dy_i\right)$; the last equation follows by arguments used to prove Theorem 2.2 of Singh and Ullah (1984). Now (3.4) follows from (4.3).

Now (4.2) and (4.3) followed by (3.1) prove that

$$\operatorname{MSE}\left(\hat{f}(\mathbf{w})\right) = O(n^{-2s/(2s+p)}) \tag{4.4}$$

which, with an application of Chebyschev's inequality, proves (3.2).

To prove (3.6), let

$$L_{nt} = \frac{1}{n}\left(\prod_{j=1}^{p} h_j^{-1}\right)\left\{K\left(\frac{\mathbf{x}_t - \mathbf{w}}{\mathbf{h}}\right) - E\left[K\left(\frac{\mathbf{x}_t - \mathbf{w}}{\mathbf{h}}\right)\right]\right\} / \left(\operatorname{var}(\hat{f}(\mathbf{w}))\right)^{\frac{1}{2}},$$

where

$$K\left(\frac{\mathbf{x}_t - \mathbf{w}}{\mathbf{h}}\right) = K\left(\frac{x_{t1} - w_1}{h_1}, \ldots, \frac{x_{tp} - w_p}{h_p}\right).$$

Then $L_{n1}, \ldots, L_{nn}$ are i.i.d. centered random variables with

$$s_n = \sum_{t=1}^{n} L_{nt} = (\hat{f}(\mathbf{w}) - E(\hat{f}(\mathbf{w}))) / (\operatorname{var}(\hat{f}(\mathbf{w})))^{\frac{1}{2}}$$

and var $(S_n) = 1$. Temporarily let $\Phi(\cdot)$ denote the distribution function of the standard normal random variable. Then by the Berry-Esseen Theorem (see Chung, 1974, Theorem 7.4.1),

$$\sup_{\xi \in R} | P(S_n \leq \xi) - \Phi(\xi) | \leq c \sum_{t=1}^{n} E | L_{nt} |^3, \tag{4.5}$$

where c is an absolute constant. But by the inequality given by Loève (1963, p. 155),

$$E | L_{nt} |^3 \leq 4 \left[\operatorname{var}(\hat{f}(\mathbf{w}))\right]^{-\frac{3}{2}} \left(n^{-1} \prod_{j=1}^{p} h_j^{-1}\right)^3 E \left| K\left(\frac{\mathbf{x}_t - \mathbf{w}}{\mathbf{h}}\right)\right|^3.$$

Since

$$\left(\prod_{j=1}^{p} h_j^{-1}\right) E \left| K\left(\frac{\mathbf{x}_t - \mathbf{w}}{\mathbf{h}}\right)\right|^3 = \int | K |^3 f(\mathbf{w} + \mathbf{h}\mathbf{y})$$
$$= f(\mathbf{w}) \int | K |^3 + o(1),$$

and $a_n \operatorname{var}(\hat{f}(\mathbf{w})) = \Lambda_0(\mathbf{w}) + o(1)$, we see that

$$\sum_{t=1}^{n} E | L_{nt} |^3 = O(a_n^{-\frac{1}{2}}).$$

Thus, we conclude that

$$\sup_{\xi \in R} \left| P\left[\frac{\hat{f}(\mathbf{w})) - E(\hat{f}(\mathbf{w})}{\left(\text{var}\,(\hat{f}(\mathbf{w}))\right)^{\frac{1}{2}}} \leq t \right] - \Phi(t) \right| = O(a_n^{-\frac{1}{2}}).$$

This result and (3.4) give (3.6).

Proof of Theorem 3.2. Throughout this proof $\hat{\ell}_0$, $\hat{\ell}_1$, $\hat{\ell}_2$, ℓ_0, ℓ_1 and ℓ_2 are evaluated at $\mathbf{w}' = (w_2, \ldots, w_p) \in R^{p-1}$ and therefore the argument $\mathbf{w}'$ in these functions will not be displayed. From the proof of Theorem 3.1, and the hypothesis on ℓ_1, it follows that

$$E(\hat{\ell}_0) = \ell_0 + O\left(\max\left\{h_2^s, \ldots, h_p^s\right\}\right),$$

and

$$\text{var}(\hat{\ell}_0) = (a_n')^{-1}\left[\ell_0 \int (K')^2 + o(1)\right]. \tag{4.6}$$

Thus with the choice of h_i's in (3.1)$'$, MSE$(\hat{\ell}_0) = O\left(n^{-2s/(2s+p-1)}\right)$, and hence

$$\hat{\ell}_0 = \ell_0 + O_p(n^{-2/(2s+p-1)}). \tag{4.7}$$

Similarly, in view of the hypothesis on ℓ_1 and ℓ_2, it follows that

$$E(\hat{\ell}_1) = \ell_1 + O\left(\max\left\{h_2^s, \ldots, h_p^s\right\}\right);$$

and

$$\text{var}(\hat{\ell}_1) = (a_n')^{-1}\left[\ell_2 \int (K')^2 + o(1)\right] \tag{4.8}$$

and, hence with (3.1)$'$,

$$\hat{\ell}_1 = \ell_1 + O_p(n^{-s(2s+p-1)}). \tag{4.9}$$

Now we evaluate the $\text{cov}(\hat{\ell}_0, \hat{\ell}_1)$. Recall that $\mathbf{x}_t' = (x_{t2}, \ldots, x_{tp})$ and $K'(y_2, \ldots, y_p) = \prod_{i=2}^{p} K_i(y_i)$. Since the summands in $\hat{\ell}_0$ are i.i.d., as are the summands in $\hat{\ell}_1$,

$$a_n' \,\text{cov}(\hat{\ell}_0, \hat{\ell}_1) = \left(\prod_{i=2}^{p} h_i^{-1}\right) \text{cov}\left[x_{11} K'\left(\frac{\mathbf{x}_1' - \mathbf{w}'}{\mathbf{h}'}\right), \; K'\left(\frac{\mathbf{x}_1' - \mathbf{w}'}{\mathbf{h}'}\right)\right], \tag{4.10}$$

where

$$K'\left(\frac{\mathbf{x}_1' - \mathbf{w}'}{\mathbf{h}'}\right) = \prod_{i=2}^{p}\left(K_i\left(\frac{\mathbf{x}_{1i} - \mathbf{w}_i}{h_i}\right)\right).$$

But the r.h.s. of (4.10) is

$$\int y_1\left((K'(\mathbf{y}))^2 f(y_1, w_2 + h_2 y_2, \ldots, w_p + h_p y_p)\, d\mathbf{y} - \left(\prod_{i=2}^{p} h_i)E(\hat{\ell}_0)E(\hat{\ell}_1\right)$$

$$= \ell_1 \int (K')^2 + O\left(\prod_{i=2}^{p} h_i\right);$$

this follows by the arguments used to prove the first part of Theorem 3.1. Thus, we conclude that

$$a_n' \operatorname{cov}(\hat{\ell}_0, \hat{\ell}_1) = \ell_1 \int (K')^2 + o(1). \tag{4.11}$$

Now, if we write

$$\begin{aligned} \hat{M} &= \hat{\ell}_1/\hat{\ell}_0 \\ &= \frac{E(\hat{\ell}_1)}{E(\hat{\ell}_0)}\left[1 + \frac{\hat{\ell}_1 - E(\hat{\ell}_1)}{E(\hat{\ell}_1)} - \frac{\hat{\ell}_0 - E(\hat{\ell}_0)}{E(\hat{\ell}_0)} + O_p\left(a_n'\right)^{-1}\right] \end{aligned} \tag{4.12}$$

and apply (4.6)–(4.9), we get (3.10). Now (4.12) followed by (4.6)–(4.9) and (4.11) gives

$$\begin{aligned} a_n' \operatorname{var}(\hat{M}) &= \frac{\ell_1^2}{\ell_0^2}\left[\frac{\ell_2 \int (K')^2}{\ell_1^2} - \frac{\int (K')^2}{\ell_0} + o(1)\right] \\ &= \Lambda_1 + o(1) \end{aligned}$$

from the definition of Λ_1 in (3.12). Thus the proof of (3.11) is complete.

Now we prove (3.13). From the arguments used to prove the asymptotic normality of $\hat{f}$, it follows that

$$(a_n')^{\frac{1}{2}}\left(\hat{\ell}_0 - E(\hat{\ell}_0)\right) \to N\left(0,\ \ell_0 \int (K')^2\right) \tag{4.13}$$

and

$$(a_n')^{\frac{1}{2}}\left(\hat{\ell}_1 - E(\hat{\ell}_1)\right) \to N\left(0,\ \ell_2 \int (K')^2\right). \tag{4.14}$$

Now (4.12) followed by (4.11), (4.13) and (4.14) give (3.13).

Proof of Theorem 3.3. As in the proof of Theorem 3.2, throughout the proof of this theorem, $\hat{\ell}_j, \hat{V}, \hat{M}, \ell_j, V$ and M are evaluated at $\mathbf{w}'$, the point considered in Theorem 3.2.

By arguments identical to those used in the proof of Theorem 3.2, it follows that

$$E(\hat{\ell}_2) = \ell_2 + O\left(\max\left\{h_2^s, \ldots, h_p^s\right\}\right),$$
$$\operatorname{var}(\hat{\ell}_2) = (a_n')^{-1}\left[\ell_4 \int (K')^2 + o(1)\right],$$

and

$$\hat{\ell}_2 = \ell_2 + O_p(n^{-s/(2s+p-1)}), \tag{4.15}$$

with the h_i's taken as in Theorem 3.1. Further, arguments applied to prove (4.11) can be used to show that

$$a_n' \operatorname{cov}(\hat{\ell}_0, \hat{\ell}_2) = \ell_2 \int (K')^2 + o(1). \tag{4.16}$$

Hence, if we write

$$\frac{\hat{\ell}_2}{\hat{\ell}_0} = \frac{E(\hat{\ell}_2)}{E(\hat{\ell}_0)}\left[1 + \frac{\hat{\ell}_2 - E(\hat{\ell}_2)}{E(\hat{\ell}_2)} - \frac{\hat{\ell}_0 - E(\hat{\ell}_0)}{E(\hat{\ell}_0)} + O_p\left(a_n'\right)^{-1}\right], \tag{4.17}$$

and use (4.6)–(4.9), (4.15) and (4.16), it follows that

$$E(\hat{\ell}_2/\hat{\ell}_0) = (\ell_2/\ell_0) + O\left(\max\left\{h_2^s, \ldots, h_p^s\right\}\right)$$

and

$$a_n' \operatorname{var}(\hat{\ell}_2/\hat{\ell}_0) = \frac{\left[\ell_4^2 - (\ell_2^3/\ell_0)\right] \int (K')^2}{\ell_0^2} + o(1)$$
$$= \Lambda_2 + o(1). \tag{4.18}$$

Hence, by Chebyschev's inequality,

$$\frac{\hat{\ell}_2}{\hat{\ell}_0} = \frac{\ell_2}{\ell_0} + O_p\left(a_n'\right)^{-\frac{1}{2}}. \tag{4.19}$$

Moreover, by the arguments used to prove the asymptotic normality of $\hat{f}$, it follows that

$$(a_n')^{\frac{1}{2}}\left(\hat{\ell}_2 - E(\hat{\ell}_2)\right) \to N\left(0,\ \ell_4 \int (K')^2\right). \tag{4.20}$$

Thus (4.17) followed by (4.13), (4.16) and (4.20) proves

$$(a_n')^{\frac{1}{2}}\left((\hat{\ell}_2/\hat{\ell}_0) - E(\hat{\ell}_2/\hat{\ell}_0)\right) \to N(0,\Lambda_2). \tag{4.21}$$

Now we obtain the results of Theorem 3.3 with regard to $\hat{V}$. From (4.12), (4.6)–(4.9) and (4.11) it follows that

$$\begin{aligned} E(M')^2 &= \frac{E(\hat{\ell}_1^2)}{E(\hat{\ell}_0)}\left[1 + \frac{\operatorname{var}(\hat{\ell}_1)}{(E(\hat{\ell}_1))^2} + \frac{\operatorname{var}(\hat{\ell}_0)}{(E(\hat{\ell}_0))^2} - \frac{2\operatorname{cov}(\hat{\ell}_1,\hat{\ell}_0)}{(E(\hat{\ell}_1))E(\hat{\ell}_0)} + O\left(a_n'\right)^{-1}\right] \\ &= (\ell_1/\ell_0)^2\left[1 + O\left(a_n'\right)^{-1}\right] \\ &= (\ell_1/\ell_0)^2 + O\left(a_n'\right)^{-1}. \end{aligned} \tag{4.22}$$

Hence from (4.17) it follows that

$$\begin{aligned} E(\hat{V}) &= (\ell_2/\ell_0) - (\ell_1/\ell_0)^2 + O\left(\max\left\{h_2^s,\ldots,h_p^s\right\}\right) \\ &= V + O\left(\max\left\{h_2^s,\ldots,h_p^s\right\}\right), \end{aligned} \tag{4.23}$$

and from (4.19) and (3.10) it follows that

$$\hat{V} = V + O_p\left(a_n'\right)^{-\frac{1}{2}}, \tag{4.24}$$

which proves (3.15). Now, from (4.12) and (4.6)–(4.9) it follows that

$$\hat{V} = (\hat{\ell}_2/\hat{\ell}_0) - \hat{M}^2 = (\hat{\ell}_2/\hat{\ell}_0) - M^2 + O_p\left(a_n'\right)^{-1}.$$

This result followed by (4.18) proves (3.16), and followed by (4.21) proves (3.18).

5. APPLICATIONS

In this section we consider the econometric applications of the estimation of densities and variances.

5.1 Estimating the Density Functions of Exact Sampling Distributions of Econometric Estimators

An important application of the kernel estimator is in estimating the density functions of the exact sampling distributions of econometric estimators and test statistics which are nonlinear functions of the endogenous data. Such estimated density functions are useful directly, such as for estimating the true size of asymptotic tests, and indirectly as input to the extended rational approximants (ERA's) of Phillips (1983). The aim of this section is to illustrate the technique with an example which is simple but which also has wide applicability and allows the production of new results.

Assume a data generating process (DGP) or joint p.d.f. of the form:

$$y_{1t} = \beta x_{1t} + \gamma y_{2t} + u_{1t}, \tag{5.1}$$

$$y_{2t} = \pi_{12} x_{1t} + \pi_{22} x_{2t} + v_{2t}; \qquad t = 1, \ldots, T. \tag{5.2}$$

Equation (5.1) is a structural equation containing the parameters of interest, β and γ, while (5.2) is a reduced form equation showing how the endogenous variable y_2 is generated as a linear combination of the two exogenous variables x_1 and x_2 plus an independent normal error v_2 which has zero mean and variance ω_{22}. The error u_1 is also assumed to be independent, normal with mean zero and variance σ_{11}. There is also a reduced form equation for y_1:

$$y_{1t} = \pi_{11} x_{1t} + \pi_{21} x_{2t} + v_{1t}, \tag{5.3}$$

where v_1 is independent $N(0, \omega_{11})$.

Because equation (5.1) is just-identified we can write the parameters of interest as functions of the reduced form parameters:

$$\gamma = \frac{\pi_{21}}{\pi_{22}} \text{ and } \beta = \pi_{11} - \frac{\pi_{12}\pi_{21}}{\pi_{22}}. \tag{5.4}$$

The normality of the errors together with the exogeneity of the x's implies that least squares (LS) applied to (5.2) and (5.3) will produce maximum-likelihood (ML) estimators of the parameters π_{ij} from which ML estimators of β and γ can be obtained using (5.4). These estimators of β and γ are also indirect-least-squares (ILS) or instrumental-variable (IV) estimators, where x_1 and x_2 are the instruments. They are consistent with asymptotically normal sampling distributions. Their exact sampling distributions were given by Basmann *et al.* (1971), who pointed out that they possess no positive integral-order moments.

Two test statistics of natural interest are:

$$t_\beta = \frac{\hat\beta - \beta}{\text{as}(\hat\beta)} \text{ and } t_\gamma = \frac{\hat\gamma - \gamma}{\text{as}(\hat\gamma)}, \tag{5.5}$$

where as($\hat\beta$) and as(β) are the (estimated) asymptotic standard errors of $\hat\beta$ and $\hat\gamma$. Asymptotically t_β and t_γ follow the standard normal distribution but their exact distribution appears to be unknown. (Richardson and Rohr (1971) derive the exact distribution for similar test statistics in the over-identified case). Although the random denominators in these ratios are always positive, the fact that the numerators lack moments makes one reluctant to assume that the ratios have moments.

Monte Carlo simulation can be used to produce samples of values of $\hat{\beta}$, $\hat{\gamma}$, t_β and t_γ. However, measures of bias, mean-squared-error and other moments computed from such samples are of little relevance, given the non-existence of the population parameters. However, nonparametric methods are well suited to estimating the density functions of $\hat{\beta}$, $\hat{\gamma}$, t_β and t_γ. Indeed, such density function estimates may well be regarded as more complete and useful than moment estimates even if the population moments did exist.

For the purposes of the Monte Carlo experiment the parameters of the DGP were set to the following values: $\beta = .6, \gamma = .2, \pi_{11} = .545455, \pi_{21} = .163636, \pi_{12} = -.272727, \pi_{22} = .818182, T = 20$. Values of x_{1t} and x_{2t} were generated such that:

$$\sum_t x_{1t} = \sum_t x_{2t} = \sum_t x_{1t}x_{2t} = 0.0 \text{ and } \sum_t x_{1t}^2 = \sum_t x_{2t}^2 = 20.0.$$

The set of x's was fixed over repeated samples. Two alternative reduced form error covariances were employed. Under the heading "loose fit", $\omega_{11} = .535537$ and $\omega_{22} = 53.5537$. When combined with the x's, these values give standard errors such that: $s(\hat{\pi}_{21}) = \pi_{21}$ and $s(\hat{\pi}_{22}) = 2\pi_{22}$, i.e., the probability of obtaining a negative value of $\hat{\pi}_{22}$ is .309. Under the heading "tight fit", $\omega_{11} = .0360331$ and $\omega_{22} = .0826446$. These values gave population goodness-of-fit measures of .90 for both reduced form equations. While this may be a realistic specification, it implies that the probability of obtaining a negative $\hat{\pi}_{22}$ is less than 1.7765×10^{-33}. In both cases the covariance between v_{1t} and v_{2t}, ω_{12}, was set to 0.0. Two experiments, each of 100 replications, were run: one loose fit and one tight fit.

The random number generator used was a version of Marsaglia's Super Duper generator as implemented by McLeod (1982).

Each experiment resulted in a frequency distribution for $\hat{\beta}$, $\hat{\gamma}$, t_β and t_γ. In addition, frequency distributions were formed for $\hat{\pi}_{21}$ and $t_\pi = (\hat{\pi}_{21} - \pi_{21})/s(\hat{\pi}_{21})$, where $s(\hat{\pi}_{21})$ is the (estimated) standard error of $\hat{\pi}_{21}$. While these frequency distributions convey some information about the underlying sampling distributions, the modest number of replications used means they are lumpy, with several empty classes.

Nonparametric estimates of the densities of $\hat{\beta}$, $\hat{\gamma}$, t_β, and t_γ were obtained from

$$\hat{f}(z) = \frac{1}{100h}\sum_{i=1}^{100} K\left(\frac{z_i - z}{h}\right), \tag{5.6}$$

where:

(i) z_i is a value of $\hat{\beta}$, $\hat{\gamma}$, t_β or t_γ, obtained by Monte Carlo simulation, standardized by subtracting its average over the 100 replications and dividing by its standard deviation over the 100 replications. This standardizing transformation alters the location and scale of the density but not its shape.

(ii) z is a value at which the density is to be estimated. Values of z were set between -5 and 5 in increments of .1.

(iii) h is the window width, $100^{-\frac{1}{5}} = .398$.

(iv) The normal kernel is used:

$$K\left(\frac{z_i - z}{h}\right) = (2\pi)^{-\frac{1}{2}} \exp\left\{-\frac{1}{2}\left(\frac{z_i - z}{h}\right)^2\right\}.$$

Of course, the estimates obtained by this procedure embody some sampling error. Therefore, when (5.6) was evaluated with z_i formed from $\hat{\pi}_{21}$, the resulting density function was not exactly standard normal: compare the standard normal density in Figure 1 to the estimated density for $\hat{\pi}_{21}$ from the tight fit experiment in Figure 2. (The estimated density of $\hat{\pi}_{21}$ from the loose fit experiment and the estimated densities of t_π for both experiments were nearly identical to Figure 2.) Although the density function in Figure 2 has its peak slightly too far right and is slightly skewed left, it is still a very good estimate of the standard normal density function in Figure 1, even though it is based on only 100 points. This gives us confidence that the estimated densities for $\hat{\beta}$, $\hat{\gamma}$, t_β and t_γ will also be close to their population counterparts.

The estimated density for $\hat{\beta}$ from the loose fit experiment is plotted in Figure 3. The analogous plot for $\hat{\gamma}$ is extremely close to that shown in Figure 3. Both densities have very high peaks and long, thin tails. The estimated density of t_β from the loose fit experiment, see Figure 5, looks very similar to Figure 2, but the estimated density of t_γ is strongly skewed to the left; see Figure 6.

The estimated density of $\hat{\beta}$ from the tight fit experiment is plotted in Figure 4. (The plot of $\hat{\gamma}$ was very similar to that for $\hat{\beta}$.) It contrasts sharply with the earlier results; the high peaks and long tails are absent. Indeed Figure 4 looks very much like Figure 2. Figures 7 and 8 show the estimated densities for t_β and t_γ when the fit was tight. Now the t_γ distribution closely resembles the t_β distribution, in contrast to the skewed distribution obtained when the fit was loose.

The estimated density functions presented in Figures 3 to 8 suffer from the disadvantage that they are point estimates. One might reasonably ask for measures of their precision or, better still, interval estimates. Asymptotic

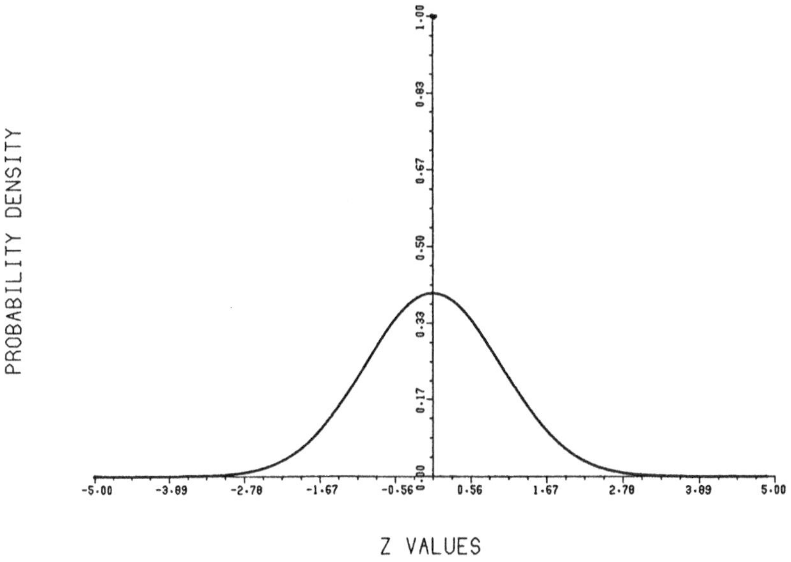

Figure 1. *Standard normal density.*

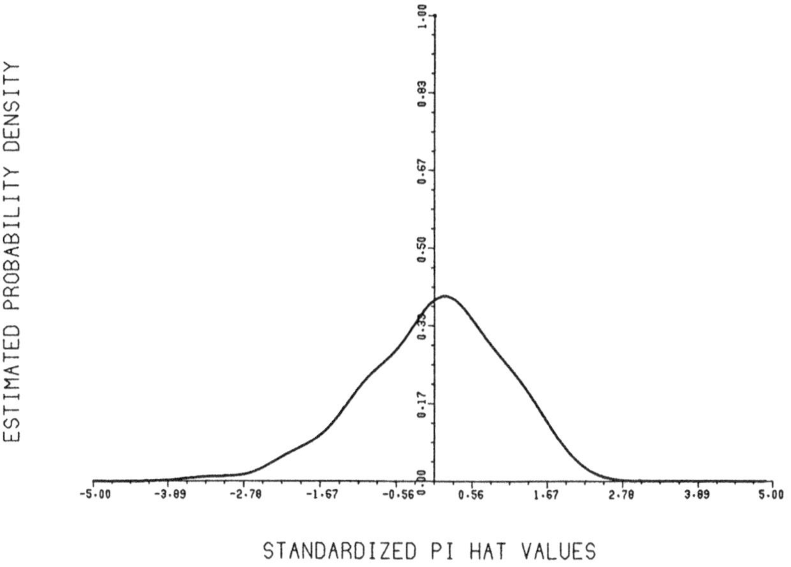

Figure 2. *Nonparametric estimate of the density of* $\hat{\pi}_{21}$, *tight fit experiment.*

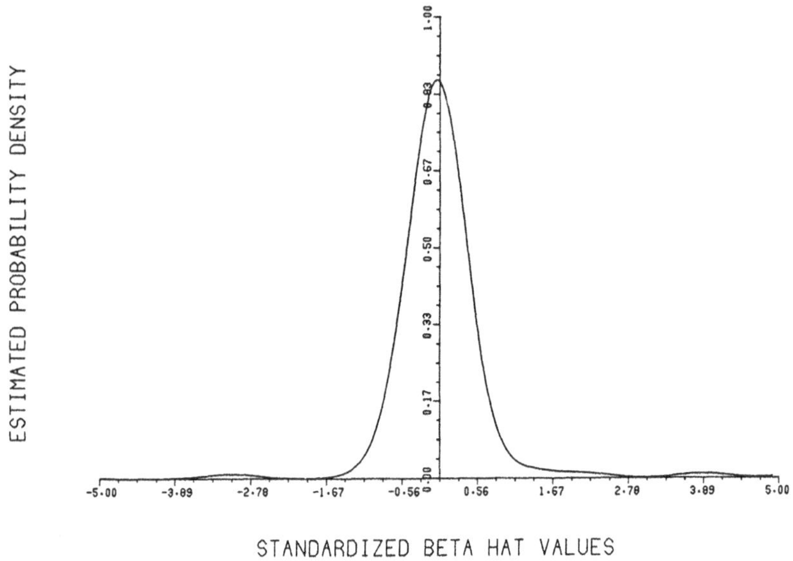

Figure 3. *Nonparametric estimate of the density of $\hat{\beta}$, loose fit experiment.*

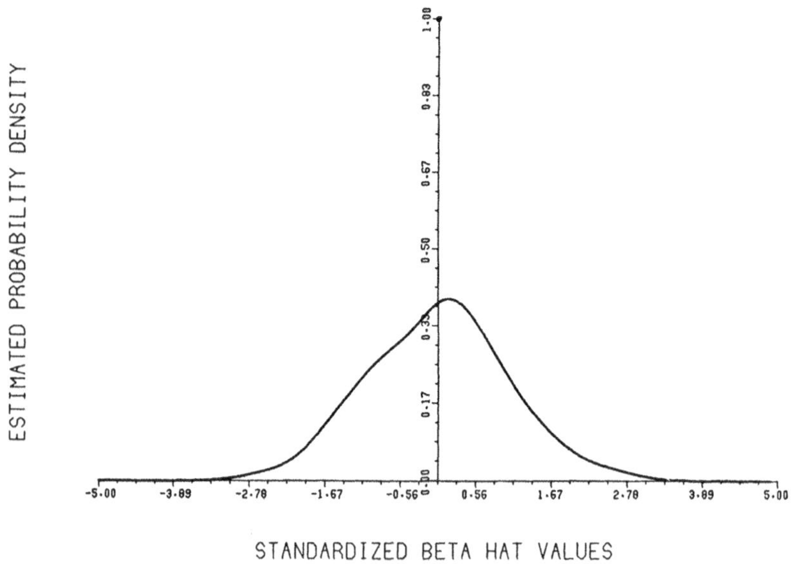

Figure 4. *Nonparametric estimate of the density of $\hat{\beta}$, tight fit experiment.*

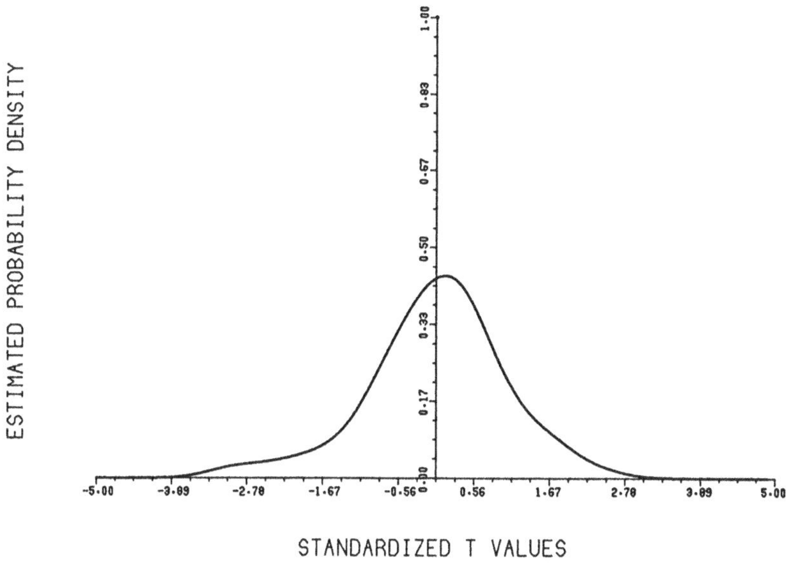

Figure 5. ***Nonparametric estimate of the density of t_β, loose fit experiment.***

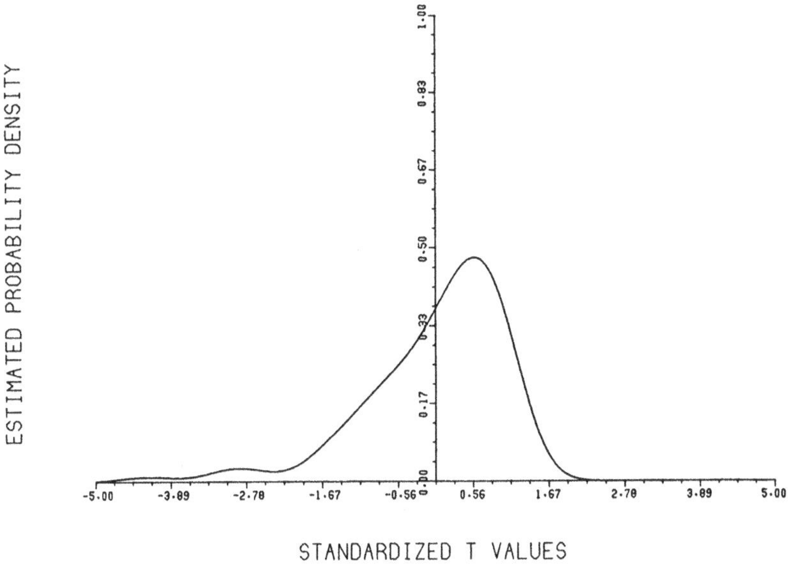

Figure 6. ***Nonparametric estimate of the density of t_γ, loose fit experiment.***

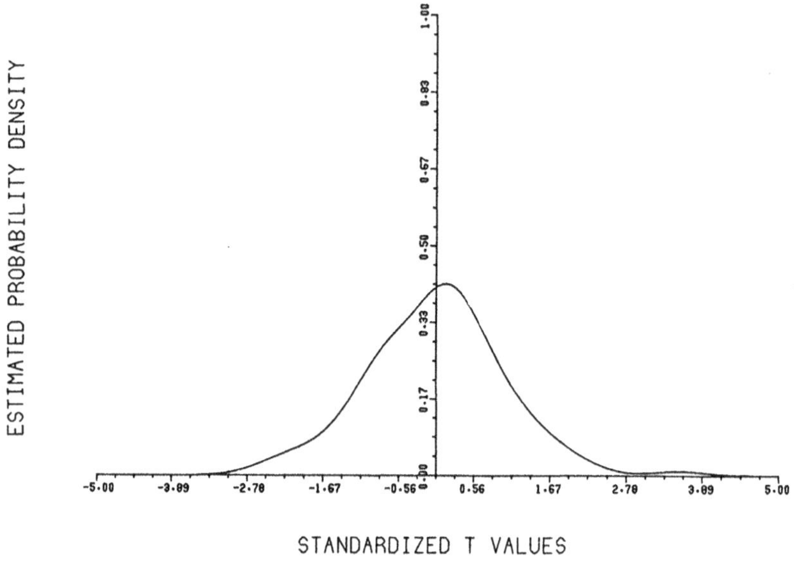

Figure 7. *Nonparametric estimate of the density of t_β, tight fit for experiment.*

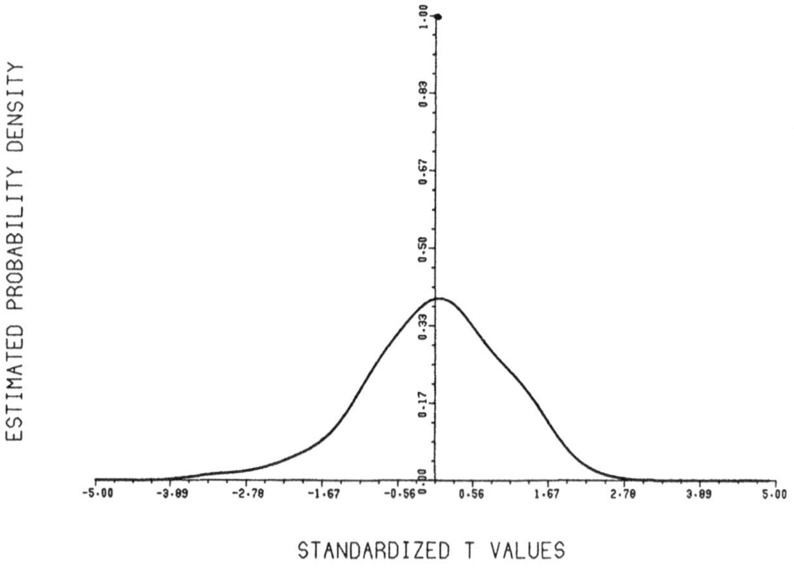

Figure 8. *Nonparametric estimate of the density of t_γ, tight fit experiment.*

95% confidence intervals for $f(z)$ are given by

$$\hat{f}(z) \pm 1.96 \left[\frac{\hat{f}(z)}{2nh\sqrt{\pi}} \right]^{\frac{1}{2}}$$

with $n = 100$, the number of replications used in the simulation. These confidence intervals are plotted for t_β and t_γ for the loose fit experiment in Figures 9 and 10. Both sets of confidence intervals from the right fit experiment closely resembled Figure 9. The standard normal density function (Figure 1) lies entirely within those confidence limits for t_β from both experiments and for t_γ from the tight fit experiment. However, it lies outside these limits for t_γ from the loose fit experiment.

The nonparametric density estimates presented in this section suggest several conclusions. First, the shape of the exact, small sample distributions of ILS/IV estimators of the structural parameters of just-identified models depends crucially upon the probability that reduced form coefficient estimates, which appear in the denominators of ratios entering the expression for structural coefficient estimates, change in sign. This probability will be high if the goodness-of-fit of the reduced form is low and/or when small samples are employed. When this probability is high the small-sample distributions of the structual coefficient estimators have high peaks and long thin tails, i.e., they are much different from their large-sample asymptotic distributions. The difference between the small and large sample distribution decreases as the probability of sign change decreases.

The second conclusion, which is of much greater operational significance, is that the small-sample distribution of t ratios depends not only on the probability of reduced form coefficient estimates changing sign, but also upon which structural coefficients enter the t ratio. Those formed from the coefficients of exogenous variables appear to have small-sample distributions which always resemble their large sample distributions. However, t ratios formed from the coefficients of endogenous variables have small-sample distributions resembling the standard normal only if the probability of sign change noted above is small, e.g., if the reduced form fits tightly. In other cases their shape is distinctly non-normal so that the use of the standard normal may yield poor inferences.

5.2 Estimation of Unknown Variances (Heteroskedasticity)

Here we analyze the conditional variance of earning (y) with respect to experience (z). For simplicity in illustration, we have assumed schooling to be constant. Our main interest is to look into the specification of the variability in earnings. For this purpose we considered Canadian data (1971

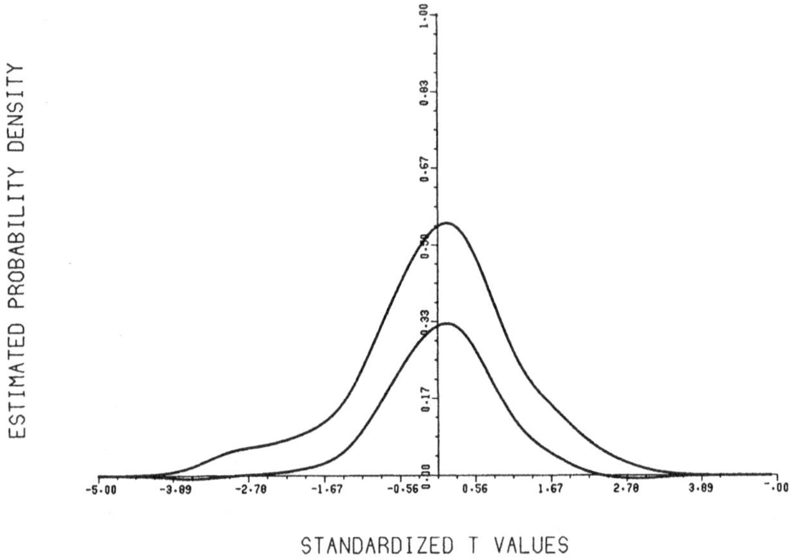

Figure 9. ***Nonparametric 95% confidence band for the density of*** t_β***, loose fit experiment.***

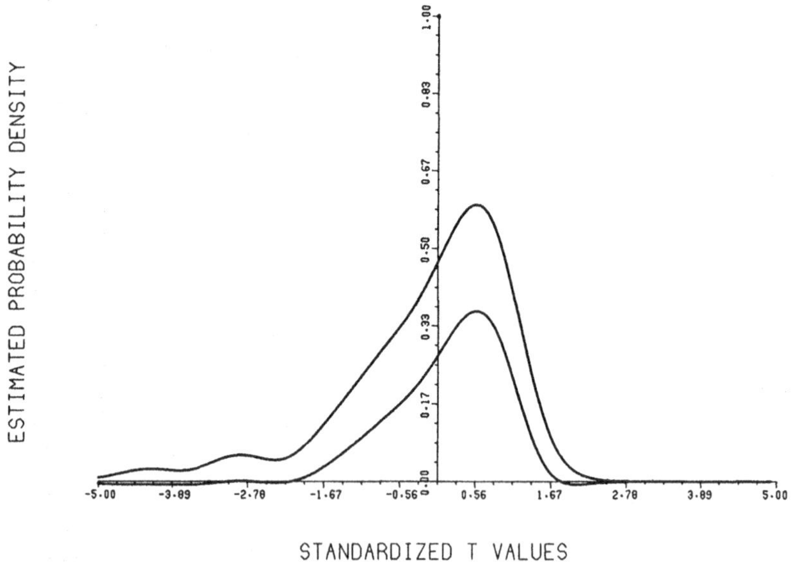

Figure 10. ***Nonparametric 95% confidence band for the density of*** t_γ***, loose fit experiment.***

Canadian Census Public Use Tapes) on 205 individuals' ages (for experience) and their earnings. These individuals were educated to Grade 13. The conditional variance, $V(y \mid z)$, in (2.4) was estimated by using the kernel function:

$$K\left(\frac{z_t - z}{h}\right) = \frac{1}{\sqrt{2\pi}} \exp\left[-\frac{1}{2}\left(\frac{z_t - z}{h}\right)^2\right], \qquad h = n^{-\frac{1}{5}}s,$$

where $s^2 = \sum (z_t - \bar{z})^2 / n$ is the sample variance of z.

It is clear from the estimate of conditional variances in Figure 11 that the true form of the variability in y with respect to $z, \hat{V}(y \mid z)$, is a second degree polynomial convex to the z axis. This is consistent with the result of Mincer (1974, p. 101). The important point to note, however, is that the variability of earnings here has been examined without using the grouped data, unlike Mincer (1974).

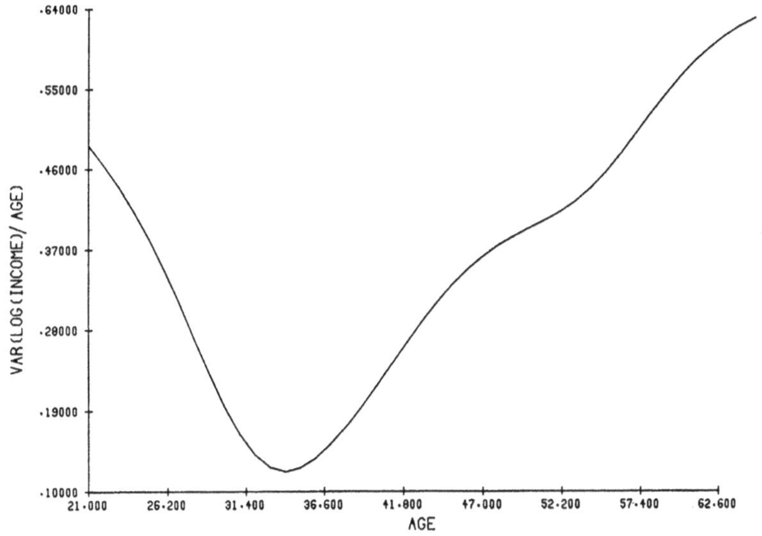

Figure 11. *Nonparametric estimate of the variance of earnings conditional on age as a function of age.*

In view of the above finding, we may conclude that $V(y \mid z)$ is negatively related with the nonparametric estimate of $E(y \mid z)$ which is, as indicated by Ullah (1985), a second degree polynomial concave to the z axis. To see that this is actually the case we estimated the regression of y on z, z^2 and $\hat{V}(y \mid z)$. The results were as follows:

$$\begin{array}{lllllll} y = 11.649 & + & .115z & - & .001z^2 & - & 1.103\hat{V}(y \mid z), \\ \quad (.987) & & (.039) & & (.005) & & \quad (.602) \end{array}$$

where the number in parentheses are standard errors. Note that the coefficient of $\hat{V}(y \mid z)$ is negative and significant indicating the negative relationship between $E(y \mid z)$ and $\hat{V}(y \mid z)$. The above result provides a possible alternative specification of the earnings equation with variability as an additional variable.

The nonparametric estimates of $V(y \mid z)$ can also be utilized to perform the generalized least squares (GLS) estimation technique in the earnings equation

$$y = \alpha + \beta z + \gamma z^2 + u = X\delta + u,$$

where $X = \left[1 \; z \; z^2\right]$ and $\delta = [\alpha \; \beta \; \gamma \,]'$. The GLS estimator is

$$\hat{\delta} = (X'\hat{\Sigma}^{-1}X)^{-1}X'\hat{\Sigma}^{-1}y,$$

where $\hat{\Sigma} = \text{ diag } [\hat{V}(\hat{u} \mid z_1), \ldots, \hat{V}(\hat{u} \mid z_n)]$. The least squares (LS) and the GLS estimates obtained are:

$$\begin{array}{lllllll} \text{LS}: \; y & = & 10.041 & + & .173z & - & .002z^2 \\ & & (.518) & & (.027) & & (.0003) \end{array}$$

and

$$\begin{array}{lllllll} \text{GLS}: \; y & = & 10.274 & + & .165z & - & .002z^2. \\ & & (.498) & & (.025) & & (.0003) \end{array}$$

It is clear that the GLS estimates outperform the LS estimates. The important point to note here is that the GLS estimates have been obtained without using any assumption about the form of heteroskedasticity. Carroll (1982) has also applied nonparametric techniques to this type of model, although his kernel and bandwidth are different to what we have used.

ACKNOWLEDGMENTS

Research supports to R. S. Singh and A. Ullah from NSERC are gratefully acknowledged. The authors are also thankful to V. Ho for research assistance.

REFERENCES

Basmann, R. L., F. L. Brown, W. S. Dawes, and G. K. Schoeplee (1971), "Exact finite sample density functions of GCL estimators of structural coefficients in a leading exactly identifiable case." *Journal of the American Statistical Association* **66**, 122–126.

Cacoullos, T. (1966), "Estimation of a multivariate density." *Annals of the Institute of Statistical Mathematics* **18**, 179–189.

Carroll, R. J. (1982), "Adapting for heteroscedasticity in linear models." *Annals of Statistics* **10**, 1224–1233.

Chung, K. L. (1974), *A Course in Probability Theory*, 2nd edition. New York: Academic Press.

Collomb, G. (1979), "Conditions necessaires et suffisantes de convergence uniforme d'un estimateur de la regression, estimation des derivatives de la regression." *Comptes Rendus Hebdomandaires des Séances. Academie des Sciences, Paris. Series A* **228**, A, 161–164.

Collomb, G. (1981), "Estimation nonparametrique de la regression: revue bibliographique. *International Statistical Review* **49**, 75–93.

Loève, M. (1963), *Probability Theory*, 3rd edition. Princeton, NJ: Van Nostrand.

McLeod, A. I. (1982), "Efficient Fortran coding of a random number generator." Department of Statistical and Actuarial Sciences, The University of Western Ontario, TR-82-08.

Mincer, J. (1974), *Schooling, Experience and Earnings*. New York: Columbia University Press.

Nadaraya, E. A. (1964), "On estimating regression." *Theory of Probability and its Applications* **9**, 141–142.

Noda, K. (1976), "Estimation of a regression function by the Parzen kernel type density estimators." *Annals of the Institute of Statistical Mathematics* **128**, 221–234

Parzen, E. (1962), "On estimation of a probability density function and mode." *Annals of Mathematical Statistics* **33**, 1065–1076.

Phillips, P. C. B. (1983), "ERA's: a new approach to small sample theory." *Econometrica* **51**, 1505–1526.

Richardson, D. H., and R. J. Rohr (1971), "Distribution of a structural *t*-statistic for the case of two included endogenous variables." *Journal of the American Statistical Association* **66**, 375–382.

Rosenblatt, M. (1956), "Remarks on some nonparametric estimates of a density function." *Annals of Mathematical Statistics* **27**, 832–837.

Rosenblatt, M. (1969), "Conditional probability density and regression estimators." In *Multivariate Analysis II*, ed. P. R. Krishnaiah, pp. 25–31. New York: Academic Press.

Singh, R. S. (1979), "On necessary and sufficient conditions for uniform strong consistency of estimators of a density and its derivatives." *Journal of Multivariate Analysis* **9**, 157–164.

Singh, R. S. (1981), "Speed of convergence in nonparametric estimation of a multivariate μ-density and its mixed partial derivatives." *Journal of Statistical Planning and Inference* **5**, 287–298.

Singh, R. S., and A. Ullah (1984), "Nonparametric recursive estimation." The University of Western Ontario, manuscript.

Ullah, A. (1985), "Specification analysis of econometric models." *Journal of Quantitative Economics* **1**, 187–209.

Watson, G. S. (1964), "Smooth regression analysis." *Sankhya*, Series A **26** 359–372.

H. D. Vinod [1]

CONFIDENCE INTERVALS FOR RIDGE REGRESSION PARAMETERS

ABSTRACT

This paper reviews various alternatives for constructing confidence intervals for ridge regression (RR) parameters, and illustrates them with an example. Among the newer alternatives are bootstrapping and those based on Stein's (1981) unbiased estimate of the mean squared error (MSE) of a biased estimator of multivariate normal mean. A simulation study supports the validity of the confidence statements based on Stein's model as modified here for the ridge regression problem. It yields confidence intervals which can be more useful and reliable than those based on other methods.

1. INTRODUCTION

Although ridge regresion has been used in econometrics by several authors, there is an impression among some econometricians that reliable confidence intervals are unavailable. The available alternatives offer a trade off between computational expense and power. This paper provides a comprehensive discussion, currently unavailable in the literature, of these alternatives. A new alternative based on the methods of Stein (1981) is discussed.

The simplest alternative advocated by Obenchain (1977) is to use confidence intervals based on ordinary least squares (OLS), even if one uses the point estimates from ridge regression. Unfortunately, the OLS interval is sometimes meaningless on *a priori* grounds. For example, Vinod and Ullah (1981, p. 12) gave an illustration of an OLS estimate of a consumption function where the OLS estimate of the marginal propensity to consume out of wage income is –0.17, thus having the wrong sign; a corresponding 95% confidence interval is (–0.26, –0.8), each point of which is invalid for substantive reasons. Such an OLS interval is centered at the wrong point; the OLS

[1] Department of Economics, Fordham University, Bronx, New York 10458

I. B. MacNeill and G. J. Umphrey (eds.), Time Series and Econometric Modelling, 279–300.

value of –0.17 and is economically meaningless. When such problems arise the best strategy is to look for better specifications. However, there are cases where, within the range of specifications consistent with economic theory, none yields reasonable OLS estimates. This is where RR is an attractive alternative to OLS, and where much of the OLS confidence interval theory may be economically meaningless. For examples where the OLS interval is meaningful, the practitioner need not consider other alternatives.

A second alternative is the approximate Bayes method, discussed near the end of Section 3. The interval is centered on the RR coefficients, rather than those for OLS, and the standard errors are based on the diagonals of the inverse of the $(X'X+kI)$ matrix in the usual notation which is defined below. Both Bayesians and frequentists can find philosophical or other reasons for rejecting this interval. In my opinion, approximate Bayes method offers a quick and reasonable alternative as a first approximation, provided the practitioner is willing to *assume* that RR is appropriate for the estimation problem and overlook philosophical questions.

A purpose of this paper is to propose a third alternative, namely a frequentist confidence interval based on Stein's (1981) use of the unbiased estimate of the MSE (UMSE) of biased estimators. If x is a normal random variable with mean ξ, its biased estimator is δx, where δ is a shrinkage factor. Stein's confidence interval based on a property of normal variables is discussed in Section 2, along with some modifications. In the remaining part of Section 1 we will develop notation, similar to Vinod and Ullah (1981), such that the ridge estimator can be thought to be a shrinkage factor δ times a normally distributed variable, denoted by $\dot{C}$ in (1.7) below. Section 3 explicitly applies the methods in Section 2 to the construction of confidence interval for RR; these intervals are illustrated by Hald's (1952) cement data in Section 4. Section 5 discusses a simulation study based on the data structure in Hald's cement data to assess the validity of confidence statements. An appendix discusses the case of stochastic k.

A fourth alternative is to use bootstrap resampling (Efron, 1982) to construct a sampling distribution. Efron's "bias corrected percentile method" can then be used to construct confidence intervals. Section 6 explains the mechanics of using the bootstrap for RR, illustrates the method with the cement data, and also discusses the "qualms" that Schenker (1985) associated with using the bootstrap intervals. We note that bootstrap intervals may be useful for understanding the sampling distribution, provided we already know that RR is a good estimate of the unknown parameter.

Let us consider the general linear regression model in the standardized form as:

$$y = X\beta + u, \ E(u) = 0, \ E(uu') = \sigma^2 I_n, \tag{1.1}$$

where y is an $n \times 1$ vector of observations on the dependent variable, X is an $n \times p$ matrix of regressors standardized in such a way that $X'X$ is non-singular correlation matrix, u is an $n \times 1$ vector of uncorrelated errors with mean zero and common unknown variance σ^2. The class of "ordinary" RR estimators of the $p \times 1$ vector of the parameters β in (1.1) is given by

$$b^k = (X'X + kI)^{-1}X'y, \tag{1.2}$$

where $k > 0$; see Hoerl and Kennard (1970a,b). When $k = 0$ we obtain the ordinary least squares estimator b^0.

The usual unbiased estimate of σ^2 is

$$s^2 = (y - Xb^0)'(y - Xb^0)/(n - p - 1). \tag{1.3}$$

For the following discussion, it is convenient to consider a canonical form of (1.1). We use the singular value decomposition (Rao, 1973, p. 42):

$$X = H\Lambda^{1/2}G', \tag{1.4}$$

where H is an $n \times p$ matrix of the coordinates of the observations along the principal axes of X, standardized in the sense that $H'H = I$. The matrix Λ is diagonal with eigenvalues $\lambda_1 \geq \lambda_2 \ldots \geq \lambda_p$, and G is the $p \times p$ matrix of eigenvectors g_i satisfying $X'X = G\Lambda G'$, and $G'G = I$.

From (1.1) and (1.4) consider the canonical model

$$y = H\Lambda^{1/2}G'\beta + u = H\Lambda^{1/2}\gamma + u, \tag{1.5}$$

which defines a parameter vector $\gamma = G'\beta$. The OLS estimate of γ is denoted by C^0. Under the normality assumption for u we have

$$C^0 \sim N_p[\gamma, \sigma^2\Lambda^{-1}], \tag{1.6}$$

a p-variate normal variable. Denote the elements of C^0 and γ by C_i^0 andγ_i respectively. For the canonical model (1.5) we have $b^{0\prime}b^0 = C^{0\prime}G'GC^0 = C^{0\prime}C^0$. Now transform C^0 to $\dot{C}$ where

$$\sigma^{-1}\Lambda^{1/2}C^0 = \dot{C} \sim N_p(\dot{\gamma}, I), \tag{1.7}$$

where $\dot{\gamma} = \sigma^{-1}\Lambda^{\frac{1}{2}}$. This follows from (1.6).

2. A PROPERTY OF THE NORMAL VARIABLE

Let $x \sim N(\xi, 1)$. The usual unbiased estimator of ξ is x. The shrinkage estimator of ξ is δx where $0 \le \delta \le 1$ is a shrinkage factor. Stein (1981) discussed shrinkage in terms of $x + f(x)$, where f is almost an arbitrary function. We choose $f(x) = (\delta - 1)x$. Straightforward calculation gives the MSE of δx as follows:

$$\begin{aligned} E(\delta x - \xi)^2 &= E(\delta^2 x^2 + \xi^2 - 2\delta x \xi) \\ &= \delta^2(\xi^2 + 1) + \xi^2 - 2\delta\xi^2 \\ &= \delta^2 + (\delta - 1)^2 \xi^2. \end{aligned} \tag{2.1}$$

An unbiased estimate of MSE of δx, denoted by UMSE (δx), is obtained by substituting $x^2 - 1$ as the unbiased estimate of ξ^2 in (2.1). Since ξ^2 is non-negative, whereas $x^2 - 1$ can be negative, we consider only a non-negative (positive part) estimate $\max[0, x^2 - 1]$ giving us two equivalent forms:

$$\mathrm{UMSE}(\delta x) = \delta^2 + (\delta - 1)^2 \max[0, x^2 - 1], \tag{2.2a}$$

and

$$\mathrm{UMSE}(\delta x) = \max[\delta^2, (\delta - 1)^2 x^2 + 2\delta - 1]. \tag{2.2b}$$

It can be shown following Baranchik (1970) that using a positive part estimate reduces the MSE of the estimator.

Stein (1981) proposed the following ingenious device for obtaining a confidence interval for the biased estimator (δx). He suggested that we consider the expectation of the squared difference between UMSE(δx) and the squared error $(\delta x - \xi)^2$. The motivation for the squared error will become clear when we define the confidence intervals in (2.9) and (2.10). In the absence of the truncation implied by the max function, we can derive this expectation by direct expansion:

$$\begin{aligned} &E\left[(\delta^2 + (\delta - 1)^2(x^2 - 1)) - (\delta x - \xi)^2\right]^2 \\ &= 8\delta^2 - 8\delta + 2 + 4(\delta - 1)^2 \xi^2 \\ &= 2(2\delta - 1)^2 + 4(\delta - 1)^2 \xi^2, \end{aligned} \tag{2.3}$$

where we use the relations

$$E(x^4) = 3 + 6\xi^2 + \xi^4, \quad E(x^3) = 3\xi + \xi^3, \tag{2.4}$$

and note that the terms involving ξ^4 and ξ^3 cancel. An alternative derivation based on Stein (1981) involves integration by parts, which is applicable for the more general case when δ is stochastic.

An unbiased estimate of the right-hand side of (2.3) is given by replacing ξ^2 by (x^2-1). Our modification lies in the suggestion that the non-negative number ξ^2 be estimated by $\max[0, x^2-1]$. This yields

$$\text{Var} = \max\left[2(2\delta-1)^2, 4\delta^2-2+4(\delta-1)^2x^2\right] \tag{2.5}$$

as an estimate of $E[\text{UMSE}(\delta x)-(\delta x-\xi)^2]^2$, which is almost unbiased.

Stein (1981) approximated a multivariate analogue of

$$\left[(\delta-1)^2x^2+2\delta-1-(\delta x-\xi)^2\right]^2 / \left[4\delta^2-2+4(\delta-1)^2x^2\right] \tag{2.6}$$

as a (central) chi-square variable with one degree of freedom, χ_1^2. Stein's approximation is obviously correct when $\delta = 1$. Note that for $\delta = 1$ the ratio in (2.6) is simply $[1-(x-\xi)^2]^2/2$. It can be verified by simulating a large number of empirical cumulative densities that the square root of this (+ or −) may be approximated by a unit normal $N(0,1)$ variable. For the χ_1^2 approximation, it is imperative that the square root of (2.6) be a real finite number for all real finite values x and ξ, and for all $|\delta| \leq 1$.

Stein's approximation can be poor when $\delta \neq 1$. For example, when $\delta = 0$ (2.6) becomes $(x^2-1-\xi^2)^2/(4x^2-2)$, which cannot be correctly approximated by a χ_1^2 variable. Its square root cannot be a unit normal variable. Note that $(4x^2-2)^{1/2}$ is imaginary for $x^2 < 1/2$. When $(4x^2-2) = 0$ and $\delta = 0$ the expression (2.6) becomes infinitely large.

Consider the case when $\delta = 1/2$ for which the square root of (2.6) becomes

$$\left[\frac{x^2}{4}-\left(\frac{x}{2}-\xi\right)^2\right] / (x^2-1)^{1/2} = \xi(x-\xi)/(x^2-1)^{1/2}. \tag{2.7}$$

Now the numerator $\xi(x-\xi)$ is a normal variable with mean zero and variance ξ^2. The term (x^2-1) in the denominator of (2.7) is simply an unbiased estimate of ξ^2. Note that when $\delta = 1/2$, Var given by (2.5) equals max $(0, x^2-1)$ which is zero for $x^2 < 1$. Hence it is obvious that we should impose a positive lower bound for Var. If we think of $(\delta-1)^2x^2+2\delta-1$ as a non-central chi-square variable with one degree of freedom and note that if a normal approximation is desired, its variance is unity; see Johnson and Kotz (1970, Ch. 28, eq. (23.2), p. 140). This suggests that Var ≥ 1 is appropriate, which is confirmed by considerable experimentation with simulated values of the ratio in (2.6). Hence we define an almost unbiased estimate as follows:

$$SD^2 = \max\left[1, 2(2\delta-1)^2, 4\delta^2-2+4(\delta-1)^2x^2\right]. \tag{2.8}$$

Further explanation of, and motivation behind the lower bound 1 in (2.8) may be obtained by regarding $U(\delta x)$ as a singly truncated normal variable; see Johnson and Kotz (1970, Table II, column 6, p. 86 of Chapter 13).

Now we write our modification of (2.6) as follows

$$\left[\mathrm{UMSE}(\delta x) - (\delta x - \xi)^2\right]^2 / (SD^2), \tag{2.9}$$

where UMSE(δx) and SD are given by (2.2b) and (2.8) respectively. Unlike (2.6), the square root of this ratio is a real and finite number for all $|\delta| \leq 1$, and for all x and ξ values.

As a further refinement one can improve the χ_1^2 approximation by considering a simulated empirical distribution function of (2.6) or (2.9) under the null hypothesis $\xi = 0$ for each δ. Since computer generation of unit normals is straightforward one can reach any desired improvement over χ_1^2 by considering a large enough simulation.

Now we develop the confidence intervals based on (2.9). For brevity denote U = UMSE(δx), M= $(\delta x - \xi)^2$, $S = SD$, and write the probability $Pr[(U - M)^2/S^2 \leq w^2]$ as follows:

$$\begin{aligned} &\Pr[|U - M| \leq wS] \\ &= \Pr[U - wS \leq M \leq U + wS] \\ &= \Pr[(U - wS)^{1/2} \leq |\delta x - \xi| \leq (U + wS)^{1/2}] \\ &= \Pr[\delta x - (U - wS)^{1/2} \leq \xi \leq \delta x + (U + wS)^{1/2}]. \end{aligned} \tag{2.10}$$

This shows that, if w is known we can determine a confidence interval for ξ centered at δx. If $\delta = 1$ the shrinkage estimator δx equals the maximum likelihood estimator x, and

$$\Pr[x - 1.96 \leq \xi \leq x + 1.96] = 0.95 \tag{2.11}$$

defines the usual confidence interval. If $\delta = 1$, we have $U = 1$ and $S^2 = 2$. Hence we can determine w so that $(U + wS)^{1/2} = 1.96$. We find that $w = 2.0092$ yields the correct 95% confidence interval for $\delta = 1$. Using this choice of w we write the upper and lower bounds for our UMSE intervals as:

$$\begin{aligned} UP(x,\delta) &= \delta x + (U + 2.0092S)^{1/2}, \\ LO(x,\delta) &= \delta x - (U + 2.0092S)^{1/2}, \end{aligned}$$

where

$$U = \max(\delta^2, (\delta - 1)^2 x^2 + 2\delta - 1)$$

and

$$S = \left[\max(1, 2(2\delta - 1)^2,\ 4\delta^2 - 2 + 4(\delta - 1)^2 x^2)\right]^{1/2}. \tag{2.12}$$

We may define the shrinkage factor δ leading to the estimator δx to be "too small" if the confidence interval centered at δx is wider than the normal confidence interval (2.11). Denote δ^{SML} to be such that for all $0 \leq \delta \leq \delta^{SML}$

$$UP(x,\delta) - LO(x,\delta) \geq 3.92. \tag{2.13}$$

Table 1 reports δ^{SML} values correct to the nearest 0.001 for a set of x values. Table 1 also reports the so-called "ridiculously small" shrinkage factors defined by

$$\delta^{RID} = 1 - 1.96(| x |)^{-1}. \tag{2.14}$$

Table 1. *'Too Small'*[1] *and 'Ridiculously Small'*[2]
Shrinkage Factors for a Unit Normal Variable

x	δ^{RID}	δ^{SML}	x	δ^{RID}	δ^{SML}
1.2	NEG[3]	0.073	7.0	0.720	0.959
1.5	NEG	0.328	8.0	0.755	0.969
1.8	NEG	0.496	9.0	0.782	0.975
1.9	NEG	0.539	10.0	0.804	0.980
2.0	0.020	0.577	11.0	0.822	0.983
2.2	0.100	0.641	12.0	0.837	0.986
2.5	0.216	0.713	13.0	0.849	0.988
2.8	0.300	0.766	14.0	0.860	0.989
2.9	0.324	0.781	15.0	0.869	0.991
3.0	0.347	0.794	16.0	0.877	0.992
3.5	0.440	0.846	18.0	0.891	0.993
4.0	0.510	0.880	20.0	0.902	0.994
5.0	0.608	0.922	22.0	0.911	0.995
6.0	0.673	0.945	25.0	0.922	0.996

[1] δ^{SML} "too small" is defined from (2.13) so that the confidence interval centered at δx is wider than the usual (OLS) interval centered at zero.

[2] $\delta^{RID} = 1 - 1.96 \mid x \mid^{-1}$ is "ridiculously small" because the point estimate δx lies outside the 95% interval for $\delta < \delta^{RID}$.

[3] NEG means negative. Thus, $\delta = 0$ is not "ridiculously small" for this $x = 1.2$. We have used the notation $\dot{C}_i$ and δ_i in the regression context, instead of x and δ respectively.

Note that the point estimate δx of ξ for all $0 \leq \delta \leq \delta^{RID}$ is outside the 95% interval centered at x. Avoidance of such extreme shrinkage may be justified by "limited translation" arguments; see Efron and Morris (1972).

The interval (2.12) is referred to as a UMSE interval, and is not intended to be used for the univariate case where the maximum likelihood estimate is known to be admissible. In Section 3 we will use (2.12) to obtain confidence intervals centered on a ridge regression estimator.

3. CONFIDENCE REGION FOR REGRESSION PARAMETERS

For the canonical model (1.5) the standardized parameter is $\dot{\gamma}_i = \sigma^{-1}\lambda^{1/2}\gamma_i$. It is estimated by

$$\dot{C}_i = \sigma^{-1}\lambda_i^{1/2}C_i^0 \sim N(\dot{\gamma}_i, 1). \tag{3.1}$$

For each i we let $\dot{C}_i$ and $\dot{\gamma}_i$ be equivalent to x and ξ respectively of Section 2 above, and rewrite the UMSE confidence interval from (2.12) as follows:

$$LO(\dot{C}_i, \delta) \leq \dot{\gamma}_i \leq UP(\dot{C}_i, \delta). \tag{3.2}$$

Now multiply all terms of (3.2) by $\sigma\lambda_i^{-1/2} > 0$ and define

$$\begin{aligned} C_i^{UP} &= \sigma\lambda_i^{-1/2}UP(\dot{C}_i, \delta) \\ &= \delta_i C_i^0 + \sigma\lambda_i^{-1/2}(U^* + 2.0092S^*)^{1/2}, \end{aligned} \tag{3.3}$$

where

$$U^* = \max\left(\delta^2, (\delta-1)^2\sigma^{-2}\lambda_i(C_i^0)^2 + 2\delta - 1\right)$$

and

$$S^* = \left[\max(1, 2(2\delta-1)^2, 4\delta^2 - 2 + 4(\delta-1)^2\sigma^{-2}\lambda_i(C_i^0)^2)\right]^{1/2}.$$

Similarly, define the lower bound of the UMSE interval as follows:

$$C_i^{LO} = \delta_i C_i^0 - \sigma\lambda_i^{-1/2}(U^* + 2.0092S^*)^{1/2}. \tag{3.4}$$

From (3.3) and (3.4) we have the approximate relation ($\alpha = 0.05$)

$$\Pr\left[C_i^{LO} \leq \gamma_i \leq C_i^{UP}\right] \doteq 1 - \alpha, \tag{3.5}$$

for all $i = 1, \ldots, p$. The probability of the joint event is approximately

$$\Pr\left[C^{LO} \leq \gamma \leq C^{UP}\right] \doteq (1-\alpha)^p, \tag{3.6}$$

because, from (1.6), C_i^0 are uncorrelated random variables.

In addition to the uncorrelatedness of C_i^0 it is important to note that we select only one finite confidence interval from each dimension i. This permits the confidence level to remain at the desired level.

From these intervals for γ we generate conservative intervals for β, the main parameter of interest in a regression model, as follows. By definition, $\beta = G\gamma$ implies that $\beta_i = \Sigma g_{ij}\gamma_j$, where g_{ij} is the (i,j)th element of G. A confidence interval for β_i can be estimated as follows. We note that the confidence set is convex, and that a pre-multiplication by a nonstochastic matrix G involves rotation of the axes. For locating the bounds with respect to the new axes we need to consider the values at the corners of the convex set. Thus we have

$$b_i^{UP} = \sum_{j=1}^{p} \max\left(g_{ij}C_j^{UP}, g_{ij}C_j^{LO}\right) \tag{3.7}$$

and

$$b_i^{LO} = \sum_{j=1}^{p} \min\left(g_{ij}C_j^{UP}, g_{ij}C_j^{LO}\right), \tag{3.8}$$

forming a confidence interval for β, centered at the RR estimator. Note that these are nonlinear equations, and the bounds are attainable. The reduction in variance often achieved by RR is reflected in narrow intervals. For $p > 6$ the intervals based on (3.7) and (3.8) can be too wide and Kabe's (1983) quadratic programming approach may be used to find confidence intervals from Stein's (1981, equation 8.6) confidence set. Otherwise it may be better to evaluate GC at say 10 equidistant values between the upper and lower bounds of C_i for each i, and find the max and min of each b_i over the ten evaluations. Then one can search for the corners of the convex set in its neighbourhood.

Obenchain (1977) advocated the use of the usual confidence intervals from normal regression theory (centered at b^0) for RR. We suggest that this might be considered when our UMSE intervals are wider than the OLS intervals, i.e., when δ_i are "too small" or "ridiculously small". More general confidence sets for linear combinations of β involving the squared length of C can be generated. Confidence intervals obtained for stochastic k (or δ_i) are derived in the appendix; however, derivatives of δ_i with respect to C_i will appear in related expressions. The appropriate correction for the fact that s^2 is a stochastic estimate of σ^2 involves a slight loss of degrees of freedom. This is discussed by Stein (1981) and explained by Vinod and Ullah (1981, p. 162). These extensions seem to be complicated from a practical viewpoint. Some of the theory is explained by Vinod (1980).

Berger (1980) suggested certain confidence ellipsoids for β. Morris (1983, p. 52) commented on Berger's ellipsoids and noted that the data analyst needs intervals for each parameter to replace the familiar t intervals, and further suggested approximate empirical Bayes confidence intervals. Morris also proposed approximate Bayes 95% confidence intervals based on

$$b_i^k \pm 1.96\left(V_{ii}^B\right)^{1/2}, \tag{3.9}$$

where V_{ii}^B are the diagonal elements of the posterior covariance matrix:

$$V^B = V^0 - V^0(V^0 + V^P)^{-1}V^0. \tag{3.10}$$

For RR we have $V^0 = \sigma^2(X'X)^{-1}$ as the sample covariance and $V^P = (\sigma^2/k)I_p$ as the prior covariance matrix from the Bayesian interpretation of ridge regression; see Vinod and Ullah (1981, p. 186). Using $X'X = G\Lambda G'$, (3.10) simplifies to $\sigma^2(X'X + kI)^{-1}$ Now the standard errors for a computational shortcut discussed by Vinod and Ullah (1981, p. 189) are precisely the ones needed for the approximate Bayes intervals. Frequentist properties of the approximate Bayesian confidence intervals are unknown for RR applications although Morris (1983, p. 52) mentioned encouraging simulation evidence for related applications.

4. AN EXAMPLE

We illustrate the methodology discussed above using a well-known example involving cement data given by Hald (1952, p. 647) having $n = 13$, $p = 4$. The computations are given in Table 2. Instead of forcing $X'X$ to be the correlation matrix, we use a similar rescaling of the data to force the sum of squares for all variables to be 12. The first row of Table 2 gives our scale factors for the four regressors. For y the scale factor is 0.066473, and s^2 is 0.026437. The eigenvalues of $X'X$ are given in the row marked λ_i. The OLS results with corresponding standard errors are given for this parameterization in the rows marked b_i^0 and SE_i respectively. The usual 95% confidence intervals using the Student's t value of 2.306 are given in the rows marked $b_i^{LO}(t)$ and $b_i^{UP}(t)$. The estimated R^2 equals 0.9824 for these data.

The elements of vectors $C^0 = G'b^0, C$, and the matrix G are also given. For UMSE, the computations of C_i^{UP} and C_i^{LO} are made for $k = 0.187$, with corresponding δ_i shown in the row marked δ_i. This does have one "too small" shrinkage, namely, $0.993 = \delta_1$, because $\dot{C}_1 = 20.9$. For $\dot{C}_i = 20$, Table 1 of the previous section shows that a $\delta_i \leq 0.994$ is "too small". However, it is not "ridiculously small".

The choice $k = 0.187$ is non-stochastic. It implies a "stable" RIDGE TRACE in terms of the so-called ISRM quantification of the stability criterion, based on the multicollinearity allowance, $m = p - \sum \delta_i$, discussed by

Table 2. *Confidence Intervals for Cement Data*

		i			
		1	2	3	4
scale factors		0.1700	0.0643	0.1561	0.0597
λ_i		26.8284	18.9128	2.2393	0.0195
b_i		0.6065	0.5277	0.0434	–0.1603
SE_i		0.2912	0.7487	0.3213	0.7889
$b_i^{UP}(t)$		1.2780	2.2543	0.7843	1.6589
$b_i^{LO}(t)$		–0.0650	–1.1988	–0.6975	–1.9795
C_1^0		0.6570	0.0083	0.3028	–0.3880
b_i^k		0.5038	0.3069	–0.0642	–0.3911
G	(1)	0.4760	–0.5090	0.6755	–0.3911
matrix	(2)	0.5639	0.4139	–0.3144	–0.6418
rows	(3)	–0.3941	0.6050	0.6377	–0.2685
	(4)	–0.5479	–0.4512	–0.1954	–0.6767
$\dot{C}_i$		20.9294	0.2220	2.7868	–0.3331
δ_i		0.9931	0.9902	0.9227	0.0941
$C_i^{UP}(UMSE)$		0.7142	0.0808	0.4805	1.7326
$C_i^{LO}(UMSE)$		0.5907	–0.0643	0.0783	–1.8056
$b_i^{UP}(UMSE)$		0.8021	1.4407	0.2630	0.9119
$b_i^{LO}(UMSE)$		0.2057	–0.8269	–0.3914	–1.6941
Relative width $b_i(UMSE/OLS)$		0.44	0.66	0.44	0.72

Vinod (1976, p. 838). The solution $k = 0.184$ corresponds to $m = 1$, and is also favored by other criteria as discussed by Vinod (1976, 1978), Obenchain (1975, 1978), and Mallows (1973). It is close to Hoerl *et al.*'s (1975) choice based on $ps^2[\sum C_i^{02}]^{-1} = 0.157$. For these data multicollinearity is a serious problem, because $\lambda_p = 0.0195$ is considerably smaller than λ_1.

The usual confidence intervals $b_i^{UP}(t)$ and $b_i^{LO}(t)$ are seen to be wider and perhaps less informative than our b_i^{UP} and b_i^{LO} for UMSE. Our nonlinear transformation (3.7) and (3.8) seems to have successfully yielded narrow intervals for β_i, from the narrow intervals for γ_i. The last row of Table 2 highlights the dramatic reduction in the width of OLS intervals by using UMSE.

5. SIMULATION TO EVALUATE APPROXIMATIONS

Because of some intrinsic difficulties associated with confidence intervals for biased estimators, we had to resort to certain approximations in deriving our UMSE intervals. For $\delta_i = 1$ these intervals coincide with the OLS intervals, and no approximation is needed. However, for $0 \leq \delta_i < 1$ it is not obvious whether our approximations will yield reliable confidence coefficients. To obtain a better understanding of the properties of our confidence intervals we have designed a simulation study as follows. The simulation involves several steps which are consecutively numbered for convenience.

(Step 1) Choose a basic data structure of the cement data of the previous section. This involves using the same eigenvalues and eigenvectors. (Step 2) Choose some arbitrary values for the true unknown γ_i. Our four sets of four γ_i values are respectively: (.12, .28, .56, −.77), (1.01, −.55, −1.02, 1.27), (.80, −2.09, 4.26, −1.35), and (5.68, .27, 6.10, 5.52). This choice implies four signal-to-noise ratios (SNR): 1, 4, 25, and 100, where $SNR = \gamma'\gamma/\sigma^2 = \beta'\beta/\sigma^2$. (Step 3) The corresponding sets of true unknown β_i are found by the definitional relation $\beta = G\gamma$. (Step 4) Specify $\sigma^2 = 1$, and the non-stochastic δ_i values from the previous section 0.9931, 0.9902, 0.9227 and 0.0941 respectively. (Step 5) Initialize the computer counters for measuring the number of times the estimated intervals for γ_i and β_i cover the respective true unknowns. Also, initialize the counters for measuring the widths of various intervals. (Step 6) Use a "super duper" generator for normal random numbers denoted by $N(0,1)$, based on a computer program developed by Marsaglia and others at McGill University in Canada. Obtain the OLS estimates of γ_i by the relation:

$$C_i^0 = \gamma_i + N(0,1)\sigma\lambda_i^{-1/2}. \tag{5.1}$$

(Step 7) Obtain $UMSE(\delta X)$ of (2.2b) for $\delta_i\dot{C}_i$ as an estimator of $\dot{\gamma}_i$. (Step 8) Obtain the UMSE confidence intervals from (2.12). Obtain OLS intervals by substituting $\delta_i = 1$ in (2.12). (Step 9) Compute the absolute values of the widths of OLS andUMSE intervals for γ_i. Verify that the OLS intervals are $2\sigma\lambda_i^{-1/2}(1.96)$ wide. (Step 10) Find the OLS and UMSE intervals for β_i by using the relations (3.7) and (3.8). (Step 11) Compute the four widths of the intervals for both β_i and γ_i, and for each of the two types of intervals considered here, namely the OLS and UMSE. These are given in Table 3. Now store the appropriate widths. Compute whether the observed intervals based on the current realization of the normal random numbers covers the true β_i, (and γ_i). If the coverage is observed, increase the appropriate counters by 1. (Step 12) Repeat steps 5 to 11 for a total of 1000 realizations. In our simulation the coverage of β_i was achieved in more than 950 cases for each i, and each SNR. (Step 13) Find the coverage proportions by dividing the coverage counts by 1000, and the average widths in a similar fashion. (Step 14) Compute an "overall" coverage proportion, as the fourth root of the product of four coverage proportions associated with the four regressors. This is to be done four times, i.e., for both γ_i and β_i, and for OLS and UMSE. (Step 15) Define the "relative width" of UMSE as the ratio of the UMSE width to the corresponding OLS width. The "overall" width of UMSE is then computed as the average of the "relative widths" over the set of the four regressors. This is to be done four times, as before. (Step 16) Summarize the results in a Table. (Step 17) Repeat Steps 4 to 16 for alternative choices of δ_i.

Table 3. *Average Width for the ith Coefficient Over 1000 Problems*

		GAMMA PARAMETER		BETA PARAMETER	
SNR	i	OLS	UMSE	OLS	UMSE
	1	.756812	.751577	9.35928	8.64175
1	2	.901381	.892573	19.64680	18.04290
	3	2.619580	2.421320	10.05430	9.27819
	4	28.082500	25.690900	20.33670	18.67270
	1	.756812	.752066	9.35928	8.73603
4	2	.901381	.892761	19.64680	18.27390
	3	2.619580	2.434470	10.05430	9.38171
	4	28.082500	26.043700	20.33670	18.91440
	1	.756812	.751881	9.35928	8.94112
25	2	.901381	.896098	19.64680	18.36500
	3	2.619580	2.739000	10.05430	9.57538
	4	28.082500	26.034500	20.33670	18.96900
	1	.756120	.767028	9.35928	9.57562
100	2	.901381	.892562	19.64680	19.64820
	3	2.619580	3.007520	10.05430	10.24900
	4	18.08250	27.89140	10.33670	20.28480

NOTE: See Steps 9 to 11 described in the text.

Table 4 gives a summary of the above simulation. The simulation does support the theory of the previous sections, and indicates that the new confidence intervals can achieve narrower widths without sacrificing the overall confidence levels. The coverage proportions (unknown to the researcher) are always larger than 95% and the averages of the "relative widths" are rarely much larger than unity. Since the width is known to the researcher, using our methods cannot mislead anyone, even if the width turns out to be slightly larger. Slightly larger width for UMSE interval when SNR is 100 in Table 4 indicates that RR may not be better than OLS in these cases. This is consistent with recent ridge literature reviewed in Chapters 7 and 8 of Vinod and Ullah (1981), and shows that UMSE intervals can discourage the use of RR in unfavorable cases. Thus wider UMSE intervals do not necessarily represent a disadvantage.

Table 4. *Confidence Interval Widths and Coverage: Simulation Over 1000 Problems*

		Confidence Level[2]		Relative Width[3]
SNR[1]		OLS	UMSE	UMSE
1	γ	0.94	0.96	0.96
	β	0.96	1.00	0.92
4	γ	0.94	0.96	0.96
	β	0.97	1.00	0.93
25	γ	0.94	0.96	0.99
	β	0.97	1.00	0.94
100	γ	0.95	0.96	1.04
	β	0.98	1.00	1.01

[1] Signal-to-noise ratio is $\gamma'\gamma/\sigma^2 = \beta'\beta/\sigma^2$.

[2] Confidence level is the fourth root of a product of four coverage proportions for four regressors over 1000 problems.

[3] Relative width is relative to the average width of OLS intervals over 1000 problems.

One can imagine a practical user of our methods trying both OLS and say, UMSE and then using the smaller of the two confidence intervals. Since the theoretical properties of such interdependent actions appear to be intractable, we discourage this approach. When the new confidence intervals are wider, RR may not be rejected completely. If ridge results are reported, both types of confidence statements should be reported. Note, for example, that a traditional user of standard normal tables may correctly consider many 95% intervals: (–2.0, 1.92), (–1.96, 1.96), (–1.88, 2.06), etc.

An optimistic assessment of our methods based on a simulation continues to hold true for additional choices of δ_i not reported here. Whenever the δ_i arise from a realistic problem they are rarely much smaller than corresponding "too small" δ_i^{SML} values; and therefore UMSE intervals are usually narrower than those for OLS. For artificial problems having $\delta_i \leq \delta_i^{SML}$, we recommend using only the OLS intervals for all coefficients.

It is interesting to note that the coverage proportions and confidence levels are usually smaller for the canonical model having γ_i as the unknowns, compared to the original model having β_i as the unknowns. This observation is reassuring to the reader who may be uncomfortable with the nonlinear transformation of equations (3.7) and (3.8), because it shows that these transformations are conservative.

6. QUALMS ABOUT BOOTSTRAP CONFIDENCE INTERVALS FOR RIDGE REGRESSION

The main purpose of this section is to show that ridge regression represents a case where one may have "qualms" (Schenker, 1985) about bootstrap confidence intervals. Since ridge regression yields biased estimators, bootstrap assumption about the existence of a pivotal quantity whose distribution does not depend on unknown parameters is not satisfied. First, we will briefly review the basic bootstrap ideas in the regression context. Next, we apply them to ridge regression illustrated by the cement data example studied before. We will also compare the results with approximate Bayes results for the same data.

The residuals from the OLS estimator, denoted by $e = y - Xb^0$, play a special role in bootstrapping. The covariance matrix of b^0 is

$$\text{Cov}(b^0) = s^2(X'X)^{-1}, \qquad s^2 = (e'e)/(n-p-1). \tag{6.1}$$

An empirical CDF of OLS residuals puts probability mass $1/n$ at each e_t, and is denoted here by F_e. Now a basic bootstrapping idea is to use F_e with mean zero and variance σ^2 as a feasible, approximate, nonparametric estimate of the CDF of the true unknown errors denoted by F_u. Let J be a suitably large number, which for the cement data is 500. We draw J sets of bootstrap samples of size n, which is 13 for the cement data, with elements denoted by e_{*jt} ($j = 1, 2, \ldots, J$ and $t = 1, \ldots, n$) from F_e using random sampling *with replacement.* One generates J sets of $n \times 1$ vectors denoted by e_{*j} having elements e_{*jt} ($t = 1, \ldots, n$). Hence the pseudo y data are obtained by:

$$y_{*j} = Xb^0 + e_{*j}, \qquad j = 1, 2, \ldots, J \tag{6.2}$$

which yields a large number, J, of regression problems to be used for bootstrap inference described below. We ensure that the probability that e_{*jt} equals e_t is $1/n$. Hence the variance is:

$$\sigma_e^2 = E\left(e_{*jt}^2\right) = \sum_{t=1}^{n}(1/n)\text{var}(e_t) = (1/n)\sum_{t=1}^{n} e_t^2 = s^2(n-p)/n. \tag{6.3}$$

This shows that when the residuals are rescaled by the square root of $n/(n-p)$, the variance of the rescaled e_{*jt} should be approximately equal to s^2. For Hald's cement data with $n = 13$ and $J = 500$, the variance (6.3) is computed to be 0.026512, which is very close to the s^2 of 0.026437; which suggests that $J = 500$ is adequate.

Efron (1982, p. 36) states that (6.2) is a "standard linear model written in unusual notation". If one applies OLS to the pseudo y data one obtains:

$$b_{*j} = (X'X)^{-1}X'y_{*j} \quad \text{and} \quad \text{Cov}(b_{*j}) = \sigma_e^2(X'X)^{-1} \tag{6.4}$$

for $j = 1, \ldots, J$. This expression for the covariance matrix is the same as $\text{Cov}(b^0)$ provided we use rescaled residuals based on (6.3). In other words, the standard errors for OLS regression coefficients calculated by the bootstrap and the conventional methods are almost identical.

Let the empirical CDF of the J estimates b_{*j} be denoted by $CDF_*(z) = \#(b_{*j} \leq z)/J$, where # denotes the number of times the condition is observed. For given α between 0 and 0.5 (without loss of generality) we define:

$$b_{LO}(\alpha) = ICDF_*(\alpha), \quad \text{and} \quad b_{UP}(\alpha) = ICDF_*(1-\alpha), \tag{6.5}$$

where $ICDF_*(z)$ denotes the inverse of $CDF_*(z)$. Now we can write the probability:

$$\text{Prob}_*\left[b_{LO}(\alpha) \leq b_{*j} \leq b_{UP}(\alpha)\right] = 1 - 2\alpha. \tag{6.6}$$

This can yield a confidence interval for β provided we can replace b_{*j} by β as an approximation.

Since the above approximation may not be good, Efron suggests the following refinement. Its basic idea is the assumption that $b^0 - \beta$ and $b_{*j} - b^0$ have approximately the same distribution, and that this distribution is symmetric about the origin. To visualize why this is a refinement note that

$$b^0 - \beta = (X'X)^{-1}X'y - \beta = (X'X)^{-1}X'u = Wu, \tag{6.7}$$

where $W = (X'X)^{-1}X'$ is a known matrix. Since (6.7) involves the unknowns u, it is approximated by:

$$\begin{aligned} b_{*j} - b^0 &= (X'X)^{-1}X'y_{*j} - b^0 \\ &= (X'X)^{-1}X'(Xb^0 + e_{*j}) - b^0, \quad \text{from (6.2)} \\ &= (X'X)^{-1}X'e_{*j} = We_{*j}. \end{aligned} \tag{6.8}$$

Note that We_{*j} can be reasonably expected to approximate Wu. To obtain the left hand side of (6.8) we subtract b^0 from each term of (6.6) and write

$$\text{Prob}_* \left[b_{LO}(\alpha) - b^0 \le b_{*j} - b^0 \le b_{UP}(\alpha) - b^0 \right] = 1 - 2\alpha. \tag{6.9}$$

Now we replace $b_{*j} - b^0$ in the middle term of (6.9) by $b^0 - \beta$ on the left hand side of (6.7), which is appropriate because the right hand side of (6.7) can be approximated by the right hand side of (6.8). This manipulation may be called a "reflection" of the confidence interval (6.6) through b^0. Thus we write

$$\text{Prob}_* \left[2b^0 - b_{UP}(\alpha) \le \beta \le 2b^0 - b_{LO}(\alpha) \right] = 1 - 2\alpha. \tag{6.10}$$

Efron's (1982, ch. 10) bias corrected percentile method makes the same "reflection" after a hypothesized standardization to $N(0,1)$, a unit normally distributed pivotal quantity denoted by θ. Let $CDF_*(\theta)$ denote the CDF based on bootstrap replications applied to the estimated pivotal quantity. Now denote

$$z_o = \Phi^{-1}[CDF_*(\theta)] \tag{6.11}$$

using the distribution function Φ of the standard normal variate.

The bias corrected percentile method leads to the following approximate $1 - 2\alpha$ confidence interval:

$$\{ICDF_*(\Phi[2z_o - z_\alpha]),\ ICDF_*(\Phi[2z_o + z_\alpha])\}, \tag{6.12}$$

where $ICDF_*$ is the inverse CDF_* as before, and z_α is the upper α point for the $N(0,1)$ normal distribution function $\Phi(z_\alpha) = 1 - \alpha$. If the bootstrap distribution is median unbiased, it can be shown that the bias correction makes no difference.

Since bootstrapping does not give better parameter estimates (Peters and Freedman, 1984), we do not use it to suggest yet another choice of k in ridge regression. The confidence interval procedure proposed in this paper accepts the mean squared error (MSE) reducing choice of k made by the investigator and provides a confidence interval conditional on that choice.

For the cement data, $k = 0.187$ as in Section 4 above. Instead of $(b_{*j} - b)$ in (6.8) we have here

$$b^k_{*j} - b^k = \left[(X'X + kI)^{-1}X'X - I\right] b^k + \left[(X'X + kI)^{-1}X'u_{*j}\right], \quad (6.13)$$

which is used to approximate the following quantity which is similar to the left hand side of (6.7):

$$b^k - \beta = \left[(X'X + kI)^{-1}X'X - I\right] \beta + \left[(X'X + kI)^{-1}X'u\right]. \quad (6.14)$$

The corresponding right-hand side in (6.7) is simply Wu, which does not depend on the unknown parameter β. In the case of OLS when $k = 0$, the first term on the right hand side of (6.14) vanishes, and the right hand side of (6.14) becomes Wu. Unfortunately, in the case of ridge based b^k equation (6.14) does depend on β, and is obviously not a pivotal quantity. If we use (6.13) to approximate (6.14), it may be incorrect to the extent that b^k does not equal β. Thus the criticism by Schenker (1985) applies, and we note that the ridge based bootstrap confidence intervals cover the regression parameters with probability $1 - 2\alpha$ only approximately. The trouble is that the coverage probability itself is affected by the value of the unknown parameters. It is tempting to use a simulation to assess the importance of this problem, except that the conclusions from the simulation will depend on the values of parameters assumed.

Table 5 reports the 95% bootstrap confidence interval for the cement data, where the column marked 1 refers to the first regression coefficient as in Table 2, which is explained as follows. Ridge estimation of 500 pseudo y data yields 500 regression coefficients. These are first ranked in an ascending order of magnitude as an approximation to the sampling distribution. Since 2.5% of 500 is 12.5 the simple average of 12th and 13th ranked regression coefficients yields 0.3730 as the lower limit. The upper limit 0.6365 is a similar average of 487th and 488th ranked values. The reflection of this interval using the ridge estimate 0.5038 instead of b^0 in (6.10) from along the 8th row of Table 2 yields 0.3711 and 0.6346. Now the 247th ranked value is 0.5033 which is closer to the ridge value 0.5038 than any other value. Hence we say that the ridge estimate is slightly median biased because the median is the average of the 250th and 251st ranked values. Now z_o of (6.11) is –0.15, using the Tables for the cumulative normal distribution. Instead of –1.96 and 1.96 we have –1.99 and 1.93 as the numbers used to find the corresponding values of the normal CDF to be 0.0233 and 0.9732, yielding 12th and 486th ranked values as the "bias corrected" percentiles reported to be 0.3724 and 0.6310 in Table 5. Other columns of Table 5 are similarly computed. Table 5 also reports approximate Bayes confidence intervals using equation (3.9),

Table 5. *Cement Data Confidence Intervals for Approximate Bayes and Bootstrap*

	i			
	1	2	3	4
Approximate Bayes Interval:				
Lower Limit:	0.2805	–0.1500	–0.2986	–0.8696
Upper Limit:	0.7273	0.7639	0.1704	0.0876
Bootstrap Interval from Quantiles:				
Lower Limit:	0.3730	0.1529	–0.2096	–0.5463
Upper Limit:	0.6365	0.4423	0.0642	–0.2539
Bootstrap Reflected Interval:				
Lower Limit:	0.3711	0.1715	–0.1926	–0.5283
Upper Limit:	0.6346	0.4609	0.0812	–0.2359
Bootstrap Bias Corrected Interval:				
Lower Limit:	0.3724	0.1471	–0.2075	–0.5463
Upper Limit:	0.6310	0.4333	0.0685	–0.2539

Notes: All are 95% intervals comparable to those of Table 2. Equation (3.9) defines the approximate Bayes method. Quantiles are based on the ranking of ridge estimators from pseudo y data for $J = 500$ replications. Reflected and Bias corrected intervals are based on equations (6.10) and (6.12) respectively, modified by using the ridge estimator instead of the OLS estimator.

which are found to be closer to the UMSE intervals of Table 2 than the bootstrap intervals, for this choice of k. In unfortunate cases where ridge regression may not be appropriate, both approximate Bayes and bootstrap intervals may be too narrow and too optimistic about the correctness of ridge estimates.

Even though one may not be satisfied with the approximate coverage probabilities, the plots of the bootstrap distributions can be useful to the economist in other ways, and may be closely studied and interpreted. They can indicate how sensitive the reported confidence interval is to the arbitrary choice of the significance level, to the standard errors based on the normality assumption, etc. These intervals are narrower than those based on Stein's

UMSE, and one may wonder about their reliability, since they require the strong assumption that b^k is a good approximation to β.

7. CONCLUSION

One should not expect the confidence interval problem to have a unique solution. This paper has studied four alternative methods of constructing confidence intervals for the regression parameter estimated by ordinary ridge regression. The OLS interval is the simplest, and if it is economically meaningful there is no need to consider any other. Often, the OLS interval is centered at the coefficient with the "wrong sign", so that much of the interval lies in a range which is economically meaningless. In these applications the practical motivation behind using ridge regression is that it yields economically meaningful estimates, such as positive marginal propensity to consume (MPC). If the economist wishes to use a confidence interval centered at the ridge estimate the choice is between three methods: approximate Bayes based on (3.9), UMSE from (3.7) and (3.8), and finally the bootstrap. The approximate Bayes may not provide sufficient penalty for using a ridge biasing parameter k that is "too large", since it uses square roots of the diagonal elements of the inverse of the $(X'X + kI)$ matrix in the standard error formulas, which means that the standard error may decrease monotonically with increasing k. Similarly, bootstrap intervals may not provide a sufficient warning against a bad choice of k.

The UMSE intervals proposed here are new and conservative in the sense that they are narrower than the OLS *only when* the reduction in variance achieved by ridge regression is sufficiently large, without using "too small" shrinkage factors. The simulation study reported here indicates that some of the approximations used here do not jeopardize the validity of the basic confidence statement. Its computational burden is lighter than that of the bootstrap. The reporting of UMSE intervals can discourage potential misuses of ridge regression. The modification of UMSE for the case of stochastic k is discussed in an Appendix. In the author's own experience, the presence of a stochastic k does not make a practical difference, but the UMSE intervals for large p are found to be too conservative.

APPENDIX: EXTENSION FOR STOCHASTIC SHRINKAGE FACTORS

For simplicity, the discussion in Section 2 assumes that the shrinkage factor δ is nonstochastic. Now we consider the general case where δ is a function of the random variable x. We use Lemma 4 of Stein (1981) to write:

$$E\left[(x+f)-\xi\right]^2 = 1 - Ef^2 + 2Ef', \tag{A.1}$$

where f is $f(x)$, and f' is the derivative with respect to x. Note that if one chooses $f = (\delta - 1)x$ one obtains (2.1) as a special case of (A.1). Now the left hand side of (2.3) is a special case of:

$$\begin{aligned} E\left[(x+f-\xi)^2 - (1+f^2+2f')\right]^2 & \\ &= E\left[(x-\xi)^2 + 2(x-\xi)f - 1 - 2f'\right]^2 \\ &= 2 + 4E[f^2 + 2f' + f'^2], \end{aligned} \tag{A.2}$$

where the expressions involving $(x-\xi)^j g(x)$ are replaced by derivatives of $g(x)$ for $j = 1, 2, 3$ by the Lemma mentioned above. The expression (A.2) is remarkably simple involving only the first derivative, thanks to a cancellation of all higher order terms.

If δ is a function of x, the f' in (A.2) will be $(\delta-1)+x\delta'$ instead of $(\delta-1)$, which leads to a modified expression for SD^2 of (2.8), especially changing the term after the last comma. When δ_i is stochastic in the regression application, one can use: the analytical expression for its derivative if it is available; or a geometrical estimate from the ridge trace; or consider a "small" change in k and approximate δ_i' by the ratio of the change in δ_i to the change in $\dot{C}_i$. Although the S^* defined after (3.3) is modified by using (A.2) to allow for stochastic δ_i, there seems to be no conceptual difficulty in deriving confidence intervals in the stochastic shrinkage case.

REFERENCES

Baranchik, A. J. (1970), "A family of minimax estimators of the mean of a multivariate normal distribution". *Annals of Mathematical Statistics* **41**, 642–645.

Berger, J. (1980), "A robust generalized Bayes estimator and confidence region for a multivariate normal mean". *Annals of Statistics* **8**, 716–761.

Efron, B. (1982), *The Jackknife, the Bootstrap and Other Resampling Plans*. Philadelphia: Society for Industrial and Applied Mathematics (SIAM), Vol. 38.

Efron, B., and C. Morris (1972), "Limiting the risk of Bayes and empirical Bayes estimators—Part II: The empirical Bayes case". *Journal of the American Statistical Association* **67**, 130–139.

Hald, A. (1952), *Statistical Theory with Engineering Applications*. New York: Wiley and Sons.

Hoerl, A. E., and R. W. Kennard (1970), "Ridge regression: biased estimation for nonorthogonal problems". *Technometrics* **12**, 55–67.

Hoerl, A. E., R. W. Kennard, and K. F. Baldwin (1975), "Ridge regression: some simulations". *Communications in Statistics* **4**, 105–123.

Johnson, N. L., and S. Kotz (1970), *Continuous Univariate Distributions 1 and 2*. New York: Wiley and Sons.

Kabe, D. G. (1983), "A quadratic programming approach to the construction of simultaneous confidence intervals". *Communications in Statistics A, Theory and Methods* **12**, 2053–2058.

Mallows, C. L. (1973), "Some comments on C_P". *Technometrics* **15**, 661–675.

Morris, C. N. (1983), "Parametric empirical Bayes inference: theory and applications". *Journal of the American Statistical Association* **78**, 47–55.

Obenchain, R. L. (1975), "Ridge analysis following a preliminary test of the shrunken hypothesis". *Technometrics* **17**, 431–441.

Obenchain, R. L. (1977), "Classical F-tests and confidence regions for ridge regression". *Technometrics* **19**, 429–439.

Peters, S. C., and D. A. Freedman (1984), "Some notes on the bootstrap in regression problems". *Journal of Business and Economic Statistics* **2**, 406–409.

Rao, C. R. (1973), *Linear Statistical Inference and Its Applications*, 2nd edition. New York: Wiley and Sons.

Schenker, N. (1985), "Qualms about bootstrap confidence intervals". *Journal of the American Statistical Association* **80**, 360–361.

Stein, C. (1981), "Estimation of the mean of a multivariate normal distribution". *Annals of Statistics* **9**, 1135–1151.

Vinod, H. D. (1976), "Application of new ridge regression methods to a study of Bell System scale economies". *Journal of the American Statistical Association* **71**, 835–841.

Vinod, H. D. (1978), "A survey of ridge regression and related techniques for improvements over ordinary least squares". *Review of Economics and Statistics* **60**, 121–131.

Vinod, H. D. (1980), "New confidence intervals for ridge regression parameters". Bell Laboratories Economics Discussion Paper No. 172, Murray Hill, N.J.

Vinod, H. D., and B. Raj (1984), "Bootstrapping confidence intervals for arbitrary functions of regression parameters". Working Paper 8413, Dept. of Economics, Fordham University, Bronx, New York 10458.

Vinod, H. D., and A. Ullah (1981), *Recent Advances in Regression Methods*. New York: Marcel Dekker.

Simon Power [1]

ASYMPTOTIC PROPERTIES OF SINGLE EQUATION ERRORS IN VARIABLES ESTIMATORS IN RATIONAL EXPECTATIONS MODELS

1. INTRODUCTION

The purpose of this paper is essentially twofold, first to introduce some new single equation errors in variables estimators for simultaneous equations models containing rational expectations variables, and second to derive their asymptotic properties. In addition, a consistency proof for a new estimator due to Fair (1984a) is presented.

Single equation errors in variables estimators have been used extensively in the applied rational expectations literature since their introduction by McCallum (1976). This popularity has been due largely to their simplicity and ease of implementation as compared to the more complex and expensive systems estimation methods explored by Hansen and Sargent (1980, 1982), Wallis (1980), and Fair and Taylor (1983) among others—avoiding, in particular, the necessity of solving the system. Additional advantages of the single equation errors in variables estimators include (i) a certain degree of robustness to misspecification elsewhere in the system, (ii) the capability of estimating a single equation from a very large system too complex for the full information methods (Fair, 1984a), and (iii) the fact that they continue to be consistent even when expectations are not in fact rational as long as the instruments used are in the information set used to form the expectations (Fair, 1984b).

The organization of the paper is as follows: Section 2 describes the types of situation which call for the estimation techniques herein considered, Section 3 outlines the existing estimators which have been proposed, Section 4 introduces the various new estimators, and Section 5 presents the asymptotic results.

[1] Department of Economics, University of Regina, Regina, Saskatchewan S4S 0A2

I. B. MacNeill and G. J. Umphrey (eds.), Time Series and Econometric Modelling, 301–315.

2. THE ESTIMATION PROBLEM

Consider the following equation from a simultaneous equations model:

$$y_t = [{}_{t-1}z_t Z_t]\delta + u_t, \tag{1}$$

where y_t is a scalar endogenous variable, u_t is a disturbance term, Z_t is a row vector containing k endogenous and $h-k-1$ predetermined variables, δ is an $h \times 1$ vector of coefficients, ${}_{t-1}z_t$ is the rational expectation of an endogenous variable not included in Z_t and not on the left hand side of the equation, and $[{}_{t-1}z_t Z_t]$ is a $1 \times h$ row vector. We consider various cases.

(a) *u_t not serially correlated*

In this case we assume that $E(u_t \mid I_{t-1}) = 0$, where I_{t-1} is the information set available at time $t-1$, which implies that u_t are not serially correlated. The errors in variables "trick" is to replace the unobservable ${}_{t-1}z_t$ with the observable z_t, yielding

$$y_t = [z_t Z_t]\delta + u_t - \delta_1 \eta_t = Q_t \delta + v_t, \tag{2}$$

where δ_1 is the first element of δ, η_t is the forecast error $z_t - {}_{t-1}z_t$ which by the rational expectations hypothesis satisfies $E(\eta_t \mid I_{t-1}) = 0$, and v_t is the composite disturbance. For estimation we now need to find some suitable instruments. Notice that using z_t as a proxy results in the composite disturbance v_t being potentially correlated with all the elements of Q_t not included in I_{t-1}, except those predetermined variables which are perfectly predictable at time $t-1$, e.g., time trends and constants. Instruments are, however, available in the form of any lagged endogenous or exogenous variables in the model as these are all included in I_{t-1} and of course, the perfectly predictable time t variables.

We may then obtain consistent estimates of δ. These estimates are termed 2SLS in the literature and we maintain this usage, although, strictly speaking, they should be referred to as instrumental variables. Moreover, the estimated standard errors will be consistent estimates of the true standard errors, this following from the fact that the composite disturbances are serially uncorrelated, given one further assumption, namely conditional homoscedasticity, i.e., $E(v_t^2 \mid F_t) = \sigma_v^2$ for all t, where F_t denotes the instrument set.

(b) *u_t serially correlated*

Consider the following example:

$$y_t = [{}_{t-1}z_t Z_t]\delta + u_t; \quad u_t = \rho u_{t-1} + w_t; \quad E(w_t \mid I_{t-1}) = 0; \quad |\rho| < 1. \tag{3}$$

Substituting for the rational expectation we obtain:

$$y_t = [z_t Z_t]\delta + u_t - \delta_1 \eta_t. \tag{4}$$

It is convenient to remove the AR part of the composite disturbance by quasi-differencing which yields:

$$y_t = [y_{t-1} z_t Z_t z_{t-1} Z_{t-1}] \begin{bmatrix} \rho \\ \delta \\ -\rho\delta \end{bmatrix} + v_t, \tag{5}$$

where $v_t = w_t - \delta_1 \eta_t + \rho \delta_1 \eta_{t-1}$.

It may now be ascertained that v_t are serially correlated, although only of first order, and therefore, by a theorem due to Anderson (1971), may be expressed as an MA(1) disturbance.

Instruments for this AR(1) example need to be lagged two periods instead of the one period as before, this following from the fact that whereas $E(v_t \mid I_{t-1}) \neq 0$, $E(v_t \mid I_{t-2}) = 0$. Having selected appropriate instruments we may then estimate as before except that having quasi-differenced we need to use nonlinear two stage least squares (NL2SLS) in place of 2SLS. Unfortunately, however, although the parameter estimates will be consistent, but not asymptotically efficient, the estimated standard errors will not even be consistent.

A similar problem arises whenever any of the following circumstances are obtained either individually or severally: (i) serial correlation in the structural disturbance, (ii) expectations formed in any period later than $t-1$, and (iii) expectations horizons exceeding the sampling interval (Cumby *et al.*, 1983; Brown and Maital, 1981; Hakkio, 1981).

Two problems therefore arise: (i) how to improve the efficiency of the parameter estimates and (ii) how to obtain consistent estimates of the standard errors.

To solve these problems, we outline some of the existing estimators which have been designed to cope with the problem. Before this, however, it will be convenient to establish a framework provided by Cumby *et al.* (1983) into which these models fall.

Essentially, our aim is to estimate parameters of models of the general form:

$$Y = Qg(\delta) + v, \tag{6}$$

where g is a one-to-one function taking elements of the parameter space into a space of equal or greater dimension, subject to the existence of an integer N and a row vector, F_t, of instruments satisfying:

$$E(v_t \mid v_{t-N}, v_{t-N-1}, \ldots, F_t, F_{t-1}, \ldots) = 0. \tag{7}$$

Clearly, our model falls into this framework as it allows for serial correlation and it assumes that the instruments are predetermined rather than strictly exogenous (Hayashi and Sims, 1983). We shall assume henceforward that a correct set of instruments is being employed at all times.

3. EXISTING ESTIMATORS

We now consider some of the existing estimators which have been put forward to deal with this type of estimation problem.

(i) McCallum (1979) suggested using both NL2SLS to estimate the parameter of (6) and a correction, provided by Hansen (1982), for the inconsistent estimates of the standard errors. Cumby *et al.* (1983) have shown that this estimator falls into the class of single equation analogues of Generalised Method of Moments (GMM) estimators (Hansen, 1982) exploiting the orthogonality conditions $E(F_t'v_t) = 0$ into which Cumby *et al.*'s two step, two stage least squares (2S2SLS) estimator (discussed next) also falls; they also have shown that within this class 2S2SLS is asymptotically efficient.

(ii) Cumby *et al.* (1983) have recently put forward the 2S2SLS estimator. This estimator may be developed in the following way. Transform model (6) by the transposed instrument matrix F' to yield:

$$F'Y = F'Qg(\delta) + F'v. \tag{8}$$

By the orthogonality conditions, $E(F_t'v_t) = 0$, nonlinear least squares will give a consistent estimate of δ. But, as a result of $E(F'vv'F)$ not being proportional to the identity matrix, we can produce a more efficient estimator. Let $\Omega_T = (1/T)E(F'vv'F)$ and assume that Ω_T is a positive definite matrix. Then Ω_T may be expressed as RR', where R is non-singular, with the result that the transformed model:

$$R^{-1}F'Y = R^{-1}F'Qg(\delta) + R^{-1}F'v \tag{9}$$

has a disturbance matrix proportional to the identity matrix. The 2S2SLS estimator may now be defined. It is the vector δ_{2S2SLS} that minimises the criterion function:

$$\Phi(\delta) = [Y - Qg(\delta)]'F\hat{\Omega}^{-1}F'[Y - Qg(\delta)], \tag{10}$$

where $\hat{\Omega}$ is a consistent estimator of $\Omega = \lim(1/T)E(F'vv'F)$. Various different estimators of Ω have been put forward by Cumby *et al.* (1983), Hansen (1982), and Newey and West (1985).

(iii) Hayashi and Sims (1983) proposed an estimator which exploits a different set of orthogonality conditions, namely $E(F'_{t-j}v_t) = 0,\ j \geq 0$. In addition, they assumed that the serial correlation in v does not depend on F, that is

$$E(v_t v_s \mid F_t, F_{t-1}, \ldots) = E(v_t v_s) \quad \text{for} \quad t \geq s. \tag{11}$$

This last assumption is necessary to evaluate the asymptotic variance-covariance matrix for this estimator. This evaluation is carried out in Section 5. Hayashi and Sims proceeded by transforming model (6) by the upper triangular inverse of Γ, where $E(vv') = \sigma^2 V$ and $V = \Gamma\Gamma'$ with Γ non-singular:

$$\Gamma^{-1}Y = \Gamma^{-1}Qg(\delta) + \Gamma^{-1}v. \tag{12}$$

By the orthogonality conditions stated above, $E(F'_t(\Gamma^{-1}v)_t) = 0$ and therefore we may use F_t as instrumental variables to obtain a consistent estimate of δ. This involves the minimization of the criterion function:

$$\Phi(\delta) = [Y - Qg(\delta)]'\Gamma^{-1'}F(F'F)^{-1}F'\Gamma^{-1}[Y - Qg(\delta)]. \tag{13}$$

Intuitively what is going on here is that we are "forward filtering" the equation (6) in such a way that we eliminate the serial correlation in the disturbance term. But, not only are we eliminating the serial correlation, we are also transforming the equation in such a way that the transformed disturbance terms are orthogonal to the instrument matrix F, which may then be used to provide consistent estimates.

To implement this technique, Hayashi and Sims (1983) proposed estimation of the matrix Γ^{-1} by one of two methods. To outline these methods, it will be convenient to make a slight digression.

We assume for simplicity that the composite disturbance v_t may be expressed as an MA(1) process, i.e., it may be written as either:

$$v_t = \epsilon_t - \theta\epsilon_{t-1} = (I - \theta L)\epsilon_t \tag{14}$$

or

$$v_t = \xi_t - \theta\xi_{t+1} = (I - \theta L^{-1})\xi_t, \tag{15}$$

the former being the standard "backward representation" the appropriate filter is $(I - \theta L)^{-1}$, while in the case of the "forward representation" and the latter being the "forward representation". To eliminate the moving average nature of the composite disturbance we need to filter v_t. In the case of the "backward representation" the appropriate filter is $(I - \theta L)^{-1}$, while in the case of the "forward representation" the appropriate filter is $(I - \theta L^{-1})^{-1}$. We now concentrate exclusively on the "forward representation", this being the relevant one for our purpose. An alternative way of representing a moving average process of finite order is as an infinite order autoregressive process. In our case this means that

$$(I - \theta L^{-1})^{-1} v_t = (I + \theta L^{-1} + \theta^2 L^{-2} + \ldots) v_t = \xi_t. \tag{16}$$

We may now outline the two methods of implementation proposed by Hayashi and Sims (1983).

The first, denoted by HS_1 hereafter, consists of the following: estimating the moving average coefficient, θ, from the residuals of a first stage estimate of (6) by some consistent technique, e.g., NL2SLS; truncating the autoregressive representation of the forward filter (16) at the sth lag, where $s = \sqrt{T}$; and transforming model (6) by the estimate of the Γ^{-1} matrix below:

$$\begin{bmatrix} 1 & \theta & \theta^2 & \cdot & \cdot & \cdot & \cdot & \theta^s & 0 & \cdot & \cdot & \cdot & 0 \\ 0 & 1 & \theta & \theta^2 & & & & & \theta^s & 0 & & & \cdot \\ \cdot & \cdot & \cdot & \cdot & \cdot & & & & & \cdot & \cdot & & \cdot \\ \cdot & & \cdot & \cdot & \cdot & \cdot & & & & & \cdot & \cdot & \cdot \\ \cdot & & & \cdot & \cdot & \cdot & \cdot & & & & & \cdot & 0 \\ 0 & \cdot & \cdot & \cdot & 0 & 1 & \theta & \theta^2 & \cdot & \cdot & \cdot & \cdot & \theta^s \end{bmatrix}. \tag{17}$$

The transformed model (6) is then estimated by instrumental variables using untransformed instruments.

The second method, denoted by HS_2 hereafter, consists of the following: estimating an unstructured autoregressive representation of order s, i.e., estimating the parameters of:

$$v_t = \alpha_1 v_{t-1} + \alpha_2 v_{t-2} + \ldots + \alpha_s v_{t-s} + \zeta_t, \tag{18}$$

where the v_{t-i}'s are the residuals from consistent estimation of (6); transforming (6) by the following estimate of the Γ^{-1} matrix:

$$\begin{bmatrix} 1 & -\hat{\alpha}_1 & -\hat{\alpha}_2 & \cdot & \cdot & \cdot & \cdot & -\hat{\alpha}_s & 0 & \cdot & \cdot & \cdot & 0 \\ 0 & 1 & -\hat{\alpha}_1 & -\hat{\alpha}_2 & & & & & -\hat{\alpha}_s & & & & \cdot \\ \cdot & \cdot & \cdot & \cdot & \cdot & & & & & \cdot & & & \cdot \\ \cdot & & & \cdot & \cdot & \cdot & & & & & \cdot & & \cdot \\ \cdot & & & & \cdot & \cdot & \cdot & & & & & \cdot & 0 \\ 0 & \cdot & \cdot & \cdot & 0 & 1 & -\hat{\alpha}_1 & -\hat{\alpha}_2 & \cdot & \cdot & \cdot & \cdot & -\hat{\alpha}_s \end{bmatrix}; \tag{19}$$

and then again estimating the transformed model (6) by instrumental variables using untransformed instruments.

It will be noted that both of the estimated Γ^{-1} matrices are of order $(T - s) \times T$; hence s observations are lost.

As we can see from the above, the estimators proposed by Cumby *et al.* (1983) and Hayashi and Sims (1983) exploit different sets of orthogonality conditions, and as a result, for a fixed instrument list, it is impossible to say which of them will be asymptotically efficient. We say "for a fixed instrument list" because Hayashi and Sims have shown that when (7) holds, as the sample size and the instrument list increase, the asymptotic variance-covariance matrices of the two estimators approach the same limiting matrix. Additional instruments come from lagging F_t. Moreover, they have demonstrated that this limiting matrix is the asymptotic variance-covariance matrix of Hansen and Sargent's (1982) optimal instrument set instrumental variables estimator. Unfortunately, one cannot determine this optimal instrument set without having details of the entire structural model.

We should also mention that standard estimation methods to cope with serial correlation in this context are not consistent, this being pointed out first by Flood and Garber (1980).

4. NEW ESTIMATORS

We now introduce our new estimators which essentially represent potential improvements on the Hayashi and Sims estimators, HS_1 and HS_2, outlined above, at least in small samples. The rationale for these new estimators comes from two sources, theoretical and empirical, and attention will be devoted to both.

It will be recalled that both HS_1 and HS_2 drop s observations from the sample. Clearly, in a small sample this procedure is wasteful and it would be preferable to retain the dropped observations. With respect to HS_1 the immediate idea which comes to mind is to fill in more of the estimate of the Γ^{-1} matrix, i.e., to transform (6) by the following matrix:

$$\begin{bmatrix} 1 & \theta & \theta^2 & \cdot & \cdot & \theta^T \\ 0 & \cdot & \cdot & & & \cdot \\ \cdot & \cdot & \cdot & \cdot & & \cdot \\ \cdot & & \cdot & \cdot & \cdot & \cdot \\ \cdot & & & \cdot & \cdot & \theta \\ 0 & \cdot & \cdot & \cdot & 0 & 1 \end{bmatrix}. \tag{20}$$

This transformation is equivalent to transforming (6) by a truncated forward filter comprising only the sample elements of y_t and Q_t. We denote this estimator by HS_3 hereafter. An equivalent representation of the transformed (6) is:

$$y_t^* = Q_t^* g(\delta) + v_t^*, \tag{21}$$

where $y_t^* = \sum \theta^i y_{t+i}$, $y_{T+1} = 0$; $Q_t^* = \sum \theta^i Q_{t+i}$, $Q_{T+1} = 0$; $v_t^* = \sum \theta^i v_{t+i}$, $v_{T+1} = 0$ and the summations are over the range $i = 0$ to $i = T-t$.

A backward form of this transformation has been used in the literature before by, inter alia, Harvey and McAvinchey (1981), Pollock (1979), and Phillips (1966). However, a variation on this transformation, which has been proposed in the backward case for the regression model with MA(1) errors by MacDonald and McKinnon (1985), and extended recently to the regression model with MA(q) errors by Ullah *et al.* (1985), and extended still more recently to the simultaneous equations case by Ullah and Power (1985), is available and constitutes our second new estimator which we denote by HS_4 hereafter. In the MA(1) case this estimator may be represented as:

$$y_t^* = Q_t^* g(\delta) + G_t^* \gamma + \xi_t, \tag{22}$$

where y_t^*, Q_t^* are defined as above; $\gamma = -\xi_{T+1}$ and $G_t^* = \theta G_{t+1}^*$; and $G_{T+1}^* = 1$. Essentially, the conversion of v_t^* to white noise represents the additional twist in HS_4 as opposed to HS_3.

Both HS_3 and HS_4 retain all T observations, and are variations on HS_1; this would seem on theoretical grounds to be a potential improvement. Some empirical evidence is also available which points in this direction. This evidence comes from Monte Carlo work on the regression model with MA(1) errors and backward transformations which has been done by Balestra (1980), Park and Heikes (1983), and Harvey and McAvinchey (1981). In general, an estimator analogous to HS_4 is found to dominate an estimator which is analogous to HS_3, which in turn is found to dominate an estimator analogous to HS_1. However, an estimator analogous to the "exact transformation" estimator, to be discussed next, is found to dominate all of the above. Before turning to this "exact transformation" estimator we mention a potential improvement on HS_2. This estimator, denoted by HS_5, is formed as HS_2 but with the s dropped observations being made up with terms analogous to the additional terms that the Prais-Winsten estimator has over the Cochrane-Orcutt estimator in the conventional regression model with AR(p) errors.

We now consider the "exact transformation" estimator. This estimator takes as its point of departure the variance-covariance matrix of the composite disturbance v_t. For convenience we shall assume here, as above, that v_t is MA(1). In this case the variance-covariance matrix is $\sigma^2 V$ where V is the well known tri-diagonal matrix with $1 + \theta^2$ on the leading diagonal and $-\theta$ on the off diagonals. The "exact transformation" estimator proposes transformation of (6) by an upper-triangular matrix H constructed such that $H'H = V^{-1}$. In the MA(1) case we have developed this matrix H explicitly.

It may be expressed as $A^{-1/2}M$, where

$$M = \begin{bmatrix} \theta^0 a_{T-1}/a_T & \theta^1 a_{T-2}/a_T & \cdot & \cdot & \cdot & \cdot & \theta^{T-1} a_0/a_T \\ 0 & \theta^0 a_{T-2}/a_{T-1} & \cdot & & & & \theta^{T-2} a_0/a_{T-1} \\ \cdot & \cdot & & \cdot & \cdot & & \cdot \\ \cdot & & & \cdot & \cdot & \cdot & \\ \cdot & & & \cdot & \cdot & & \cdot \\ 0 & & \cdot & \cdot & \cdot & 0 & \theta^0 a_0/a_1 \end{bmatrix}, \quad (23)$$

$A = \text{diag}(a_{T-1}/a_T, a_{T-2}/a_{T-1}, \ldots, a_0/a_1)$, and $a_j = 1 + \theta^2 + \ldots + \theta^{2j}$.

In the case of higher order MA disturbances, the development of the explicit form of the H matrix appears tedious and indeed has not even been developed for the general case, i.e., the case of H not being restricted to be an upper-triangular matrix. In this case, we propose numerical inversion and decomposition of V. In practice, of course, to implement this exact estimator we need to estimate θ—as indeed we do to implement HS_1, HS_3, and HS_4. For this, we propose one of the following four methods: (i) estimate (6) by some consistent procedure such as NL2SLS and then use Box-Jenkins methods on the residuals, (ii) use a grid search over the range of possible values of θ, (iii) use a new method initially put forward by Ullah *et al.* (1985) for the regression case and recently extended to the simultaneous equations case by Ullah and Power (1985), and (iv) estimate (6) by some consistent procedure such as NL2SLS and then use a method of moments estimator due to Fuller (1976).

An alternative estimation method which is also based on the variance-covariance matrix $\sigma^2 V$ has been used by Fair (1984a). This method estimates the matrix $\sigma^2 V$ by a tri-diagonal matrix in which the elements on the leading diagonal are $(1/T)\sum v_t v_t$, where the v_t's are the residuals from consistent estimation of (6), and the off-diagonal elements are $(1/(T-1))\sum v_t v_{t-1}$. Having constructed this matrix estimate one then inverts and decomposes it as above to provide a suitable upper-triangular transformation matrix for (6).

All of the above single equation errors in variables estimators may be extended easily to the general case of an ARMA(p,q) structural disturbance u_t and a general ${}_{t-1}z_{t+j}$ or ${}_t z_{t+j}$ rational expectation, or indeed to the case where more than one rational expectation appears in the equation. The procedure is to replace all the rational expectations terms with their realized values, to quasi-difference the resulting equation until the composite disturbance is free of any AR component, to establish the order of the MA process describing the new composite disturbance, and to proceed as above.

5. ASYMPTOTIC PROPERTIES

In this section we derive the asymptotic properties of the various new single equation errors in variables estimators that we have outlined in Section 4. We begin by showing the asymptotic properties of the Hayashi and Sims (1983) estimator and then prove that our various new estimators are asymptotically equivalent. Finally, we present a consistency proof for the Fair (1984a) estimator.

Following Hayashi and Sims (1983), we restrict ourselves to the linear case; that is, we shall be considering the estimation of:

$$Y_t = Q_t\delta + v_t \tag{24}$$

under the identifying assumption

$$E[v_t \mid v_{t-N}, v_{t-N-1}, \ldots, F_t, F_{t-1}, \ldots] = 0 \tag{25}$$

and exclude the case of unit roots in the composite disturbance. In addition, we make the following two general assumptions: (i) there exists $H \geq h$ instrumental variables, F_t, and a positive integer N satisfying the identifying assumption (25), and (ii) Y_t, Q_t, and F_t are jointly covariance-stationary and ergodic processes.

We begin by considering the Hayashi and Sims (1983) estimator. Recall that above we have used the notation $E(vv') = \sigma^2 V$, where $V = \Gamma\Gamma'$, and that the transformation matrix was Γ^{-1}.

Proposition 1. Let $Q^+ = \Gamma^{-1}Q$, $Y^+ = \Gamma^{-1}Y$, and $v^+ = \Gamma^{-1}v$. Assume:
(a) $E[F_t'F_t]$ is positive definite.
(b) $E[Q_t^{+\prime}F_t]$ is of full row rank.
(c) $\lim(1/T)E[F'v^+v^{+\prime}F] = \Delta$ exists and is positive definite.
Then the Hayashi and Sims estimator $\delta_{HS} = (Q^{+\prime}F(F'F)^{-1}F'Q^+)^{-1}Q^{+\prime}F (F'F)^{-1}F'Y^+$, with known V, is consistent and $\sqrt{T}(\delta_{HS} - \delta)$ converges in distribution to a normal random variable with mean 0 and variance-covariance matrix $\sigma^2 \text{plim}[(Q^{+\prime}F/T)(F'F/T)^{-1}(F'Q^+/T)]^{-1}$.

Proof. *Consistency.* $\delta_{HS} = [Q^{+\prime}F(F'F)^{-1}F'Q^+]^{-1}Q^{+\prime}F(F'F)^{-1}F'Y^+$ implies that $\delta_{HS} = \delta + [Q^{+\prime}F(F'F)^{-1}F'Q^+]^{-1}Q^{+\prime}F(F'F)^{-1}F'v^+$. This, in turn, implies that $\text{plim}\,\delta_{HS} = \delta + \text{plim}[Q^{+\prime}F(F'F)^{-1}F'Q^+]^{-1}Q^{+\prime}F(F'F)^{-1} F'v^+$. By ergodicity, $\text{plim}(F'v^+/T) = E(F_t'v_t^+)$ and by the identifying assumption (25), $E(F_t'v_t^+) = 0$. Hence, if one uses this result and assumptions (a) and (b), then one can show that $\text{plim}\,\delta_{HS} = \delta$.

Asymptotic Normality. Consider $\sqrt{T}(\delta_{HS} - \delta) = [Q^{+\prime}F(F'F)^{-1}F'Q^+]^{-1} Q^{+\prime}F(F'F)^{-1}(F'v^+/\sqrt{T})$. By a central limit theorem used by Hansen

(1982), $(F'v^+/\sqrt{T}) \stackrel{\cdot}{\sim} N(0,\Delta)$ where $\Delta = \lim(1/T)E[F'v^+v^{+\prime}F]$. By this result and assumptions (a), (b), and (c), $\sqrt{T}(\delta_{HS} - \delta) \stackrel{\cdot}{\sim} N(0, \sigma^2 \text{plim}$ $[(Q^{+\prime}F/T)(F'F/T)^{-1}(F'Q^+/T)]^{-1}$.

We now consider the case where V is unknown and consequently has to be estimated. In this case it is not sufficient for asymptotic equivalence that V be estimated consistently. Before we can make such a statement we need to prove that: (i) $\text{plim}\{[(\hat{Q}^{+\prime} - Q^{+\prime})F]/T\} = 0$ and (ii) $\text{plim}\{[F'(\hat{\Gamma}^{-1} - \Gamma^{-1})]/\sqrt{T}\} = 0$, where $\hat{Q}^+$ is the estimated Γ^{-1} multiplied by Q and $\hat{\Gamma}^{-1}$ is the estimate of Γ^{-1}. Hayashi and Sims (1983) have proved that these conditions are satisfied for their implementations of the δ_{HS} estimator, i.e., HS_1 and HS_2, under the following assumptions:

I.1 Y_t, Q_t and F_t are jointly stationary, ergodic, of finite variance, linearly regular and of maximal rank.

I.2 $E[v_t \mid F_{t-s}] = 0$ for all $s \geq 0$.

I.3 $E[v_t v_s \mid F_{t-s}] = E[v_t v_s]$ for all $s \geq 0$.

I.4 For $Q_t^s = a(L^{-1})Q_t$, $E[Q_t^{s\prime}F_t]$ is of full row rank.

I.5 The "backwards innovation" ξ_t in v_t is independent of F_t.

II. $T^{1/2} \max_s\{\mid g_s^T - a_s \mid /B(s)\}$ is bounded in probability, where $B(s)$ is positive and absolutely summable.

We now consider our new estimators and the Fair (1984a) estimator.

Proposition 2. The asymptotic properties of HS_3 are identical to those of HS_1, i.e. plim $\sqrt{T}(HS_3 - HS_1) = 0$; hence they are also identical to those of δ_{HS}.

Proof. Compare the transformation matrix involved in HS_3 with that for HS_1. This matrix has s additional rows and, in addition, the zero elements in the upper triangular part of the matrix are replaced by additional terms as described above. As T goes to infinity, we may, following standard arguments, ignore the s additional observations corresponding to the additional s rows. Similarly, we may argue that as T goes to infinity, the additional upper triangular non-zero terms tend to zero and may be ignored.

Proposition 3. The asymptotic properties of HS_4 are identical to those for HS_3, i.e. plim $\sqrt{T}(HS_4 - HS_3) = 0$; hence they are also identical to those of δ_{HS}.

Proof. Compare estimators HS_4 and HS_3. The only difference between them lies in the fact that HS_4 includes an artificial regressor, G_t^*, in the MA(1) example given in Section 4. As T tends to infinity, the terms in this artificial regressor tend to zero and hence asymptotically this regressor may be ignored.

Proposition 4. The asymptotic properties of HS_5 are identical to those for HS_2, i.e. plim $\sqrt{T}(HS_5 - HS_2) = 0$; hence they are also identical to those of δ_{HS}.

Proof. Compare the transformation matrices for HS_5 and HS_2. The only difference is the additional s rows. However, as T tends to infinity, the s observations corresponding to these additional s rows may be ignored.

We now come to the "exact transformation" estimator and the Fair (1984a) estimator. These estimators, it will be recalled, involve estimation of the matrices V and $\sigma^2 V$ respectively. Given the fact that Hayashi and Sims (1983) have shown that the two sufficient conditions (i) and (ii) are satisfied for their implementations, it follows that it will be sufficient for us to show the consistency of these estimators' estimates of V and $\sigma^2 V$ respectively to ensure that their asymptotic properties are equivalent to those of δ_{HS}.

In the case of the "exact transformation" estimator, the elements of the V matrix are functions of the moving average parameters, and therefore the V matrix will be estimated consistently as long as the moving average parameters are estimated consistently. The consistency of the moving average parameter estimates follows from the fact that we use Box-Jenkins procedures on the consistently estimated residuals of (24), i.e., Box-Jenkins procedures are either maximum likelihood or asymptotically equivalent to maximum likelihood. Alternatively, in the case where we estimate the moving average parameters using the method of Ullah and Power (1985), consistency may be proved as an extension of the results in Ullah *et al.* (1985).

Finally, we consider the Fair (1984a) estimator which estimates the $\sigma^2 V$ matrix slightly more formally.

Proposition 5. Assume (a), of Proposition 1 and also the following:
(d) $E[Q_t' F_t]$ is of full row rank.
(e) $E[Q_t' v_t]$ is positive definite.
Then the Fair (1984a) estimator of the $\sigma^2 V$ matrix is consistent.

Proof. We consider the MA(1) case for simplicity. Recall that the disturbance term of equation (24) is denoted v_t, that the backward representation of v_t is $v_t = \epsilon_t - \theta\epsilon_{t-1}$, and that we have referred above to the consistently estimated residuals from this equation as v_t, where the consistent estimator was 2SLS. The proof will be in four parts. Together they are sufficient.

(I.) We first prove: $\text{plim}[(1/T)v'v] = (1+\theta^2)\sigma^2$.

$\text{plim}[(1/T)v'v] = \text{plim}[(1/T)\sum v_t^2] = \text{plim}[(1/T)\sum \epsilon_t^2] + \theta^2 \text{plim} [(1/T)\sum \epsilon_{t-1}^2] - 2\theta\, \text{plim}[(1/T)\sum \epsilon_t\epsilon_{t-1}] = (1+\theta^2)\sigma^2$.

(II.) We next prove: $\text{plim}[(1/(T-1))\sum v_t v_{t-1}] = -\theta\sigma^2$.

$\text{plim}[(1/(T-1))\sum v_t v_{t-1}] = \text{plim}[(1/(T-1))\sum \epsilon_t\epsilon_{t-1}] - \theta\ \text{plim}[(1/(T-1))\sum \epsilon_t\epsilon_{t-2}] - \theta\ \text{plim}[(1/(T-1))\sum \epsilon_{t-1}^2] + \theta^2\text{plim}[(1/(T-1))\sum \epsilon_{t-1}\epsilon_{t-2}] = -\theta\sigma^2$.

(III.) We next prove: $\text{plim}[(1/T)v'v] = \text{plim}[(1/T)v'v]$.
If $M = [I - Q(Q'F(F'F)^{-1}F'Q)^{-1}Q'F(F'F)^{-1}F']$ then $T^{-1}v'v = T^{-1}v'Mv = T^{-1}v'v - (v'Q/T)[(Q'F/T)(F'F/T)^{-1}(F'Q/T)]^{-1}(Q'F/T)(F'F/T)^{-1}(F'v/T)$. This implies that $\text{plim}[(1/T)v'v] = \text{plim}[(1/T)v'v] - \text{plim}\{(v'Q/T)[(Q'F/T)(F'F/T)^{-1}(F'Q/T)]^{-1} \times (Q'F/T)(F'F/T)^{-1}(F'v/T)\}$.
Consider $\text{plim}(F'v/T)$. By ergodicity $\text{plim}(F'v/T) = E[F_t'v_t]$ and by the identifying assumption, $E[F_t'v_t] = 0$. Therefore these results and the assumptions (a), (d), and (e) imply that $\text{plim}[(1/T)v'v] = \text{plim}[(1/T)v'v]$.

(IV.) We finally prove: $\text{plim}[(1/(T-1))\sum v_t v_{t-1}] = \text{plim}[(1/(T-1))\sum v_t v_{t-1}]$.
Let B be a $T \times T$ matrix whose first row consists of zeros and whose i^{th} row ($i \geq 2$) has a 1 in the $(i-1)^{th}$ position and zeros elsewhere. Then $\text{plim}[(1/(T-1))\sum v_t v_{t-1}] = \text{plim}[(1/T)v'Bv] = \text{plim}[(1/T)v'MBMv] = \text{plim}[(1/T)v'Bv] = \text{plim}[(1/(T-1))\sum v_t v_{t-1}]$, where the third equality follows by arguments similar to those used in (III).

ACKNOWLEDGMENTS

This paper constitutes part of my Ph.D. thesis work at the University of Western Ontario. As such, I should like to thank my thesis committee, Aman Ullah, R. A. L. Carter, and Allan Gregory, for all their help and encouragement. I should also like to thank Mike Veall and Vicky Zinde-Walsh for many useful conversations. In addition, I am very grateful to an editor and an anonymous referee for valuable comments regarding the exposition of this paper.

REFERENCES

Anderson, T. W. (1971), *The Statistical Analysis of Time Series*. New York: Wiley and Sons.

Balestra, P. (1980), "A note on the exact transformation associated with the first-order moving average process". *Journal of Econometrics* **14**, 381–394.

Brown, B. W., and S. Maital (1981), "What do economists know? An empirical study of experts' expectations". *Econometrica* **49**, 491–504.

Cumby, R. E., J. Huizinga, and M. Obstfeld (1983), "Two-step two-stage least squares estimation in models with rational expectations". *Journal of Econometrics* **21**, 333–355.

Fair, R. C. (1984a), "Effects of expected future government deficits on current economic activity". NBER Working Paper No. 1293.

Fair, R. C. (1984b), "The use of expected future variables in macroeconometric models". NBER Working Paper No. 1445.

Fair, R. C., and J. Taylor (1983), "Solution and maximum likelihood estimation of dynamic rational expectations models". *Econometrica* **51**, 1169–1185.

Flood, R. P., and P. M. Garber (1980), "A pitfall in estimation of models with rational expectations". *Journal of Monetary Economics* **6**, 433–435.

Fuller, W. A. (1976), *Introduction to Statistical Time Series*. New York: Wiley and Sons.

Hakkio, C. (1981), "Expectations and the forward exchange rate". *International Economic Review* **22**, 663–678.

Hansen, L. P. (1982), "Large sample properties of generalised method of moments estimators". *Econometrica* **50**, 1029–1054.

Hansen, L. P., and T. J. Sargent (1980), "Formulating and estimating dynamic linear rational expectations models". *Journal of Economic Dynamics and Control* **2**, 7–46.

Hansen, L. P., and T. J. Sargent (1982), "Instrumental variables procedures for estimating linear rational expectations models". *Journal of Monetary Economics* **9**, 263–296.

Harvey, A. C., and I. D. McAvinchey (1981), "On the relative efficiency of various estimators of regression models with moving average disturbances". In *Proceedings of the Econometric Society Meetings*, 1979, ed. E. G. Charatsis, pp. 105–118. Amsterdam: North-Holland.

Hayashi, F., and C. Sims (1983), "Nearly efficient estimation of time series models with predetermined, but not exogenous instruments". *Econometrica* **51**, 783–798.

McCallum, B. T. (1976), "Rational expectations and the natural rate hypothesis: some consistent estimates". *Econometrica* **44**, 43–52.

McCallum, B. T. (1979), "Topics concerning the formulation, estimation and use of macroeconomic models with rational expectations". *American Statistical Association Proceedings of the Business and Economics Section*, 65–72.

MacDonald, G. M., and J. G. MacKinnon (1985), "Convenient methods for estimation of linear regression models with MA(1) errors". *Canadian Journal of Economics* **18**, 106–116.

Newey, W. K., and K. D. West (1985), "A simple, positive definite, heteroskedasticity and autocorrelation consistent covariance matrix". Princeton University Discussion Paper 92.

Park, C. Y., and R. G. Heikes (1983), "A note on Balestra's (1980) approximate estimator for the first-order moving average process". *Journal of Econometrics* **21**, 387–388.

Phillips, A. W. (1966), "The estimation of systems of difference equations with moving average disturbances". Paper read to the Econometric Society Meetings, San Francisco, reprinted in *Stability and Inflation*, 1978, ed. A. R. Berstrom *et al.* New York: Wiley and Sons.

Pollock, D. S. G. (1979), *The Algebra of Econometrics*. New York: Wiley and Sons.

Ullah, A., and S. Power (1985), "Estimation of linear models with moving average processes". Paper presented at the Second International Symposium on Probability and Information Theory, McMaster University.

Ullah, A., H. D. Vinod, and R. S. Singh (1985), "Estimation of linear models with moving average disturbances". *Journal of Quantitative Economics*, forthcoming.

Wallis, K. F. (1980), "Econometric implications of the rational expectations hypothesis". *Econometrica* **48**, 49–73.

Jean-Marie Dufour [1]

LINEAR WALD METHODS FOR INFERENCE ON COVARIANCES AND WEAK EXOGENEITY TESTS IN STRUCTURAL EQUATIONS

ABSTRACT

Inference about the vector of covariances between the stochastic explanatory variables and the disturbance term of a structural equation is an important problem in econometrics. For example, one may wish to test the independence between stochastic explanatory variables and the disturbance term. Tests for the hypothesis of independence between the full vector of stochastic explanatory variables and the disturbance have been proposed by several authors. When more than one stochastic explanatory variable is involved, it can be of interest to determine whether all of them are independent of the disturbance and, if not, which ones are. We develop simple large-sample methods which allow us to construct confidence regions and test hypotheses concerning any vector of linear transformations of the covariances between the stochastic explanatory variables and the disturbance of a structural equation. The main method described is a generalized Wald procedure which simply requires two linear regressions. No nonlinear estimation is needed. Consistent tests for weak exogeneity hypotheses are derived as special cases.

1. INTRODUCTION

Inference about the vector of covariances between the stochastic explanatory variables and the disturbance term of a structural equation is an important problem in econometrics. For example, one may wish to test whether a set of stochastic explanatory variables are statistically independent of the

[1] Département de sciences économiques, Université de Montréal, C.P. 6128, succursale A, Montréal, Québec, H3C 3J7

I. B. MacNeill and G. J. Umphrey (eds.), Time Series and Econometric Modelling, 317–338.

disturbance of a structural equation, i.e., whether the stochastic explanatory variables considered can be treated as "exogenous" (or "predetermined"). In particular, it is well known that independence between explanatory variables and disturbances is usually needed to ensure that standard inference procedures, like ordinary least squares or F-tests, are appropriate in linear models. Furthermore, a number of economic hypotheses can be formulated in terms of the independence between stochastic explanatory variables and disturbances.[2]

Tests for the hypothesis of independence between a vector of stochastic explanatory variables and a disturbance term were proposed by several authors; see Durbin (1954), Wu (1973, 1974), Revankar and Hartley (1973), Farebrother (1976), Hausman (1978), Revankar (1978), Kariya and Hodoshima (1980), Richard (1980), and Holly and Sargan (1982).[3] These articles deal especially with the problem of testing whether the full vector of stochastic explanatory variables is independent of the disturbance. When more than one stochastic explanatory variable are involved, it is often necessary to determine whether all of them are independent of the disturbances and, if not, which ones are. This can be useful, for example, to check the specification of a simultaneous equation model (e.g., block recursiveness assumptions) and to get more efficient estimators for such models.

Tests for the hypothesis of independence between a subset of stochastic explanatory variables and the disturbance in a structural equation have been proposed by a number of authors: Hwang (1980) and Smith (1984) studied likelihood ratio (LR) tests, Hausman and Taylor (1981a), Spencer and Berk (1981) and Wu (1983b) proposed extensions of the tests previously studied by Wu (1973) and by Hausman (1978), while Engle (1982) derived Lagrange multiplier (LM) tests.

Each of these procedures has important drawbacks, either practical or theoretical. Some of them require nonlinear estimation, e.g., LR tests and certain forms of the LM tests. All of them require a separate estimation for each null hypothesis tested. It is difficult to construct confidence intervals for the covariances of interest because covariance estimates or their standard errors are not typically produced.

[2] See Wu (1973). For an example of a structural equation where the stochastic explanatory variables can be treated as "exogenous", see Zellner *et al.* (1966).

[3] Further useful discussions and extensions of these tests are provided by Bronsard and Salvas-Bronsard (1984), Engle (1982, 1984), Gouriéroux and Trognon (1984), Hausman and Taylor (1980, 1981a,b), Holly (1980, 1982a,b, 1983), Holly and Monfort (1982), Nakamura and Nakamura (1980, 1981), Plosser *et al.* (1982), Reynolds (1982), Riess (1983), Ruud (1984), Tsurumi and Shiba (1984), Turkington (1980), White (1982), and Wu (1983a).

Hausman-type tests are better viewed as consistency tests. By comparing an efficient estimator under the null hypothesis with a consistent estimator under the alternative hypothesis, one checks whether the constrained estimator is consistent (see Holly, 1982a,b; Hausman and Taylor, 1980, 1981b). When testing exogeneity, this is not equivalent to testing independence between possible endogenous variables and the disturbance term: the condition tested is weaker (unless special assumptions hold) and the test may not be consistent. (This is easy to see from Hausman and Taylor (1980, 1981b) and Wu (1983b).) Even though this condition may be sufficient to ensure the consistency or the efficiency of the constrained estimator, it is not generally sufficient to guarantee the validity of inferences obtained from the model by treating the regressors whose exogeneity is in doubt as being exogenous: tests and confidence intervals pertaining to the various coefficients of the model may not have the correct levels, even asymptotically.[4] In many if not most practical situations, the relevant hypothesis is whether one can treat some stochastic explanatory variables as being exogenous for all purposes of inference (i.e., the independence assumption).

In this paper, we consider a single linear structural equation and develop a class of linear Wald-type procedures which allow us to construct confidence regions as well as to test any set of linear restrictions on the vector of covariances between the stochastic explanatory variables and the disturbance term in the equation. Besides a set of instrumental regressions, all that is needed is a simple linear regression which yields consistent estimates of both the structural coefficients in the equation and the relevant vector of covariances. The asymptotic covariance matrix of the coefficients is then easily obtained. Using these results, one can test any set of linear restrictions on the covariances and construct confidence regions. Cross-restrictions between the structural coefficients and the covariances may also be tested. Special cases of this family of tests include tests of zero restrictions on the covariances, either for individual covariances or subvectors of covariances. In particular, one can compute in a routine way asymptotic "t-values" for each covariance, an especially convenient instrument to explore the recursiveness properties of a model. All the tests suggested are consistent.

Because they are based on consistent asymptotically normal estimators different from the maximum-likelihood estimators (Wald, 1943), the tests developed here should be viewed as generalized Wald tests rather than Wald tests in the usual sense (see Stroud, 1971; Szroeter, 1983). We will not need the information matrix associated with the maximum likelihood estimators. As we shall see below, the tests proposed can be obtained as a byproduct of

[4] See White (1982, p. 16), and Breusch and Mizon (comment to Ruud, 1984, p. 249).

the estimation of a structural equation by any instrumental-variable method (including two-stage least squares). They thus have a natural complementary with the latter estimation method.

In Section 2, we formulate the model considered and the assumptions used. In Section 3, we describe the procedures proposed and formulate the theorems underlying them. In Section 4, we examine three important special situations: the case where we want to test independence between the full vector of stochastic explanatory variables and the disturbance term, the one where a subset of stochastic explanatory variables is taken *a priori* as being exogenous and the case where the matrix of instruments includes all the fixed (or exogenous) regressors in the equation considered. In Section 5, we discuss econometric applications. Finally, in Section 6, we provide the proofs of the theorems.

2. FRAMEWORK

We consider the model described by the following assumptions.

ASSUMPTION 1:

$$y = Y\beta + Z_1\gamma + u, \tag{2.1}$$

where y is a $T \times 1$ random vector, u is a $T \times 1$ vector of disturbances, Y is a $T \times G$ matrix of stochastic explanatory variables, Z_1 is a $T \times K_1$ non-stochastic matrix of rank K_1, β and γ are $G \times 1$ and $K_1 \times 1$ vectors of coefficients.

ASSUMPTION 2:

$$Y = Z\Pi + V, \tag{2.2}$$

where Z is a $T \times K$ non-stochastic matrix of rank K, Π is a $K \times G$ matrix of coefficients and V is a $T \times G$ matrix of disturbances. Furthermore, we will denote by y_k, Π_k and w_k the kth columns of the matrices Y, Π and V respectively $(1 \leq k \leq G)$:

$$Y = [y_1, \ldots, y_G],\ \Pi = [\Pi_1, \ldots, \Pi_G],\ V = [w_1, \ldots, w_G]. \tag{2.3}$$

ASSUMPTION 3: The rows (u_t, v_t'), $t = 1, \ldots, T$, of the matrix $[u : V]$ are independent and normally distributed with mean zero and non-singular covariance matrix

$$\Sigma = \begin{bmatrix} \sigma_{00} & \delta' \\ \delta & \Sigma_{22} \end{bmatrix}, \tag{2.4}$$

where

$$\delta = (\sigma_{01}, \sigma_{02}, \ldots, \sigma_{0G})',\ \Sigma_{22} = [\sigma_{jk}]_{\ j,\ k=1,\ldots,G}. \tag{2.5}$$

ASSUMPTION 4: Let $Z = [Z_{11} : Z_2]$ and $\Pi = [\Pi_{11}' : \Pi_2']'$, where Z_2 is the $T \times K_2$ matrix of non-stochastic variables excluded from equation (2.1), Π_2 is the $G \times K_2$ corresponding matrix of coefficients, Z_{11} is a set of variables included in Z_1, so that $Y = Z_{11}\Pi_{11} + Z_2\Pi_2 + V$, $\text{rank}(\Pi_2) = G$ and $T > 2G + K_1$.
[This condition ensures identification of the coefficients of equation (2.1); see Fisher (1966, p. 53). Note also that Z_1 is not constrained to be a submatrix of Z.]

ASSUMPTION 5: The matrix $\frac{1}{T}Z'Z$ converges, as $T \to \infty$, to a positive definite matrix Q_z.

ASSUMPTION 6: The matrices $\frac{1}{T}Z_1'Z_1$ and $\frac{1}{T}Z'Z_1$ converge, as $T \to \infty$, to the matrices Q_{11} and Q_1 respectively, where Q_{11} is positive definite.

We want to test some set of linear restrictions on the parameter vector δ, i.e., a hypothesis of the type

$$H_0 : H\delta = d_0, \tag{2.6}$$

where H is an $r \times G$ matrix of rank $r \leq G$ and d_0 is a fixed $r \times 1$ vector. Since the vectors $(u_t, v_t')'$, $t = 1, \ldots, T$, are i.i.d. normal, we obtain by regressing u_t on v_t:

$$u = Va + e, \tag{2.7}$$

where $a = \Sigma_{22}^{-1}\delta$ and the vector e is $N[0, \sigma_e^2 I_T]$ independent of all the elements of V.[5] Then, substituting (2.7) into (2.1), we get

$$y = Y\beta + Z_1\gamma + Va + e, \tag{2.8}$$

where the disturbance vector e is independent of all the regressors. The latter formulation illustrates clearly that the existence of correlation between some of the regressors and the disturbance term in an econometric relationship, as generated, for example, by simultaneous equations, may be viewed as a problem of omitted variables. If the matrix V were observed, we would test any set of linear restrictions on the coefficients β, γ and a in equation (2.8) by standard F-tests, and these tests would be exact in small samples. In particular, linear hypotheses regarding the parameter vector a could be tested by using the least squares estimate $\hat{a}$ obtained from (2.8). Furthermore, if Σ_{22} were also known, the transformation $\delta = \Sigma_{22}a$ would allow one to test $H_0 : H\delta = d_0$ by a standard F-test. The difficulty, of course, is that neither V nor Σ_{22} are known. We also note that, although hypotheses regarding δ have relatively direct and intuitive interpretations (e.g., in

[5] This transformation is also used by Revankar (1973), Revankar and Hartley (1973) and Reynolds (1982).

terms of independence), the auxiliary parameter vector $a = \Sigma_{22}^{-1}\delta$ itself may be of interest. One may wish to test linear restrictions on a directly in the reparameterized model (2.8). In any event, we will deal with both problems.

We will first consider the problem of testing arbitrary linear restrictions on the parameter vector $\alpha = (\beta', \gamma', a')'$ and then restrictions on the covariance vector δ. In each case, we will first define a vector of linear consistent asymptotically normal estimators, derive the asymptotic covariance matrix and propose generalized Wald tests. In particular, we will derive the asymptotic distribution of the covariance estimator $\hat{\delta}$ under both the null and the alternative hypotheses. As a special case, it will be straightforward to test zero restrictions on δ, for example, $H_0 : \delta_1 = 0$ where $\delta = (\delta_1', \delta_2')'$. In the context of the model considered here, the hypothesis $\delta_1 = 0$ is equivalent to the independence between Y_1 and u, where $Y = [Y_1 \vdots Y_2]$, or the weak exogeneity of Y_1 inside equation (2.1).[6] Further, from the same results, it is easy to construct a confidence region for any element or subvector of δ or α.

3. DESCRIPTION OF THE TESTS

In equation (2.8), replace the disturbance matrix V by the corresponding ordinary least squares (OLS) residuals

$$\hat{V} = Y - Z\hat{\Pi}, \tag{3.1}$$

where $\hat{\Pi} = (Z'Z)^{-1}Z'Y$. We obtain in this way the equivalent equation

$$y = Y\beta + Z_1\gamma + \hat{V}a + e^* = X\alpha + e^*, \tag{3.2}$$

where $X = [Y : Z_1 : \hat{V}]$, $\alpha = (\beta', \gamma', a')'$ and

$$e^* = Z(\hat{\Pi} - \Pi)a + e. \tag{3.3}$$

Also let

$$\hat{\Sigma}_{22} = \hat{V}'\hat{V}/T. \tag{3.4}$$

Under Assumptions 2 through 6, we have

$$\text{plim}\frac{Z'V}{T} = 0, \quad \text{plim}\ \hat{\Sigma}_{22} = \Sigma_{22} \tag{3.5}$$

---

[6] For a general discussion of exogeneity and related notions, see Engle *et al.* (1983).

(where plim refers to the probability limit as $T \to \infty$), hence

$$Q_x = \text{plim}\frac{X'X}{T} = \begin{bmatrix} \Pi' Q_z \Pi + \Sigma_{22} & \Pi' Q_1 & \Sigma_{22} \\ Q_1' \Pi & Q_{11} & 0' \\ \Sigma_{22} & 0 & \Sigma_{22} \end{bmatrix} \tag{3.6}$$

and

$$Q_{zx} = \text{plim}\frac{Z'X}{T} = [Q_z \Pi : Q_1 : 0], \tag{3.7}$$

where $\text{rank}(Q_x) \equiv L = 2G + K_1$. Consider the OLS estimate of α obtained from (3.2):

$$\hat{\alpha} = (X'X)^{-1} X'y. \tag{3.8}$$

Under the assumptions made, this estimate is unique with probability one. Further, the asymptotic distribution of $\hat{\alpha}$ is given by the following theorem. (The proofs of the theorems are given in Section 5.)

Theorem 1. Suppose that Assumptions 1 through 6 are satisfied, and let matrix Q_x defined in (3.6) be non-singular. Then the estimator $\hat{\alpha}$ given in (3.8) is consistent for α and $\sqrt{T}(\hat{\alpha} - \alpha)$ has a normal limiting distribution with mean zero and covariance matrix

$$\begin{aligned} \Sigma_\alpha &= Q_x^{-1} \left[\sigma_e^2 Q_x + \rho Q_{zx}' Q_z^{-1} Q_{zx} \right] Q_x^{-1} \\ &= \sigma_e^2 Q_x^{-1} + \rho Q_x^{-1} Q_{zx}' Q_z^{-1} Q_{zx} Q_x^{-1}, \end{aligned} \tag{3.9}$$

where Q_{zx} is given by (3.7) and

$$\rho = a' \Sigma_{22} a = \delta' \Sigma_{22}^{-1} \delta. \tag{3.10}$$

Further, the statistics

$$\hat{\sigma}_e^2 = (y - X\hat{\alpha})'(y - X\hat{\alpha})/T \tag{3.11}$$

and

$$\hat{\Sigma}_\alpha = \left(\frac{X'X}{T}\right)^{-1} \left[\hat{\sigma}_e^2 \left(\frac{X'X}{T}\right) + \hat{\rho} \left(\frac{X'Z}{T}\right) \left(\frac{Z'Z}{T}\right)^{-1} \left(\frac{Z'X}{T}\right) \right] \left(\frac{X'X}{T}\right)^{-1} \tag{3.12}$$

are consistent estimators of σ_e^2 and Σ_α, where $\hat{\rho} = \hat{a}' \hat{\Sigma}_{22} \hat{a}$, $\hat{\Sigma}_{22} = \hat{V}'\hat{V}/T$ and $\hat{a}$ is the estimate of a from $\hat{\alpha}$.

We can test any set of linear restrictions on the vector α, such as $M\alpha = m_0$, where M is a $\nu \times L$ matrix of rank $\nu \leq L$ and m_0 is a $\nu \times 1$ fixed vector, by using a critical region of the form $\{S(M, m_0) \geq c\}$, where

$$S(M, m_0) = T(M\hat{\alpha} - m_0)'(M\hat{\Sigma}_\alpha M')^{-1}(M\hat{\alpha} - m_0) \tag{3.13}$$

and c is a constant which depends on the level of the test. The asymptotic distribution of the test statistic $S(M, m_0)$ is chi-square with ν degrees of freedom under the null hypothesis.

Since the coefficient a is of special interest here, it will be useful to summarize the asymptotic properties of $\hat{a}$ by the following corollary.

Corollary 1.1. Under the assumptions of Theorem 1, the subvector $\hat{a}$ of $\hat{\alpha} = (\hat{\beta}', \hat{\gamma}', \hat{a}')'$ is a consistent estimator of a and $\sqrt{T}(\hat{a} - a)$ has a normal limiting distribution with mean zero and covariance matrix:

$$\Sigma_a = A_2 \left[\sigma_e^2 Q_x + \rho Q'_{zx} Q_z^{-1} Q_{zx}\right] A'_2, \tag{3.14}$$

where $A_2 = \text{plim}(C_2)$ and C_2 is the $G \times (2G + K_1)$ matrix such that

$$\left(\frac{X'X}{T}\right)^{-1} = \begin{bmatrix} C_1 \\ \\ C_2 \end{bmatrix}. \tag{3.15}$$

Further, the submatrix

$$\hat{\Sigma}_a = C_2 \left[\hat{\sigma}_e^2 \left(\frac{X'X}{T}\right) + \hat{\rho} \left(\frac{X'Z}{T}\right) \left(\frac{Z'Z}{T}\right)^{-1} \left(\frac{Z'X}{T}\right)\right] C'_2 \tag{3.16}$$

in (3.12) is a consistent estimator of Σ_a.

Of course, tests of linear restrictions on a are special cases of the tests given by (3.13). However, if our interest lies in δ rather than $a = \Sigma_{22}^{-1}\delta$, the estimator directly relevant to us is not $\hat{a}$. We need an estimator of δ. Since $\hat{a}$ and $\hat{\Sigma}_{22}$ are consistent estimators of a and Σ_{22}, $\hat{\delta} = \hat{\Sigma}_{22}\hat{a}$ is a consistent estimator of δ. The asymptotic distribution of δ is given by the following theorem.

Theorem 2. Under the assumptions of Theorem 1, the estimator $\hat{\delta} = \hat{\Sigma}_{22}\hat{a}$ is consistent for δ and the vector $\sqrt{T}(\hat{\delta} - \delta)$ has a normal limiting distribution, as $T \to \infty$, with mean zero and covariance matrix

$$\Sigma_\delta = \Sigma_{22}\Sigma_a\Sigma_{22} + \rho\Sigma_{22} + \delta\delta', \tag{3.17}$$

where Σ_a is given by (3.16). Further, a consistent estimator of Σ_δ is provided by

$$\hat{\Sigma}_\delta = \hat{\Sigma}_{22}\hat{\Sigma}_a\hat{\Sigma}_{22} + \hat{\rho}\hat{\Sigma}_{22} + \hat{\delta}\hat{\delta}', \tag{3.18}$$

where $\hat{\rho}$ and $\hat{\Sigma}_{22}$ are defined in Theorem 1.

Consequently, we can test the hypothesis $H_0 : H\delta = d_0$, where H is a $r \times G$ matrix of rank $r \leq G$ and d_0 is a fixed $r \times 1$ vector, by using a critical region of the form $\{W(H, d_0) \geq c\}$, where

$$W(H, d_0) = T(H\hat{\delta} - d_0)'(H\hat{\Sigma}_\delta H')^{-1}(H\hat{\delta} - d_0) \tag{3.19}$$

and c depends on the level of the test. The asymptotic distribution of the statistic $W(H, d_0)$, under H_0, is chi-square with r degrees of freedom. Again, this test is valid for large samples.

Concerning the power of the above tests, we can make the important observation that they are consistent whenever $M\alpha \neq m_0$ or $H\delta \neq d_0$ (see Section 6.5).[7] Besides, by considering complements of the critical regions described above, we can obtain confidence regions for $M\alpha$ or $H\delta$, for example confidence intervals for the individual covariances in δ.

3. SPECIAL CASES

We will now examine three cases of special interest. First, consider the situation where the null hypothesis is $H_0 : \delta = 0$ or equivalently, $H_0' : a = 0$. Under H_0, we can rewrite equation (3.2) as

$$y = Y\beta + Z_1\gamma + \hat{V}a + e, \tag{4.1}$$

where e follows a $N[0, \sigma_e^2 I_T]$ distribution and is independent of both Y and $\hat{V}$. Then the standard F-statistic for testing $a = 0$ is

$$F = \frac{\hat{a}'(\hat{V}'M_1\hat{V})\hat{a}/G}{\hat{e}'\hat{e}/(T - K_1 - 2G)}, \tag{4.2}$$

where $M_1 = I_T - X_1(X_1'X_1)^{-1}X_1'$ and $X_1 = [Y : Z_1]$; under H_0, F follows a Fisher distribution with $(G, T - K_1 - 2G)$ degrees of freedom. The resulting test is exact rather than asymptotic.[8] It is not equivalent (even asymptotically) to the test of $a = 0$ based on the statistic $S_0 = T\hat{a}'\hat{\Sigma}_a^{-1}\hat{a}$, obtained

[7] This property is especially important in view of Holly's (1982a) discussion of Hausman-type tests.

[8] One can see easily that this test is equivalent to a test suggested by Wu (1973, T_2 statistic) and, in a different form, by Hausman (1978, eq. 2.23), except that Z_1 is not necessarily a submatrix of Z. On alternative forms of the Wu-Hausman test, see Nakamura and Nakamura (1981).

from (3.13). The main difference is that $\hat{\rho}$ is set to zero in the estimator of Σ_a in (3.16). This restriction is justified under H_0, for then $\rho = a'\Sigma_{22}a = 0$. If we write F in the form

$$F = \frac{1}{G}\left(T\hat{a}'\tilde{\Sigma}_a^{-1}\hat{a}\right), \tag{4.3}$$

where

$$\tilde{\Sigma}_a = \hat{\sigma}_0^2(\hat{V}'M_1\hat{V}/T)^{-1}, \;\; \hat{\sigma}_0^2 = \hat{e}'\hat{e}/(T - K_1 - 2G), \tag{4.4}$$

we see easily that the statistics F and S_0/G are asymptotically identical under H_0 (since $\hat{\rho} \to 0$). Nevertheless, under the alternative, this equivalence does not hold because $\hat{\rho}$ does not, in general, converge to zero.

The second problem we wish to examine is to test whether a subset of the variables in Y are independent of u, conditional on the assumption that the others are independent of u. More precisely, given $Y = [Y_1 : Y_2]$, we want to test whether Y_1 and u are independent, knowing that Y_2 and u are independent. To do this, we can simply include Y_2 in Z_1 and reshape equation (2.1) accordingly:

$$y = Y_1\beta_1 + Z_3\gamma_3 + u, \tag{4.5}$$

where $Z_3 = [Y_2 : Z_1]$, $\gamma_3 = (\beta_2', \gamma')'$, and $\beta = (\beta_1', \beta_2')'$ is the partition of β corresponding to $[Y_1 : Y_2]$. We then proceed as previously on the transformed model.

Finally, consider the important case where the matrix Z_1 is a submatrix of Z, say $Z = [Z_1 : Z_2]$. This is probably the most frequent situation when (2.1) is viewed as a "structural equation" (presumably inside some system of equations) and (2.2) represents the "reduced-form equation" for the endogenous variables appearing on the right-hand side of (2.1). In this case, the estimates $\hat{\beta}$ and $\hat{\gamma}$, obtained from the regression given by (3.3), are the two-stage least squares (2SLS) estimates of β and γ. To see this, rewrite equation (3.2) as

$$y = \hat{Y}\beta + Z_1\gamma + \hat{V}a^* + e^*, \tag{4.6}$$

where $a^* = a + \beta$. By the orthogonality relations $\hat{V}'\hat{Y} = 0$ and $\hat{V}'Z_1 = 0$, the estimates of β and γ obtained by OLS from (4.6) are identical to those obtained from the regression

$$y = \hat{Y}\beta + Z_1\gamma + e^{**}. \tag{4.7}$$

They are thus identical to the 2SLS estimates of β and γ, showing clearly that the linear Wald tests described above have a natural complementary with the estimation of a structural equation by 2SLS.

In the same special case, the estimate $\hat{\delta}$ used in Theorem 2 may be derived in a second interesting manner. Using again the orthogonality relations, we see that

$$\hat{a} = \hat{\Sigma}_{22}^{-1}\left(\frac{1}{T}\hat{V}y\right),$$

hence

$$\hat{a} = \hat{a}^* - \hat{\beta}$$

and

$$\hat{\delta} = \frac{1}{T}\hat{V}'y - \hat{\Sigma}_{22}\hat{\beta}.$$

Further, substitute (2.2) into (2.1) to get the reduced-form equation for y:

$$y = Z\Pi\beta + Z_1\gamma + v_0,$$

where $v_0 = V\beta + u$. If we denote the tth element of v_0 by $v_{0t} = v_t'\beta + u_t$ and define $\omega_0 = E[v_t v_{0t}]$, we have

$$\delta = \omega_0 - \Sigma_{22}\beta.$$

Since ω_0 can be consistently estimated by $\hat{\omega} = \frac{1}{T}\hat{V}'\hat{v}_0$, where $\hat{v}_0$ is the vector of residuals from the regression of y on Z, this suggests the following estimate of δ:

$$\tilde{\delta} = \hat{\omega}_0 - \hat{\Sigma}_{22}\tilde{\beta},$$

where $\tilde{\beta}$ is a consistent estimate of β. Then, if we take $\tilde{\beta} = \hat{\beta}$, the 2SLS estimate of β, we see that

$$\tilde{\delta} = \frac{1}{T}\hat{V}'y - \hat{\Sigma}_{22}\hat{\beta} = \hat{\delta}, \tag{4.8}$$

which shows that the estimator $\hat{\delta}$ can be generated in a second natural manner.

5. ECONOMETRIC APPLICATIONS

As previously indicated, assumptions concerning the independence of various stochastic explanatory variables in a structural equation and the disturbance term can have important implications for the appropriate choice of method of inference. On the one hand, if all the stochastic explanatory variables are correlated with the disturbance term, OLS does not usually provide consistent estimates of the structural coefficients in the equation and, even more generally, standard inference techniques (like F-tests) are not valid; one should use a simultaneous equations technique (e.g., instrumental variables). On the other hand, if they are all independent of the disturbance term, standard linear regression techniques (OLS, F-tests) are appropriate. Furthermore, between these two extremes, several intermediate cases are possible. If some but not all stochastic explanatory variables are independent of the disturbance term, standard inference techniques are not generally valid. However, we can exploit this information to get a more efficient method. In particular, if we split the matrix of stochastic explanatory variables into two submatrices $Y = [Y_1 : Y_2]$, where Y_2 is independent of u, we can get more efficient consistent estimators and more powerful tests by treating Y_2 as exogenous: in particular, this can be done by using Y_2 as an additional set of instruments or, at least, by not replacing Y_2 by $\hat{Y}_2$ (see Maddala, 1977, pp. 477–478).

The procedures developed above allow one to test the exogeneity of each stochastic explanatory variable included in a given equation by looking at asymptotic t-values. It is easy to compute these in a routine way while estimating the equation by an instrumental-variable method. In this manner, one can get automatic indications on the simultaneity properties of a model and possible ways of improving estimation efficiency.

Finally, we may observe that a number of economic hypotheses can be formulated in terms of the independence between certain stochastic explanatory variables and the error term in an equation. Wu (1973) described a number of such cases, such as the permanent income hypothesis, the expected profit maximization hypothesis, and the recursiveness hypothesis in simultaneous equation models.

6. PROOFS

6.1 Proof of Theorem 1

First, from (3.2), (3.3) and (3.8), we have the identity:

$$\sqrt{T}(\hat{\alpha} - \alpha) = \left(\frac{X'X}{T}\right)^{-1} e_T, \tag{6.1}$$

where

$$e_T = \frac{1}{\sqrt{T}} X'e + \frac{X'Z}{T}\sqrt{T}(\hat{\Pi} - \Pi)a.$$

Moreover, we can see that

$$(\hat{\Pi} - \Pi)a = (Z'Z)^{-1}Z'Va;$$

hence $E[(\hat{\Pi} - \Pi)a] = 0$ and

$$E[(\hat{\Pi} - \Pi)aa'(\hat{\Pi} - \Pi)'] = \rho(Z'Z)^{-1},$$

where $\rho = a'\Sigma_{22}a$, for, from Assumption 3, it is easily verified that $E[Vaa'V'] = \rho I_T$. Consequently,

$$\sqrt{T}(\hat{\Pi} - \Pi)a \sim N\left[0, \rho\left(\frac{Z'Z}{T}\right)^{-1}\right]. \tag{6.2}$$

Also, since e is independent of Y, the distribution of $\frac{1}{\sqrt{T}}X'e$, conditional on Y, is $N[0, \sigma_e^2(\frac{X'X}{T})]$.

Consider the characteristic function of e_T,

$$\begin{aligned}\phi_T(\tau) &\equiv E\{\exp[i\tau' e_T]\}\\ &= E\left\{\exp\left[i\tau'\frac{1}{\sqrt{T}}X'e + i\tau'\left(\frac{X'Z}{T}\right)\sqrt{T}(\hat{\Pi} - \Pi)a\right]\right\},\end{aligned}$$

where $\tau \in R^{2G+K_1}$ and $i = \sqrt{-1}$. In order to get an explicit expression for $\phi_T(\tau)$, we first compute the expected value of $\exp\{i\tau' e_T\}$ conditional on Y. Since $\frac{1}{\sqrt{T}}X'e$ is normal for given Y, we have

$$R_T(\tau) \equiv E[\exp(i\tau' e_T) \mid Y] = R_T^{(1)}(\tau)R_T^{(2)}(\tau),$$

where

$$R_T^{(1)}(\tau) = \exp\left[-\frac{\sigma_e^2}{2}\tau'\left(\frac{X'X}{T}\right)\tau\right]$$

and

$$R_T^{(2)}(\tau) = \exp\left[i\tau'\left(\frac{X'Z}{T}\right)\sqrt{T}(\hat{\Pi} - \Pi)a\right].$$

Then, using (3.6) and (6.2), we see that

$$\text{plim } R_T^{(1)}(\tau) = \exp\left\{-\frac{1}{2}\sigma_e^2\tau' Q_x\tau\right\}.$$

Also, from (3.7), (6.2) and Assumption 5, we have

$$R_T^{(2)}(\tau) \xrightarrow{L} \exp\{i\tau' B\},$$

where $B \sim N[0, \rho Q'_{zx} Q_z^{-1} Q_{zx}]$ and $\xrightarrow{L}$ refers to convergence in distribution as $T \to \infty$. Consequently,

$$R_T(\tau) \xrightarrow{L} \exp\left(-\frac{1}{2}\sigma_e^2 \tau' Q_x \tau\right) \exp(i\tau' B);$$

hence, by the Helly-Bray Theorem,

$$\begin{aligned}\lim_{T\to\infty} E\{R_T(\tau)\} &= \exp\left(-\frac{1}{2}\sigma_e^2 \tau' Q_x \tau\right) E\{\exp(i\tau' B)\} \\ &= \exp\left\{-\frac{1}{2}\tau' \left(\sigma_e^2 Q_x + \rho Q'_{zx} Q_z^{-1} Q_{zx}\right)\tau\right\},\end{aligned}$$

for all τ. Since $\phi_T(\tau) = E\{R_T(\tau)\}$, it follows that $\phi_T(\tau)$ converges to the characteristic function of the $N[0, \sigma_e^2 Q_x + \rho Q'_{zx} Q^{-1} Q_{zx}]$ distribution. Therefore,

$$e_T \longrightarrow N[0, \sigma_e^2 Q_x + \rho Q'_{zx} Q_z^{-1} Q_{zx}] \tag{6.3}$$

and, using (6.1),

$$\sqrt{T}(\hat{\alpha} - \alpha) \longrightarrow N[0, \Sigma_\alpha], \tag{6.4}$$

where Σ_α is given by (3.9). The consistency of $\hat{\alpha}$ follows from (6.4). Concerning the estimator $\hat{\sigma}_e^2$, we can write

$$\hat{\sigma}_e^2 = \frac{e^{*\prime} e^*}{T} - \frac{1}{T}\left(\frac{X' e^*}{\sqrt{T}}\right)' \left(\frac{X'X}{T}\right)^{-1} \left(\frac{X' e^*}{\sqrt{T}}\right),$$

hence

$$\text{plim } \hat{\sigma}_e^2 = \text{plim } \frac{e^{*\prime} e^*}{T}.$$

Moreover, by the definition of e^* in (3.3), we have

$$\frac{e^{*\prime} e^*}{T} = \frac{e'e}{T} + 2a'(\hat{\Pi} - \Pi)' \frac{Z'e}{T} + a'(\hat{\Pi} - \Pi)' \left(\frac{Z'Z}{T}\right)(\hat{\Pi} - \Pi)a;$$

hence, since plim $(Z'e/T) = 0$ and plim $(\hat{\Pi} - \Pi)a = 0$,

$$\text{plim } \hat{\sigma}_e^2 = \text{plim } \frac{e'e}{T} = \sigma_e^2,$$

which shows that $\hat{\sigma}_e^2$ is a consistent estimator of σ_e^2. Finally, we can see that $\hat{\Sigma}_\alpha$ is a consistent estimator of Σ_α by considering the definitions of Q_x and Q_{zx}, and by noting that $\hat{\rho}$ and $\hat{\sigma}_e^2$ are consistent for ρ and σ_e^2. Q.E.D.

6.2 Proof of Corollary 1.1

The consistency of $\hat{a}$ follows from the consistency of $\hat{\alpha}$. The asymptotic distribution of $\sqrt{T}(\hat{a}-a)$ follows from the identity

$$\sqrt{T}(\hat{a}-a) = C_2\left(\frac{1}{\sqrt{T}}X'e^*\right) \tag{6.5}$$

and from (6.3). The consistency of $\hat{\Sigma}_a$ follows from the consistency of $\hat{\Sigma}_\alpha$ and the definition of A_2. Q.E.D.

6.3 Lemma

In order to obtain the asymptotic distribution of $\hat{\delta} = \hat{\Sigma}_{22}\hat{a}$, we will need the following lemma.

Lemma 1. Suppose that Assumptions 2, 3 and 5 are satisfied. Let σ_i' and $\hat{\sigma}_i'$ be the ith rows of Σ_{22} and $\hat{\Sigma}_{22}$, respectively ($i = 1, \ldots, G$), and

$$\sigma = (\sigma_1', \sigma_2', \ldots, \sigma_G')', \quad \hat{\sigma} = (\hat{\sigma}_1', \hat{\sigma}_2', \ldots, \hat{\sigma}_G')'.$$

Then, the vector $\sqrt{T}(\hat{\sigma}-\sigma)$ has a normal limiting distribution, as $T \to \infty$, with mean zero and covariance matrix

$$\Sigma_\sigma = \begin{bmatrix} V_{11} & \cdots & V_{1G} \\ \vdots & & \vdots \\ V_{G1} & \cdots & V_{GG} \end{bmatrix}, \tag{6.6}$$

where $V_{ij} = \sigma_{ij}\Sigma_{22} + \sigma_j\sigma_i'$. Furthermore, the vector $\sqrt{T}(\hat{\Sigma}_{22}-\Sigma_{22})c$, where c is any fixed $G \times 1$ vector, has a normal limiting distribution with mean zero and covariance matrix

$$\psi_c = (c'\Sigma_{22}c)\,\Sigma_{22} + (\Sigma_{22}c)\,(\Sigma_{22}c)'. \tag{6.7}$$

Proof. Let $\hat{w}_i = M_z w_i$, where w_i is the ith column of the matrix V and $M_z = I_T - Z(Z'Z)^{-1}Z'$. The (i,j)th element of $\hat{\Sigma}_{22}$ has the form $\hat{\sigma}_{ij} = \hat{w}_i'\hat{w}_j/T$; hence

$$\hat{\sigma}_{ij} = \frac{w_i'w_j}{T} - \frac{1}{T}\left(\frac{Z'w_i}{\sqrt{T}}\right)'\left(\frac{Z'Z}{T}\right)^{-1}\left(\frac{Z'w_j}{\sqrt{T}}\right).$$

Let $\tilde{\sigma}_{ij} = w_i'w_j/T, \tilde{\sigma}_i = (\tilde{\sigma}_{i1}, \tilde{\sigma}_{i2}, \ldots, \tilde{\sigma}_{iG})'$, $i,\ j = 1, \ldots, G$, and $\tilde{\sigma} = (\tilde{\sigma}_1', \tilde{\sigma}_2', \ldots, \tilde{\sigma}_G')'$. Then, using Assumptions 3 and 5, we get

$$\operatorname{plim}\left[\sqrt{T}\,(\hat{\sigma}_{ij}-\sigma_{ij}) - \sqrt{T}\,(\tilde{\sigma}_{ij}-\sigma_{ij})\right] = 0, \quad i,\ j = 1, \ldots, G.$$

Thus, the vectors $\sqrt{T}(\hat{\sigma} - \sigma)$ and $\sqrt{T}(\tilde{\sigma} - \sigma)$ have the same limiting distribution.

Let w_{it} be the tth element of w_i and define $S_{it} = (w_{it}w_{1t}, w_{it}w_{2t}, \ldots, w_{it} w_{Gt})'$, $i = 1, \ldots, G$, $S_t = (S'_{1t}, S'_{2t}, \ldots, S'_{Gt})'$, $t = 1, \ldots, T$. It is clear that the vectors S_t, $t = 1, \ldots, T$, are independent and identically distributed with mean σ. Furthermore, since

$$\begin{aligned} E\left[(S_{it} - \sigma_i)(S_{jt} - \sigma_j)'\right] &= \left[\sigma_{ij}\sigma_{k\ell} + \sigma_{i\ell}\sigma_{jk}\right]_{k,\ \ell=1,\ldots,G} \\ &= \sigma_{ij}\Sigma_{22} + \sigma_j\sigma_i', \ i, j = 1, \ldots, G, \end{aligned}$$

for all t (see Anderson, 1958, p. 39), the covariance matrix of S_t is Σ_σ, as given in (6.6). Thus, since

$$\sqrt{T}(\tilde{\sigma} - \sigma) = \frac{1}{\sqrt{T}} \sum_{t=1}^{T} (S_t - \sigma),$$

and using the Multivariate Central Limit Theorem (see Anderson, 1958, Theorem 4.2.3), we can conclude that the limiting distribution of $\sqrt{T}(\tilde{\sigma} - \sigma)$ is $N[0, \Sigma_\sigma]$. Furthermore, for any $G \times 1$ fixed vector c,

$$\sqrt{T}(\hat{\Sigma}_{22} - \Sigma_{22})c = (I_G \otimes c')\sqrt{T}\,(\hat{\sigma} - \sigma),$$

where I_G is the identity matrix of order G and $\otimes$ refers to the Kronecker product. Since $\sqrt{T}(\hat{\sigma} - \sigma)$ is asymptotically $N[0, \Sigma_\sigma]$, we can conclude that the vector $\sqrt{T}(\hat{\Sigma}_{22} - \Sigma_{22})c$ is asymptotically normal with mean zero and covariance matrix

$$\begin{aligned} \psi_c &\equiv (I_G \otimes c')\,\Sigma_\sigma\,(I_G \otimes c) \\ &= \left[c'V_{ij}c\right]_{i,\ j=1,\ldots,G}. \end{aligned}$$

We see easily that ψ_c reduces to the expression in (6.7). Q.E.D.

6.4 Proof of Theorem 2

First, note that the vector $\sqrt{T}(\hat{\delta} - \delta)$ can be decomposed as follows:

$$\begin{aligned} \sqrt{T}(\hat{\delta} - \delta) &= \sqrt{T}\hat{\Sigma}_{22}(\hat{a} - a) + \sqrt{T}(\hat{\Sigma}_{22} - \Sigma_{22})a \\ &= \hat{\Sigma}_{22}C_2\left[\frac{1}{\sqrt{T}}X'e + \left(\frac{X'Z}{T}\right)\sqrt{T}(\hat{\Pi} - \Pi)a\right] \\ &\quad + \sqrt{T}\left(\hat{\Sigma}_{22} - \Sigma_{22}\right)a, \end{aligned}$$

where we have used (6.5) and (3.3). Let

$$W_T \equiv \Sigma_{22}A_2\left[\frac{1}{\sqrt{T}}X'e + Q'_{zx}\sqrt{T}(\hat{\Pi} - \Pi)a\right] + \sqrt{T}\left(\hat{\Sigma}_{22} - \Sigma_{22}\right)a.$$

Since $\text{plim}(\hat{\Sigma}_{22}C_2) = \Sigma_{22}A_2$ and $\text{plim}(X'Z/T) = Q'_{zx}$, we have $\text{plim}[\sqrt{T}(\hat{\delta} - \delta) - W_T] = 0$; $\sqrt{T}(\hat{\delta} - \delta)$ and W_T must have the same asymptotic distribution (see Billingsley, 1968, p. 25, Theorem 4.1). Consider now the characteristic function of W_T,

$$\phi_T(\tau) = E\left\{\exp\left[i\tau' W_T\right]\right\},$$

where $\tau \in R^G$. Since e is independent of Y, $\frac{1}{\sqrt{T}}X'e \sim N[0, \sigma_e^2(\frac{X'X}{T})]$ for Y fixed and, by taking the expected value of $\exp(i\tau' W_T)$ conditional on Y, we get

$$S_T(\tau) \equiv E\left\{\exp\left(i\tau' W_T\right) \mid Y\right\} = S_T^{(1)}(\tau) S_T^{(2)}(\tau),$$

where

$$S_T^{(1)}(\tau) = \exp\left\{-\frac{1}{2}\sigma_e^2 \tau' \Sigma_{22} A_2 \left(\frac{X'X}{T}\right) A_2' \Sigma_{22} \tau\right\}$$

and

$$S_T^{(2)}(\tau) = \exp\left\{i\tau' \Sigma_{22} A_2 Q'_{zx} \sqrt{T}(\hat{\Pi} - \Pi)a + i\tau' \sqrt{T}\left(\hat{\Sigma}_{22} - \Sigma_{22}\right) a\right\}.$$

Furthermore, using (3.6),

$$\text{plim}\ S_T^{(1)}(\tau) = \exp\left\{-\frac{1}{2}\sigma_e^2 \tau' \Sigma_{22} A_2 Q_x A_2' \Sigma_{22} \tau\right\} \equiv S^{(1)}(\tau). \tag{6.8}$$

Consequently, by the Helly-Bray Theorem,

$$\lim_{T\to\infty} E[S_T(\tau)] = S^{(1)}(\tau) \lim_{T\to\infty} E\left[S_T^{(2)}(\tau)\right], \tag{6.9}$$

where the expectation E is taken over Y.

Each column $\hat{\Pi}_j$ of $\hat{\Pi}$ is independent of each column $\hat{w}_k$ of $\hat{V}$, since $E[(\hat{\Pi}_j - \Pi_j)\hat{w}_k'] = 0$, $j, k = 1, \ldots, G$. Therefore, $\hat{\Pi}$ and $\hat{\Sigma}_{22}$ are independent and

$$\begin{aligned} E\left\{S_T^{(2)}(\tau)\right\} &= E\left\{\exp\left[i\tau' \Sigma_{22} A_2 Q'_{zx} \sqrt{T}(\hat{\Pi} - \Pi)a\right]\right\} \\ &\quad \times E\left\{\exp\left[i\tau' \sqrt{T}\left(\hat{\Sigma}_{22} - \Sigma_{22}\right) a\right]\right\} \\ &= \exp\left[-\frac{1}{2}\rho\tau' \Sigma_{22} A_2 Q'_{zx} \left(\frac{Z'Z}{T}\right)^{-1} Q_{zx} A_2' \Sigma_{22} \tau\right] \\ &\quad \times E\left\{\exp\left[i\tau' \sqrt{T}\left(\hat{\Sigma}_{22} - \Sigma_{22}\right) a\right]\right\}, \end{aligned} \tag{6.10}$$

where the second identity comes from (6.2). By Lemma 1,

$$\sqrt{T}\left(\hat{\Sigma}_{22} - \Sigma_{22}\right) a \xrightarrow{L} N[0, \psi],$$

where $\psi = (a'\Sigma_{22}a)\Sigma_{22} + (\Sigma_{22}a)(\Sigma_{22}a)'$. Since $\rho = a'\Sigma_{22}a$ and $\delta = \Sigma_{22}a$, we see that $\psi = \rho\Sigma_{22} + \delta\delta'$. Thus, the limit characteristic function of $\sqrt{T}(\hat{\Sigma}_{22} - \Sigma_{22})a$ is

$$\lim_{T\to\infty} E\left\{\exp\left[i\tau'\sqrt{T}\left(\hat{\Sigma}_{22} - \Sigma_{22}\right)a\right]\right\} = \exp\left[-\frac{1}{2}\tau'\psi\tau\right]. \tag{6.11}$$

Using (6.8)–(6.11), we obtain

$$\begin{aligned}&\lim_{T\to\infty} E[S_T(\tau)]\\ &= \exp\left\{-\frac{1}{2}\tau'\left[\Sigma_{22}A_2\left(\sigma_e^2 Q_x + \rho Q'_{zx}Q_z^{-1}Q_{zx}\right)A'_2\Sigma_{22} + \psi\right]\tau\right\},\end{aligned}$$

which implies that the asymptotic distribution of $\sqrt{T}(\hat{\delta} - \delta)$ is normal with mean zero and covariance matrix $\Sigma_\delta = \Sigma_{22}\Sigma_a\Sigma_{22} + \psi$, where Σ_a is given by (3.14). Finally, the consistency of $\hat{\delta} = \hat{\Sigma}_{22}\hat{a}$ follows from that of $\hat{\Sigma}_{22}$ and $\hat{a}$ for Σ_{22} and a respectively, and the consistency of $\hat{\Sigma}_\delta$ follows from the consistency of $\hat{\Sigma}_{22}$, $\hat{\Sigma}_a$ and $\hat{\psi}$ for Σ_{22}, Σ_a and ψ respectively. Q.E.D.

6.5 Asymptotic Power

We will now show that the tests discussed above are consistent. The statistic $S(M, m_0)$ used to test $M\alpha = m_0$, where M is a $\nu \times L$ matrix of rank ν, can be decomposed in the following way:

$$S(M, m_0) = S_1(M, \alpha) + \sqrt{T}S_2(M, \alpha, m_0),$$

where

$$S_1(M, \alpha) = T(M\hat{\alpha} - M\alpha)'(M\hat{\Sigma}_\alpha M')^{-1}(M\hat{\alpha} - M\alpha)$$

converges to a chi-square distribution with ν degrees of freedom and

$$\begin{aligned}&S_2(M, \alpha, m_0)\\ &= \left[2\sqrt{T}M(\hat{\alpha} - \alpha) + \sqrt{T}(M\alpha - m_0)\right]'\left(M\hat{\Sigma}_\alpha M'\right)^{-1}(M\alpha - m_0).\end{aligned}$$

We will show that plim $S_2(M, \alpha, m_0) = +\infty$, whenever $M\alpha \neq m_0$.

Consider first the case where all the elements of the vector $M\alpha - m_0$ are different from zero. In the sum $[2\sqrt{T}M(\hat{\alpha} - \alpha) + \sqrt{T}(M\alpha - m_0)]$, the second term always dominates as $T \to \infty$ and $\sqrt{T}(\hat{\alpha} - \alpha)$ has a limiting distribution. Consequently

$$\text{plim } S_2(M, \alpha, m_0) = \text{ plim } \sqrt{T}(M\alpha - m_0)'\left(M\hat{\Sigma}_\alpha M'\right)^{-1}(M\alpha - m_0) = +\infty,$$

where the fact that plim $(M\hat{\Sigma}_\alpha M')^{-1} = (M\Sigma_\alpha M')^{-1}$ is positive definite has been used. Second, for the case where $M\alpha \neq m_0$ but some elements of $(M\alpha - m_0)$ are zero, we can assume without loss of generality that these constitute the lower vector of $(M\alpha - m_0)$:

$$M\alpha - m_0 = (d_1', 0')',$$

where all the elements of the $\nu_1 \times 1$ vector d_1 are different from zero. Furthermore, let us partition $(\hat{\alpha} - \alpha)$ and $(M\hat{\Sigma}_\alpha M')^{-1}$ conformably with $(d_1', 0')'$:

$$\hat{\alpha} - \alpha = \begin{bmatrix} (\hat{\alpha} - \alpha)_1 \\ (\hat{\alpha} - \alpha)_2 \end{bmatrix}, \quad (M\hat{\Sigma}_\alpha M')^{-1} = \begin{bmatrix} A_{11} & A_{12} \\ A_{21} & A_{22} \end{bmatrix},$$

where $(\hat{\alpha} - \alpha)_1$ is a $\nu_1 \times 1$ vector and A_{11} is a $\nu_1 \times \nu_1$ positive definite matrix. Then

$$S_2(M, \alpha, m_0) = 2\sqrt{T}(\hat{\alpha}-\alpha)_1' M' A_{11} d_1 + 2\sqrt{T}(\hat{\alpha}-\alpha)_2' M' A_{21} d_1 + \sqrt{T} d_1' A_{11} d_1.$$

Since plim (A_{11}) is a positive definite matrix and $\sqrt{T}(\hat{\alpha} - \alpha)$ has a limiting distribution, we have

$$\text{plim } S_2(M, \alpha, m_0) = +\infty. \tag{6.12}$$

Thus, (6.12) holds whenever $M\alpha \neq m_0$,

$$\text{plim } S(M, m_0) = \text{ plim } \left[S_1(M, \alpha) + \sqrt{T} S_2(M, \alpha, m_0)\right] = +\infty,$$

whenever $M\alpha \neq m_0$, and

$$\lim_{T\to\infty} P\left[S(M, m_0) \geq c\right] = \begin{cases} \varepsilon, & \text{if } M\alpha = m_0 \\ 1, & \text{if } M\alpha \neq m_0, \end{cases} \tag{6.13}$$

where ε is the level of the test. This proves the consistency of the tests proposed for linear hypotheses regarding α. The consistency of tests regarding δ can be shown in a similar way.

ACKNOWLEDGMENTS

This work was supported by grants from the Natural Sciences and Engineering Research Council of Canada, the Social Sciences and Humanities Research Council of Canada, Foundation FCAR (Government of Quebec) and the Centre de recherche et développement en économique (Université de Montréal). The author thanks Arnold Zellner, Marcel Dagenais, Alberto Holly, Nicholas Kiefer, De-Min Wu and participants to the University of Chicago Econometrics and Statistics Colloquium for several very helpful comments. This paper is a revised version of Dufour (1979, Chapter 4; 1980).

REFERENCES

Anderson, T. W. (1958), *An Introduction to Multivariate Statistical Analysis*. New York: Wiley and Sons.

Billingsley, P. (1968), *Convergence of Probability Measures*. New York: Wiley and Sons.

Bronsard, C., and L. Salvas-Bronsard (1984), "On price exogeneity in complete demand systems." *Journal of Econometrics* **24**, 235–247.

Dufour, J.-M. (1979), "Methods for specification errors analysis with macroeconomic applications." Ph.D. Thesis, Department of Economics, University of Chicago.

Dufour, J.-M. (1980), "Tests of exogeneity." Cahier 8026, Département de sciences économiques, Université de Montréal.

Durbin, J. (1954), "Errors in variables." *Review of the International Statistical Institute* **22**, 23–32.

Engle, R. F. (1982), "A general approach to Lagrange multiplier diagnostics." *Journal of Econometrics* **20**, 83–104.

Engle, R. F. (1984), "Wald, likelihood ratio, and Lagrange multiplier tests in econometrics." In *Handbook of Econometrics*, Volume 2, ed. Z. Griliches and M. Intrilligator, pp. 775–826. Amsterdam: North-Holland.

Engle, R. F., D. F. Hendry, and J.-F. Richard (1983), "Exogeneity." *Econometrica* **51**, 277–304.

Farebrother, R. W. (1976), "A remark on the Wu test." *Econometrica* **44**, 475–477.

Fisher, F. M. (1966), *The Identification Problem in Econometrics*. New York: McGraw-Hill.

Gouriéroux, C., and A. Trognon (1984), "Specification pre-test estimation." *Journal of Econometrics* **25**, 15–28.

Hausman, J. (1978), "Specification tests in econometrics." *Econometrica* **46**, 1251–1272.

Hausman, J., and W. E. Taylor (1980), "Comparing specification tests and classical tests." Mimeo, Massachusetts Institute of Technology, Cambridge, MA.

Hausman, J., and W. E. Taylor (1981a), "Panel data and unobservable individual effects." *Econometrica* **49**, 1377–1398.

Hausman, J., and W. E. Taylor (1981b), "A generalized specification test." *Economics Letters* **8**, 239–245.

Holly, A. (1980), "Testing recursiveness in a triangular simultaneous equation model." Mimeo, Université de Lausanne.

Holly, A. (1982a), "A remark on Hausman's specification test." *Econometrica* **50**, 749–760.

Holly, A. (1982b), "Tests de spécification." *Cahiers du séminaire d'économétrie* **24**, 151–173.

Holly, A. (1983), "Une présentation unifiée des tests d'exogénéité dans les modèles à équations simultanées." *Annales de l'INSEE* **50**, April–June, 3–24.

Holly, A., and A. Monfort (1982), "Some useful equivalence properties of Hausman's test." Cahier 8201, Université de Lausanne.

Holly, A., and D. Sargan (1982), "Testing of exogeneity within a limited information framework." Cahier 8204, Université de Lausanne.

Hwang, H.-S. (1980), "Test of independence between a subset of stochastic regressors and disturbances." *International Economic Review* **21**, 749–760.

Kariya, T., and H. Hodoshima (1980), "Finite sample properties of the tests for independence in structural systems and the LRT." *The Economic Studies Quarterly* **31**, 45–56.

Maddala, G. S. (1977), *Econometrics*. New York: McGraw-Hill.

Nakamura, A., and M. Nakamura (1980), "On the usefulness of the Wu-Hausman test for detecting the least squares bias problem." Mimeo, University of Alberta.

Nakamura, A., and M. Nakamura (1981), "On the relationships among several specification error tests presented by Durbin, Wu and Hausman." *Econometrica* **49**, 1583–1588.

Plosser, C. J., G. W. Schwert, and H. White (1982), "Differencing as a test of specification." *International Economic Review* **23**, 535–552.

Rao, C. R. (1973), *Linear Statistical Inference and its Applications*, 2nd edition. New York: Wiley and Sons.

Revankar, N. S. (1978), "Asymptotic relative efficiency analysis of certain tests in structural systems." *International Economic Review* **19**, 165–179.

Revankar, N. S., and M. J. Hartley (1973), "An independence test and conditional unbiased predictions in the context of simultaneous equation systems." *International Economic Review* **14**, 625–631.

Reynolds, R. A. (1982), "Posterior odds for the hypothesis of independence between stochastic regressors and disturbances." *International Economic Review* **23**, 479–490.

Richard, J.-F. (1980), "Models with several regimes and changes in exogeneity." *Review of Economic Studies* **17**, 1–20.

Riess, P. (1983), "Alternative interpretations of Hausman's test." Mimeo, Yale University.

Ruud, P. A. (1984), "Tests of specification in econometrics." *Econometric Reviews* **3**, 211–242. [with discussion by T. S. Breusch and G. E. Mizon, J. Hausman, L.-F. Lee and H. White, 243–276].

Smith, R. J. (1984), "A note on likelihood ratio tests for the independence between a subset of stochastic regressors and disturbances." *International Economic*

Review **25**, 263–269.

Spencer, D. E., and K. N. Berk (1981), "A limited-information specification test." *Econometrica* **49**, 1079–1085. [Erratum, **50**, 1087].

Stroud, T. W. F. (1971), "On obtaining large-sample tests for asymptotically normal estimators." *Annals of Mathematical Statistics* **42**, 1412–1424.

Szroeter, J. (1983), "Generalized Wald methods for testing nonlinear implicit and overidentifying restrictions." *Econometrica* **51**, 335–353.

Theil, H. (1971), *Principles of Econometrics*. New York: Wiley and Sons.

Thursby, J. G. (1982), "The common structure of regression-based specification error tests." Mimeo, Department of Economics, Ohio State University.

Tsurumi, H., and T. Shiba (1984), "Tests of exogeneity using restricted reduced form coefficients." Mimeo, Rutgers University.

Turkington, D. A. (1980), "A note on Hausman's test for the limited information model." Mimeo, Department of Economics, University of Western Australia (Nedlands, Australia).

Wald, A. (1943), "Tests of statistical hypotheses concerning several parameters when the number of observations is large". *Transactions of the American Mathematical Society* **54**, 426–482.

White, H. (1982), "Maximum likelihood estimation of misspecified models." *Econometrica* **50**, 1–26.

Wu, D.-M. (1973), "Alternative tests of independence between stochastic regressors and disturbances." *Econometrica* **41**, 733–750.

Wu, D.-M. (1974), "Alternative tests of independence between stochastic regressors and disturbances: finite sample results." *Econometrica* **42**, 529–546.

Wu, D.-M. (1983a), "Tests of causality, predeterminedness and exogeneity." *International Econometric Review* **24**, 547–558.

Wu, D.-M. (1983b), "A remark on a generalized specification test." *Economics Letters* **11**, 365–370.

Zellner, A., J. Kmenta, and J. Drèze (1966), "Specification and estimation of Cobb-Douglas production functions." *Econometrica* **34**, 784–795.

Thomas A. Peters [1]

THE FINITE SAMPLE MOMENTS OF OLS IN DYNAMIC MODELS WHEN DISTURBANCES ARE SMALL

1. INTRODUCTION

For many years, econometric researchers have been using Least Squares methods (LS) to estimate the coefficients in dynamic single equation econometric models. These methods are usually justified by citing research that proves LS estimators have desirable asymptotic properties when the errors are assumed to be independent and identically distributed. Seminal articles by Mann and Wald (1943), White (1958) and Anderson (1959) are among those cited. These articles prove that under the assumption of independent and identically distributed errors, LS is asymptotically unbiased, asymptotically efficient and consistent in the context of dynamic econometric models. However, in small samples typical of econometric research involving time series data, the distribution of the LS estimator in dynamic models is much less certain and is difficult to obtain. As a result, researchers who ignore these finite sample problems may sacrifice the accuracy of their conclusions.

In this paper, I shall present both theoretical and numerical results on the bias and the Mean Squared Error (MSE) of the LS estimator of the autoregressive coefficient in a first order stochastic difference equation with white noise normal errors. The theoretical results are divided into two groups: Exact Results and Small Disturbance Approximate Results. The exact formulae rely heavily on the assumption of a normal error structure while the approximate formulae only require the existence of the first four moments of the error distribution. Both sets of results are dependent on the characteristics of the model and hence are functions of all the unknown parameters and the exogenous data series that exist in the model. Consequently, these formulae are very general and are valid for any specification of the model parameters. However, in this general form, they are unable to provide any quantitative information about the sign direction and/or size of

[1] Department of Economics, Trent University, Peterborough, Ontario K9J 7B8

I. B. MacNeill and G. J. Umphrey (eds.), Time Series and Econometric Modelling, 339–358.

the bias and MSE of the LS estimator. Thus, numerical evaluation of these formulae is necessary.

Through the use of computers, the theoretical formulae are evaluated under alternative specifications for the unknown parameters and exogenous data. These results are presented in this paper and comparisons are made, not only among alternative parameter scenarios but also between the approximate results and the exact results. Although the numerical results are not exhaustive, they do provide some limited, yet important, information about the nature of the distribution of the LS estimator in a stochastic difference equation.

The plan of this paper is as follows: Section 2 outlines the model and the estimators considered in this paper; Sections 3 and 4 present the theoretical and numerical results respectively; while Section 5 summarizes the general findings.

2. MODEL AND ESTIMATOR

In this paper, I consider a first-order stochastic difference equation with normally distributed, white noise errors. Specifically,

$$y_t = \mu + \gamma y_{t-1} + X_t\beta + u_t, \tag{2.1}$$

where y_t is the tth observation on the dependent variable. The parameter μ is an intercept that may or may not be included with X_t which is a $(1 \times K)$ vector of observations on K exogenous variables at time t. The u_t is a random error term at time t with distribution $N(0, \sigma_u^2)$. The parameter γ is the scalar autoregressive coefficient on the lagged dependent variable and β is a $(K \times 1)$ vector of coefficients corresponding to the K exogenous variables in the model.

Following the example of Evans and Savin (1984) it is convenient to use a transformed version of (2.1). Subtraction of the initial value, y_0, from both sides of (2.1) results in the following model:

$$z_t = \gamma z_{t-1} + \delta + X_t\beta + \sigma w_t, \tag{2.2}$$

where $z_{t-i} = (y_{t-i} - y_0)$ for $i \neq t$, $z_0 = 0$, $u_t/\sigma = w_t \sim N(0,1)$, $\delta = y_0(\gamma - 1) + \mu$, and $y_0 = c \geq 0$. This transformed version of model (2.1) is used because it allows us to write model (2.1) in matrix notation as follows:

$$z = \gamma Lz + \tilde{X}\tilde{\beta} + \sigma w, \tag{2.3}$$

where

$$Lz = z_{-1},$$

$$z_{-1} = [z_0, z_1, \ldots, z_{T-1}]',$$

$$z = [z_1, z_2, \ldots, z_T]',$$

$$L_{T\times T} = \begin{bmatrix} 0\ 0 \cdots\cdots 0 & \vert & 0 \\ ----- & \vert & -- \\ & \vert & 0 \\ I_{(T-1)\times(T-1)} & \vert & \vdots \\ & \vert & 0 \end{bmatrix},$$

$$\tilde{X}\tilde{\beta} = [\imath : X] \begin{bmatrix} \delta \\ \cdots \\ \beta \end{bmatrix},$$

and

$$\imath = [1, 1, \ldots, 1]'.$$

Rearrangment of (2.3) reveals that

$$z = H\tilde{X}\tilde{\beta} + \sigma H w, \tag{2.4}$$

where $H = (I - \gamma L)^{-1}$, and $\tilde{X}$ is $(T \times K + 1)$.

We have decomposed model (2.1) into its stochastic and non-stochastic components without making any assumptions about the stability of the autoregressive portion of the model. That is, as long as $(I - \gamma L)$ is invertible, the representation of model (2.1) in (2.4) is valid.

In Section 4 it will be necessary to know the general form of the distribution for y and/or z. Thus, using the decomposed form of z in (2.4), the mean of z is seen to be

$$\bar{z} = E(z) = H\tilde{X}\tilde{\beta}, \tag{2.5}$$

and the variance-covariance matrix is

$$\begin{aligned} V(z) &= E[(z - E(z))(z - E(z))'] \\ &\quad \times E[(H\tilde{X}\tilde{\beta} + \sigma H w - H\tilde{X}\tilde{\beta})(H\tilde{X}\tilde{\beta} + \sigma H w - H\tilde{X}\tilde{\beta})'] \\ &= E[(\sigma H w)(\sigma H w)'] \\ &= E(\sigma^2 H w w' H') \\ &= \sigma^2 H H' = \Omega. \end{aligned} \tag{2.6}$$

Therefore, due to the normality of w and to expressions (2.5) and (2.6), the distribution of z is seen to be

$$z \sim N(H\tilde{X}\tilde{\beta}, \Omega). \tag{2.7}$$

Having completely specified the model and derived the distribution of the dependent variable, we can now focus our attention on the form of the least squares estimator of γ.

The least squares estimator of γ is derived by minimizing the sum of the squared errors with respect to the true parameters. Thus, the LS estimator for γ is

$$\hat{\gamma} = \frac{z'Az}{z'Bz} = \frac{z'_{-1}\widetilde{M}z}{z'_{-1}\widetilde{M}Z_{-1}} = \frac{y'_{-1}My}{y'_{-1}My_{-1}}, \tag{2.8}$$

where

$$\begin{aligned} M &= I - X(X'X)^{-1}X', \\ \widetilde{M} &= I - \tilde{X}(\tilde{X}'\tilde{X})^{-1}\tilde{X}', \\ A &= L'\widetilde{M}, \\ B &= L'\widetilde{M}L, \end{aligned}$$

and z, z_{-1}, y and y_{-1} are defined in (2.1) and (2.4). M and $\widetilde{M}$ are idempotent matrices of rank $K+1$. It is obvious that if $y_0 = 0$ then $M = \widetilde{M}$ with rank $K+1 < T$ and $z = y$ and $z_{-1} = y_{-1}$. Finally, it can be noted that B is idempotent of rank K and A is nonsymmetric.

The model as specified in (2.3)—(2.8) will be used to derive the small disturbance asymptotic results in Section 3. However, when deriving the exact results, it will be much more convenient to use the canonical form of model (2.3) and estimator (2.8). Thus the following adjustments to the notation are necessary. Notice that $B = L'\widetilde{M}L$ is idempotent of rank K. Thus, there exists a $T \times T$ orthogonal matrix P such that

$$PBP' = I^* = \left[\begin{array}{c|c} I_n & 0 \\ -- & -- \\ 0 & 0 \end{array}\right],$$

where I_n is an $n \times n$ identity matrix and $n = T - K$. If one uses P to transform model (2.4) one obtains

$$Pz = s^* = PH\tilde{X}\tilde{\beta} + \sigma PHw, \tag{2.9}$$

where $s^* \sim N(\bar{s}^*, \Omega^*)$, $\Omega^* = P\Omega P'$, and $\bar{s}^* = PH\tilde{X}\tilde{\beta}$. Thus, the LS estimator in (2.8) becomes

$$\hat{\gamma} = \frac{z'Az}{z'Bz} = \frac{s^{*\prime}A^*s^*}{s^{*\prime}I^*s^*}, \tag{2.10}$$

where $A^* = P\tilde{A}P'$, $\tilde{A} = (A + A')/2$, $P'P = I_T$, and $\tilde{A}$ is now symmetric. $\tilde{A}$ has been introduced for convenience since $z'Az = z'\tilde{A}z$ (for examples of this usage, see Carter and Ullah, 1979; Hoque, 1980). This form of the estimator is easier to work with when deriving exact results. Furthermore, we can make these derivations even easier by standardizing z in the following manner. Recall that the distribution of s^* in (2.9) is

$$s^* \sim N(\bar{s}^*, \Omega^*). \tag{2.11}$$

Thus s^* can be standardized as follows:

$$\Omega^{*-1/2}s^* = \Omega^{*-1/2}\bar{s}^* + \sigma\Omega^{*-1/2}PHw, \tag{2.12}$$

where

$$\Omega^{*-1/2}\bar{s}^* \sim N(\Omega^{*-1/2}\bar{s}^*, I_T),$$

and $\Omega^{*-1/2}$ is a standardizing $T \times T$ transformation derived from the variance-covariance matrix of s^*. This transformation is derived by an eigenvalue-eigenvector decomposition of Ω^*. That is, there exists a $T \times T$ orthogonal matrix R such that

$$R\Omega^*R' = \Lambda_{T\times T} = \text{diag}\{\lambda_i;\ i = 1, \ldots, T\} \tag{2.13}$$

and $R'R = I_T$. Therefore,

$$R'\Lambda R = \Omega^*.$$

Hence, if

$$R'\Lambda RR'\Lambda R = R'\Lambda^2 R,$$

where

$$\Lambda^2 = \text{diag}\{\lambda_i^2;\ i = 1, \ldots, T\}$$

then

$$R'\Lambda^{(-j)}R = \Omega^{*(-j)}, \qquad j \in \mathcal{R}_1. \tag{2.14}$$

This allows us to write

$$\Omega^{*-1/2} = R'\Lambda^{-1/2}R, \tag{2.15}$$

where

$$\Lambda^{-1/2} = \text{diag}\{\lambda_i^{-1/2};\ i = 1, \ldots, T\}.$$

This allows us to write the estimator given in (2.10) as follows:

$$\hat{\gamma} = \frac{s'Cs}{s'Ds}, \tag{2.16}$$

where

$$s = R\Omega^{*-1/2}s^* \sim N(\bar{s}, I_T),$$
$$\bar{s} = R\Omega^{*-1/2}PH\tilde{X}\tilde{\beta},$$
$$C = R\Omega^{*1/2}A^*\Omega^{*1/2}R',$$

and

$$D = \begin{bmatrix} \Lambda_n & | & 0 \\ -- & | & -- \\ 0 & | & 0 \end{bmatrix};$$

$D = \Lambda$ when the last K diagonal elements are equal to zero.

The model and estimator expressed in (2.16) are equivalent to the form in (2.3) and (2.10). However, the alternative forms are required because they greatly simplify the theoretical derivations in Section 3 and the programming of the computer algorithm for the numerical results in Section 4.

3. THEORETICAL RESULTS

The results presented in the following subsections are derived using exact analytical techniques and small disturbance asymptotic techniques. The exact results are based on the work of Sawa (1972, 1978), Phillips (1980) and Hoque (1980). The small disturbance approximate results and the asymptotic results are based on work by Kadane (1970, 1971), Carter and Ullah (1979) and Evans and Savin (1984). These articles have been instrumental in developing the analytical tools needed to solve the technical problems in this section.

3.1 Exact Results

We write (2.16) as follows:

$$s = \bar{s} + \eta, \tag{3.1.1}$$

where

$$\eta = \sigma R\Omega^{*-1/2}PHw \sim N(0, I),$$
$$s \sim N(\bar{s}, I_T);$$

the LS estimator for γ is

$$\hat{\gamma} = \frac{s'Cs}{s'Ds}, \tag{3.1.2}$$

where $D = \text{diag}\{d_i;\ i = 1, \ldots, T\ ]$, d_i is non-zero for $i = 1, \ldots, T - h$, and is zero otherwise, $h = K$, and $T - h$ is the degrees of freedom.

The distribution of s and the formula for $\hat{\gamma}^2$ can be used to derive the first and second moments[2] of $\hat{\gamma}$. (See "Endnotes" section after the Conclusions.)

The following theorems, without proof[3], contain the formulae for these exact moments.

Theorem 3.1. *The exact mean of the Least Squares estimator of the autoregressive coefficient in a first-order stochastic difference equation with white noise, normal errors is,*

$$E(\hat{\gamma}) = \int_0^\infty \exp\left\{\frac{\bar{s}'Q^*\bar{s}}{2}\right\} \mid I + 2tD \mid^{-1/2} \{\bar{s}'QCQ\bar{s} + \operatorname{tr}(QC)\}\, dt,$$

where

$$Q^* = Q - I$$

and

$$Q = (I + 2td)^{-1} = \operatorname{diag}\{1/(1+2td_i);\ i = 1, \ldots, T\}.$$

Theorem 3.2. *The exact second moment of the Least Squares estimator of the autoregressive coefficient for a first-order stochastic difference equation with white noise, normal errors is*

$$E(\hat{\gamma}^2) = \int_0^\infty \exp\left\{\frac{\bar{s}'Q^*\bar{s}}{2}\right\} \mid I + 2tD \mid^{-1/2} \times \left\{(\bar{s}'QCQ\bar{s} + \operatorname{tr}QC)^2 + 2\operatorname{tr}QCQC + 4\bar{s}'QCQCQ\bar{s}\right\} t\, dt,$$

where

$$Q^* = Q - I$$

and

$$Q = (I + 2tD)^{-1} = \operatorname{diag}\{1/(1+2td_i);\ i = 1, \ldots, T\}.$$

3.2 Small Disturbance Approximate Results

The purpose of this subsection is to present formulae for the first and second moments of $\hat{\gamma}$ in (2.10) that are approximations to the exact formulae in subsection (3.1). The analytical technique used is the small disturbance asymptotic approximation popularized by Kadane.

The following theorems summarize the small disturbance approximate bias and mean squared error for $\hat{\gamma}$ in (2.10).

Theorem 3.3. *The small disturbance approximate bias of the Least Squares estimator of the autoregressive coefficient in a first-order stochastic difference equation with white noise, normal errors up to $O(\sigma^2)$ in probability is*

$$E_a(\hat{\gamma}-\gamma)=\sigma^2\left(\frac{\mathrm{tr}H'A}{\bar{z}'B\bar{z}}\right)-2\sigma^2\left(\frac{\bar{z}'BHA'\bar{z}}{(\bar{z}'B\bar{z})^2}\right),$$

where $\bar{z}$, A, B and H are defined in Section 2.[4]

Theorem 3.4. *The small disturbance approximate Mean Squared Error of the Least Squares estimator of the autoregressive coefficient in a first-order stochastic difference equation with white noise, normal errors up to $O(\sigma^4)$ in probability is given by*

$$E_a(\hat{\gamma}-\gamma)^2=\sigma^2(\bar{z}'B\bar{z})^{-1}+\sigma^4(\bar{z}'B\bar{z})^{-2}\tilde{G},$$

where

$$\begin{aligned}\tilde{G}&=4(\bar{z}'B\bar{z})^{-1}\left[\bar{z}'BHH'B\bar{z}+6(\bar{z}'B\bar{z})^{-1}(\bar{z}'BHA'\bar{z})^2\right]\\&\quad-8(\bar{z}'B\bar{z})^{-1}\left[\bar{z}'BHA'\bar{z}\cdot\mathrm{tr}H'A+\bar{z}'AH'AH'B\bar{z}+\frac{1}{2}\bar{z}'AH'BHA'\bar{z}\right]\\&\quad+(\mathrm{tr}H'A)^2+\mathrm{tr}H'AH'A-\mathrm{tr}H'BH,\end{aligned}$$

and where H, A, B, $\bar{z}$ are defined in Section 2.

The exact results in this section present general formulae for the first and second-order moments (hence bias and MSE) of the distribution of the LS estimator of γ in model (2.1). They are general in the sense that they are valid for any prior setting of the unknown nuisance parameters y_0, σ, and X matrix as well as the true parameters γ and B. However, the formulae are not general in the sense that we have assumed normality of the error distribution which is crucial for the exact results.

The small disturbance results are also general, in the sense that they are valid for all combinations of y_0, σ, γ, β, and X matrix that will render the small disturbance expansion valid.[5] This will inevitably restrict the admissible parameter space and limit the broad generality of the approximation. The consequences of violating the admissibility condition show up vividly in the numerical calculations section.

4. NUMERICAL RESULTS

The purpose of this section is to evaluate the formulae developed in the previous section under alternative scenarios for the unknown nuisance parameters and the exogenous data series. Based on these alternative prior beliefs about the location in the parameter space, comparative tables are presented featuring numerical evaluations of the general moment formulae with emphasis on the sign and size of biases, on the size of the MSE, and on the performance of the approximation formulae compared to the exact results.

The choice of the parameter space was governed by the availability of computing resources.[6] Three types of data for the X matrix were used, each being used as an exclusive case and not together. Thus, $\tilde{X}$ is a $T \times 2$ matrix or $K = 1$. This is done to see the effect of alternative types of data. The choices for x_t are:

1) TRENDING DATA

$$x_t = e^{.1t}, \qquad t = 1, \ldots, T.$$

2) NONTRENDING DATA

x_t is randomly drawn from a uniform distribution in the interval $[0, 1]$.

3) AR(2) DATA

$$x_t = \rho_1 x_{t-1} + \rho_2 x_{t-2} + \varepsilon_t,$$

where $\rho_1 = .75$, $\rho_2 = -.5$, $\varepsilon_t \sim N(0, 1)$, and x_t follows a stationary AR(2) process.[7]

The rest of the parameters of the model also will take on various values. In all tables, $\gamma = [.4, .8, 1.0, 1.01, 1.025]$, $\sigma^2 = [.5, 1.0]$, $y_0 = 1.0$, $T = 10$, $\mu = 0$ or 1.0. In one table the effect of an increasing sample size is observed by setting $T = 20$. The specification of the parameter space is far from exhaustive; however, from the analysis of this limited number of scenarios some interesting observations can be made.

Using the specifications of the model parameters and the exogenous data series, the $\{z_t\}$ series was generated, allowing us to obtain the elements of the vector $\bar{z}$ that are present in all the formulae. Therefore $\bar{z}$ is generated by starting with model (2.2),

$$z_t = \delta + \gamma z_{t-1} + x_t \beta + \sigma w_t,$$

then

$$z_t = \gamma^t z_0 + \sum_{i=0}^{t-1} \gamma^i \delta + \sum_{i=0}^{t-1} \gamma^i x_{t-i} \beta + \sigma \sum_{i=0}^{t-1} \gamma^i w_{t-i}, \tag{5.1}$$

where $z_0 = 0$, and $\delta = y_0(\gamma - 1) + \mu$. Therefore

$$E(z_t) = \bar{z}_t = \sum_{i=0}^{t-1} \gamma^i \delta + \sum_{i=0}^{t-1} \gamma^i x_{t-i} \beta. \tag{5.2}$$

Table 1. *Trending Data*
$x_t = \exp\{.1t\}$, $t = 1, \ldots, T$, $T = 10$. $\mu = 0$.

		Exact		σ-Approximate		Asymptotic
σ^2	γ	BIAS	MSE	BIAS	MSE	MSE
.5	.400	–.0590	.0859	–.7729	.8313	.2722
1.0	.400	–.1280	.1088	–1.5460	2.7810	.5444
.5	.800	–.0666	.0297	–.1270	.0561	.0313
1.0	.800	–.1344	.0616	–.2540	.1619	.0626
.5	1.000	–.0240	.0063	–.0234	.0068	.0060
1.0	1.000	–.0490	.0139	–.0468	.0152	.0121
.5	1.010	–.0226	.0058	–.0210	.0061	.0055
1.0	1.010	–.0454	.0126	–.0421	.0135	.0110
.5	1.025	–.0206	.0050	–.0179	.0053	.0048
1.0	1.025	–.0403	.0107	–.0357	.0114	.0096

The variance-covariance matrix for vector z can be deduced from (5.1):

$$E\{[z_t - E(z_t)][z_s - E(z_s)]\} = \begin{cases} \sigma^2 \sum_{i=0}^{t-1} \gamma^{2i} & t = s \\ \sigma^2 \gamma^{|t-s|} \sum_{i=0}^{t-1} \gamma^{2i} & t \neq s, \end{cases} \tag{5.3}$$

which are the elements of $\sigma^2 HH' = \Omega$ when $t = 1, 2, \ldots, T$. Thus with the specification of the parameter space and the construction of the distribution of z completed, running the computer algorithm[8] for various parameter values produced Tables 1–5.

The numerical results presented in Tables 1–5 reveal some interesting information about the behaviour of the LS bias and MSE in finite samples. After thorough examination of this tabular information, the following groups of observations can be made.

1. The exact theoretical results in Section 4 show some interesting tendencies with respect to σ^2, T, γ and x_t, based on the numbers presented in

Table 2. *Nontrending Data*

x_t is randomly selected from a uniform distribution in the interval $[0, 1]$.
$T = 10, \quad \mu = 0.$

		Exact		σ-Approximate		Asymptotic
σ^2	γ	BIAS	MSE	BIAS	MSE	MSE
.5	.400	–.1310	.1057	–2.562	.1868	1.206
1.0	.400	–.1508	.1126	–5.125	–1.6650	2.412
.5	.800	–.2588	.1432	–.7862	1.308	.2648
1.0	.800	–.2651	.1488	–1.5720	4.702	.5296
.5	1.000	–.1222	.0675	–.0750	.0485	.0218
1.0	1.000	–.1859	.1045	–.1499	.1505	.0436
.5	1.010	–.1133	.0631	–.0658	.0414	.0193
1.0	1.010	–.1786	.1009	–.1315	.1271	.0387
.5	1.025	–.1003	.0567	–.0538	.0329	.0162
1.0	1.025	–.1672	.0951	–.1077	.0991	.0324

Tables 1–5. First, it is obvious from viewing Tables 2 and 4 that the absolute value of the bias and MSE are decreasing functions of the sample size, T, at all values of σ^2, γ and x_t. Also, all tables predict that the absolute bias and MSE are decreasing as σ decreases for all values of T, γ and x_t, except when x_t is AR(2) and $\gamma = .4$ or $\gamma = .8$. These contrary cases are not easily explained but could have something to do with common factors in the autoregressive portions of the model. Further, Table 5 suggests that the inclusion of an intercept in the model together with the AR(2) series seems to eliminate these counter-intuitive observations. Hence, with few exceptions, the expected asymptotic tendencies of the LS bias and MSE are observed with respect to both σ and T.

Secondly, while holding the sample size (T), standard deviation (σ) and data series (x_t) constant, and letting γ vary, some interesting conclusions

Table 3. *Autoregressive Process of Order 2*

$$x_t = \rho_1 x_{t-1} + \rho_2 x_{t-2} + v_t, \quad T = 10$$
$$\rho_1 = .75, \quad \rho_2 = -.5, \quad v_t \sim N(0,1), \quad \mu = 0$$

		Exact		σ-Approximate		Asymptotic
σ^2	γ	BIAS	MSE	BIAS	MSE	MSE
.5	.400	.2582	.0833	−.0380	.0227	.0202
1.0	.400	.2065	.0758	−.0760	.0503	.0404
.5	.800	.0295	.0239	−.0661	.0325	.0142
1.0	.800	−.0117	.0407	−.1322	.1017	.0284
.5	1.000	−.0716	.0390	−.0671	.0343	.0095
1.0	1.000	−.0945	.0542	−.1343	.1182	.0190
.5	1.010	−.0755	.0399	−.0668	.0342	.0093
1.0	1.010	−.0971	.0544	−.1335	.1183	.0185
.5	1.025	−.0812	.0415	−.0661	.0341	.0089
1.0	1.025	−.1008	.0549	−.1322	.1184	.0178

can be drawn from these tables. For all values of σ, T and x_t, the absolute bias first tends to increase as γ increases from .4 to .8 then decreases as γ reaches unity and beyond. Again, Table 3 shows a result contrary to this generalization. When x_t is an AR(2) series without an intercept, absolute bias increases as γ approaches unity and beyond. The MSE results outline an identical pattern for all x_t series, including the AR(2) series.

An interesting observation concerning the behaviour of the LS estimator, $\hat{\gamma}$, is that it confirms other numerical work done in the literature. The increasing bias and MSE as γ approaches the unit circle (without reaching it) confirms the findings of Phillips (1977), Sawa (1978), Tse (1981) and Tanaka (1982) for the stationary case. Also, the findings of Evans

Table 4. *Nontrending Data*

x_t randomly selected from a uniform distribution in the interval $[0, 1]$.
$T = 20, \quad \mu = 0$

		Exact		σ-Approximate		Asymptotic
σ^2	γ	BIAS	MSE	BIAS	MSE	MSE
.5	.400	−.0469	.0468	−1.146	.4633	.4276
1.0	.400	−.0709	.0503	−2.292	.9979	.8552
.5	.800	−.1476	.0508	−.3702	.1440	.0759
1.0	.800	−.1522	.0550	−.7404	.4244	.1518
.5	1.000	−.0295	.0068	−.0196	.0037	.0024
1.0	1.000	−.0618	.0162	−.0392	.0100	.0048
.5	1.010	−.0243	.0056	−.0158	.0029	.0020
1.0	1.010	−.0553	.0150	−.0316	.0077	.0039
.5	1.025	−.0169	.0024	−.0112	.0020	.0014
1.0	1.025	−.0437	.0076	−.0223	.0053	.0029

and Savin (1984) suggest that anything that increases the signal-to-noise ratio (or non-centrality parameter) will cause the absolute bias and MSE to decrease. From (5.1) of this section, the expression for $\bar{z}_t$ shows that γ is an increasingly important parameter of the model especially when γ reaches the unit circle. Thus as γ increases beyond unity, it is obvious that the non-centrality parameter $(\bar{z}'\bar{z}/(2\sigma))$ is increasing rapidly and hence causing the damping effect on the bias and MSE.

Finally, the tables present information about the behaviour of the bias and MSE when alternative exogenous data series are assumed to be present. At all values of σ and γ, the bias and MSE tend to be smaller in the case of trending data than in the case of AR(2) data. Also, from Tables 2 and 3, the AR(2) data yield results that are smaller in absolute value than the results for nontrending data. Therefore, it is obvious that trending data scenarios provide LS estimators that have a smaller bias and MSE than in the case of nontrending data.

Table 5. *Autoregressive Process of Order 2*

Intercept, $\mu = 1.0$, $T = 10$

σ^2	γ	Exact BIAS	Exact MSE	σ-Approximate BIAS	σ-Approximate MSE	Asymptotic MSE
.5	.400	−.0796	.0604	−.1410	.0510	.0948
1.0	.400	−.1119	.0851	−.2821	.0146	.1895
.5	.800	−.1110	.0598	−.1263	.0712	.0387
1.0	.800	−.1752	.1032	−.2526	.2075	.0774
.5	1.000	−.0428	.0209	−.0333	.0142	.0111
1.0	1.000	−.0968	.0546	−.0665	.0346	.0222
.5	1.010	−.0396	.0191	−.0303	.0130	.0103
1.0	1.010	−.0918	.0514	−.0605	.0315	.0206
.5	1.025	−.0350	.0164	−.0261	.0114	.0092
1.0	1.025	−.0843	.0463	−.0522	.0274	.0183

An additional point can be raised upon the comparison of Tables 3 and 5. The inclusion of an intercept into the model (in the AR(2)) case seems to cause the absolute bias and MSE to increase when $\gamma > 1$, but when $\gamma < 1$ the values tend to be smaller. The intuitive explanation for this result might be that when γ reaches unity or beyond the increased uncertainty (noise) of an additional nuisance parameter is completely overwhelmed by the rapidly increasing noncentrality parameter (signal). On the other hand, the increased uncertainty of an additional regressor in the model (nuisance parameter) is too large to be offset when $\gamma < 1$. This result was also found by Sawa (1978) in a simpler context and under stationary conditions.

The exact results on the LS estimator for γ have revealed some valuable conclusions. Generally, the results show that the distribution of $\hat{\gamma}$ in finite samples is very sensitive to the portion of the parameter space considered and to the exogenous data-generating series. Finally, the tendencies of bias and MSE are linked to the signal-to-noise ratio in an inverse way.

2. The small disturbance approximate results are also presented along with the small disturbance asymptotic MSE in Tables 1 through 5. It can be noted first, just as with the exact results, that the approximate bias and MSE shrink as σ decreases and T increases for all values of γ and exogenous data. This is also seen with the asymptotic MSE for all γ and x_t. Thus, the expected asymptotic tendencies of the LS bias and MSE are observed using the approximate formulae.

   Secondly, the numerical values for the approximate bias and MSE reveal some strong conclusions about tendencies as γ varies. If one considers the bias, there is a tendency for the absolute approximate bias to decrease as γ increases from $\gamma = .4$ to $\gamma = 1.025$ for all data except the AR(2) case. In that case, the bias first increases as γ increases from .4 to .8 and then declines as γ increases further. The approximate MSE behaves consistently when $\gamma \geq 1$ for all x_t; however, it behaves erratically at values of γ less than unity. Specifically, when $\gamma \geq 1$, the approximate MSE decreases as γ increases for all x_t. In situations such as are illustrated in Tables 1 and 4, the MSE decreases as γ increases towards unity. Tables 2, 3 and 5 show the MSE increasing as γ increases from .4 to .8, and, in Table 3, as γ increases from .4 to 1.0. Further, from all tables, the asymptotic MSE is decreasing as γ increases from .4 to 1.025.

   As in the exact results, the approximate results are sensitive to the exogenous data series generating $\overline{z}_t$. Identical conclusions about the size of the bias and MSE in trending data scenarios, relative to nontrending situations, can be made. That is, that absolute bias and MSE are larger for nontrending x_t data as opposed to AR(2) and trended data. Also, the validity of the observation about the inclusion of an extra regressor (intercept) in the model is evident using the approximate formula. That is, the bias and MSE increase with an extra nuisance parameter when $\gamma < 1$ but decrease when $\gamma \geq 1$. This evidence suggests that the stronger the mean (signal) the smaller the values of the bias and MSE.

   The evidence presented by Tables 1–5 concerning the approximate bias, MSE and asymptotic MSE, essentially corroborate the general conclusions drawn on the exact results. However, caution must be observed in situations where the signal-to-noise ratio is small.

3. Having shown that most of the proper conclusions about the exact bias and MSE can be found through the approximate results ($\gamma \geq 1$), the absolute accuracy of the approximation as opposed to the order 1 asymptotic approximation should be assessed. In light of this concern, the

results show that for all x_t, σ and T, the approximate results perform very well when $\gamma \geq 1$. Also, the approximation formulae outperform the asymptotic formulae for all $\gamma \geq 1$, σ, T and x_t series. Earlier, it was suggested that the signal-to-noise ratio played a large role in determining the size of bias and MSE. Again, we suggest that a large signal-to-noise ratio is the main reason for increased accuracy of approximation. For example, any of our tables will reveal that the approximation to the exact results is much better when $\sigma^2 = .5$ than when $\sigma^2 = 1.0$; also, consider the approximation when $\gamma \geq 1$ and x_t is trending, compared to when $\gamma \geq 1$ and x_t is nontrending. Further, comparison of the results when $\gamma < 1$ leads to the conclusion that the approximate formulae perform very poorly. In fact, the asymptotic results of unbiasedness and asymptotic MSE outperform the approximate results in most cases when $\gamma < 1$ (specifically $\gamma \leq .8$).

In situations where the signal of the model dominates the noise, the approximate formulae will behave very well in terms of predictions about tendencies with respect to nuisance parameters and will outperform the asymptotic formulae in terms of numerical accuracy.

The major findings in this section are summarized in the three distinct groups above; however, a common theme seems to be clear. It was found that the exact, approximate and asymptotic moments are very sensitive to the type of exogenous data and parameter specification used to generate the mean of z_t. Ultimately, the stronger the signal ($\overline{z}_t$), the smaller the risk in using LS in this context. The changes in the factors σ, T, and γ that increased the signal-to-noise ratio caused the absolute bias and MSE to shrink and improved the performance of the small disturbance approximations.

5. CONCLUSIONS

It was the objective of this paper to analyze the distribution, in terms of its first and second moments, of the LS estimator of the autoregressive coefficient in a first-order stochastic difference equation under the assumption of white noise, normal errors. Under the condition that the moments of this estimator exist up to a specific order, theoretical formulae were derived, using some well established techniques, for the bias and MSE of the estimator. Unfortunately, these formulae were not able to provide any quantitative information about the size and sign of the bias and size of the MSE in their raw form. Hence they were evaluated numerically under alternative scenarios for the nuisance parameters. After analysis of the computational results, conclusions were drawn about the behavioural tendencies of the bias and MSE of LS across a defined portion of the nuisance parameter space.

The conclusions reached in this paper are that the bias and MSE vary inversely with the signal-to-noise ratio, and that the σ-approximation's performance improves with increasing signal-to-noise ratio. Also in the context of the model studied here, the σ-approximation performs well when the true parameter is around or outside the unit circle. This isn't surprising, since the signal blows up quickly when $\gamma \geq 1$.

In most cases, it was found that the σ-approximation outperformed the asymptotic results; however, in certain portions of the parameter space, researchers would be better off using the asymptotic formulae for MSE. This conclusion is related to the validity of the σ-expansion itself under specific scenarios for the nuisance parameters. That is, when the signal-to-noise ratio is low (i.e., σ is large and γ is small), the condition under which a σ-expansion is admissible is violated. Therefore, the expansion is invalid and the numerical values for the formula make little sense.[9]

The information contained in this paper could be of some benefit to empirical researchers. Once they are able to decide on a prior (or null hypothesis) for the true parameter, they could determine the risk (in terms of MSE) of using LS estimators by applying the above exact formulae and/or the σ-approximate formulae when appropriate ($\gamma \geq 1$). The MSE numbers could be used to specify corrected standard errors for the t-statistic and improve the accuracy of the statistical inference.

ENDNOTES

2. The moment formulae presented in this paper, both exact and approximate, are defining approximating moments that actually exist. Evans and Savin (1981) and Mariano (1972) proved that the rth moment exists under the condition that the degrees of freedom in the model are greater than twice the order of the moment. Their estimators are similar to that under study here. Additional proof of this condition for existence can be obtained from the author.

3. Proofs for the theorems in this section can be obtained from the author. The proofs are derived using the results given by Sawa (1978), Phillips (1980), and Hoque (1980).

4. This notation is used to distinguish the exact bias $E(\hat{\gamma} - \gamma)$ from the truncated series approximation, $E_a(\hat{\gamma} - \gamma)$.

5. In deriving the small disturbance approximation, the expansion is only valid under certain conditions. For example, the LS sampling error can

be expanded as follows:

$$(\hat{\gamma} - \gamma) = \sigma(\overline{z}'B\overline{z})^{-1}\left[1 + \left(\frac{2\sigma\overline{z}'BHw + \sigma^2 w'H'BHw}{(\overline{z}'B\overline{z})}\right)\right]^{-1}$$

$$\times\,[\overline{z}Aw + \sigma w'H'Aw]$$

$$= a_0 + a_1 + a_2 + \cdots,$$

where the a_i's are of successively higher orders in σ. This expansion is valid if and only if

$$\left|\frac{2\sigma\overline{z}'BHw + \sigma^2 w'H'BHw}{\overline{z}'B\overline{z}}\right| < 1.$$

As can be seen, this ratio is random and will obviously take on values greater than one in absolute value in some parts of the parameter space.

6. Numerical approximation of an integral is a difficult task to do cheaply. Thus, I must thank the Department of Economics at The University of Western Ontario and ultimately the University itself for allowing me enough resources to complete my work.
7. This process was generated with an initial value of $x_0 = 1$. The first 20 values were dropped which allows the value for x_{20} to be our initial value. This is a method suggested by Phillips (1980).
8. The algorithm used to compute the exact results was based on directly approximating the area under the function. Using Simpson's Rule to calculate this approximation, the values in the tables are accurate up to 1.0×10^{-4}. The algorithm used was obtained from John L. Knight. Professor Knight spent a good deal of his own time helping me program this algorithm for which I am greatly in his debt. Any unseen errors, however, are mine alone.
9. See endnote 5.

ACKNOWLEDGMENTS

Many thanks to A. Ullah, R. A. L. Carter, M. Veall, A. Gregory, D. E. A. Giles, P. C. B. Phillips and J. L. Knight for their helpful suggestions and comments. The errors, of course, remain my own.

REFERENCES

Anderson, T. W. (1959), "On asymptotic distributions of estimates of parameters of stochastic difference equations". *Annals of Mathematical Statistics* **30**, 676–687.

Anderson, T. W., and T. Sawa (1973), "Distributions of estimates of coefficients of a single equation in a simultaneous system and their asymptotic expansions". *Econometrica* **41**, 683–714.

Carter, R. A. L., and A. Ullah (1979), "The finite sample properties of OLS and IV estimators in special rational distributed lag models". *Sankhyā*, Series D **41**, 1–18.

Dickey, D. A., and W. A. Fuller (1979), "Distribution of the estimators for autoregressive time series with a unit root". *Journal of the American Statistical Association* **74**, 427–431.

Evans, G. B. A., and N. E. Savin (1981), "Testing for unit roots: 1". *Econometrics* **49**, 753–779.

Evans, G. B. A., and N. E. Savin (1984), "Testing for unit roots: 2". *Econometrica* **52**, 1241–1269.

Giles, D. E. A., and M. Beattie (1985), "Autocorrelated pre-test estimation in models with a lagged dependent variable". In *Specification Analysis in the Linear Model: In Honour of Donald Cochrane*, ed. M. L. King and D. E. A. Giles. London: Routledge, and Kegan Paul.

Hoque, A. (1980), "Finite sample analysis of three dynamic models". Ph.D. Thesis, The University of Western Ontario.

Kadane, J. B. (1971), "Comparison of k-class estimators when disturbances are small". *Econometrica* **39**, 723–737.

Kendall, M. G. (1954), "Note on bias in estimation of autocorrelation". *Biometrika* **41**, 403–404.

Maekawa, K. (1983), "An approximation to the distribution of the least squares estimators in an autoregressive model with exogenous variables". *Econometrica* **51**, 229–238.

Mann, H. B., and A. Wald (1943), "On the statistical treatment of linear stochastic difference equations". *Econometrica* **11**, 173–200.

Mariano, R. S. (1972), "The existence of moments of the ordinary least squares and two-stage least squares estimators". *Econometrica* **40**, 643–652.

Phillips, P. C. B. (1977), "Approximations to some finite sample distributions associated with a first-order stochastic difference equation". *Econometrica* **45**, 463–486.

Phillips, P. C. B. (1980), "Finite sample theory and the distribution of alternative estimators of the marginal propensity to consume". *Review of Economic Studies*, **47**, 183–224.

Phillips, P. C. B. (1983), "Small sample distribution theory in econometric models of simultaneous equations". In *Handbook of Econometrics* Volume 1, ed. D. M. Intrilligator and Z. Griliches. New York: North-Holland.

Sargan, J. D. (1974), "The validity of Nagar's expansion for the moments of econometric estimators". *Econometrica* **42**, 169–176.

Satchell, S. E. (1984), "Approximations to the finite sample distribution for the nonstable first-order stochastic difference equation". *Econometrica* **52**, 1271–1290.

Sawa, T. (1972), "Finite sample properties of k-class estimators". *Econometrica* **40**, 653–680.

Sawa, T. (1978), "The exact moments of the least squares estimator for the autoregressive model". *Journal of Econometrics* **8**, 159–172.

Tanaka, K. (1983), "Asymptotic expansions associated with the AR(1) model with unknown mean". *Econometrica* **51**, 1221–1231.

Tse, Y. K. (1982), "Edgeworth approximations in first-order stochastic difference equations with exogenous variables". *Journal of Econometrics* **20**, 175–195.

Ullah, A. (1980), "The exact, large sample and small disturbance conditions of dominance of biased estimators in linear models". *Economic Letters* **6**, 339–344.

White, J. S. (1958), "The limiting distribution of the serial correlation coefficient in the explosive case". *Annals of Mathematical Statistics* **29**, 1188–1197.

Esfandiar Maasoumi [1]

THE APPROXIMATE MOMENTS OF THE 3SLS REDUCED FORM ESTIMATOR AND A MELO COMBINATION OF OLS-3SLS FOR PREDICTION

1. INTRODUCTION

This paper develops the small sample approximations to the moments of the 3SLS *reduced form* estimator in a general linear simultaneous equations model under the classical assumptions. These approximations easily specialize to the case of k-class reduced form estimators, and they may be used to evaluate conditional forecasts based on the k-class or 3SLS estimators.

The method of derivation and its interpretation are discussed in Section 2. It can be seen that the techniques are more generally applicable and may be used to derive the approximate moments of any standard reduced form estimator.

Sections 2 and 3 of the paper discuss examples of how these approximate expressions have been used, for instance, to evaluate estimators or mixtures of estimators under quadratic loss and other moment-based criteria. In particular, an approximate Minimum Expected Loss (MELO) mixture of unrestricted OLS and 3SLS reduced form estimators is described.

Derivation of asymptotic expansions for the moments of econometric estimators began with Nagar (1959) who dealt with the k-class estimators of the *structural* coefficients in a linear simultaneous equations model. Nagar also showed how his formula for the bias may be used to develop "almost unbiased" estimators. Sawa (1973a) developed the "almost unbiased" application of these formulae, and (e.g.) Nagar and Carter (1976), Sawa (1973b) and Zellner and Vandaele (1975) have utilized these approximations to develop Minimum Variance, Minimum Mean Squared Error (MMSE) and MELO estimators. Such expansions to estimator biases have more recently been

[1] Department of Economics, Indiana University, Bloomington, Indiana 47405, also University of Southern California.

I. B. MacNeill and G. J. Umphrey (eds.), Time Series and Econometric Modelling, 359–371.

used by Rothenberg (1984) and others for correction of ML, GLS and other statistics in order to discuss questions of second and higher order efficiency.

Considerably less attention has been given to estimators and forecasts in the *reduced form* context. While asymptotic properties of the reduced form estimators can be evaluated simply by the methods of Goldberger *et al.* (1961) and Dhrymes (1973), higher order approximations do not seem to have been developed for many reduced form estimators. One exception is the work of Nagar and Sahay (1978), who were concerned with the Partially Restricted Reduced Form and the 2SLS estimators. The existence of the moments of several reduced form estimators is discussed by McCarthy (1972), Sargan (1976b), Maasoumi (1977, 1978, 1985, 1986), and Knight (1977).

2. APPROXIMATE MOMENTS OF THE 3SLS

Following Rothenberg (1984), we consider a random sample of size T from a population with a continuous density function which depends on an unknown parameter θ. Let $\hat{\theta}_T$ be the standardized estimator of θ and $F_T(k) = Pr[\hat{\theta}_T \leq k]$. For estimators which are Best Asymptotic Normal (BAN) and admit an Edgeworth expansion to order T^{-1}, we can write:

$$F_T(k) = F(k) + O(T^{-1}), \tag{1}$$

$$F(k) = \eta(k) + \frac{\mu(k)}{\sqrt{T}} + \frac{\gamma(k)}{T}. \tag{2}$$

In (2), $\eta(\cdot)$ is the standard normal distribution function and $\mu(\cdot)$ and $\gamma(\cdot)$ are usually polynomials multiplied by $\eta(k)$. η is referred to as a "first order" approximation, and F a "second order" approximation to F_T. Consequently, estimators that satisfy (1)–(2) are first-order efficient and so may be compared on the basis of the moments of F, i.e., the "second-order" moments. In this section we derive the moments of $F(\cdot)$ for the case of the 3SLS reduced form estimator. These approximate moments are thus well defined as the moments of the distributions which approximate the exact distribution, $F_T(\cdot)$. [The moments of η are first-order approximate moments of F_T in exactly the same sense. The only distinction with the traditional "asymptotic theory" is therefore one of degree of approximation and not of interpretation.]

The validity of Edgeworth approximations to F_T is discussed by Sargan (1976a) and generally does not depend on the *existence* of the moments. When the moments of F_T do not exist, however, care must be taken in interpreting the moments of F (or η) as approximations to the moments of F_T. In this situation the value of the approximate moments derives from the value of F (or η) as a representation of F_T. To the extent that this is done adequately by F, the study of its moments is of value in characterizing both F *and* F_T. We emphasize that, once again, there is no distinction between the interpretations of the moments of the asymptotic distribution of estimators and those of higher order approximate distributions.

The Linear Simultaneous Equations Model (SEM) is defined for Y $(T \times n)$ endogenous variables and Z $(T \times m)$ non-stochastic exogenous variables as follows:

$$AX' = BY' + \Gamma Z' = U', \tag{3}$$

where $X = (Y, Z)$ is a $T \times (n+m)$ matrix of all observations, $A = (B, \Gamma)$ is the $(n \times (n+m))$ matrix of the unknown coefficients, and U is the $(T \times n)$ matrix of the random disturbances such that each row, U_t, satisfies the following assumption:

$A1:\ U_t \sim \text{i.i.d.}(0, \Sigma)$.

Further assumptions of the classical SEM are:

$A2:\ \lim_{T\to\infty} T^{-1} Z'Z = M$, a constant matrix of rank m;

$A3:\ B$ is non-singular.

The reduced form of (3) is given by:

$$Y' = PZ' + V', \tag{4}$$

where $P = -B^{-1}\Gamma$, $V' = B^{-1}U'$ and, from $A1$, rows of V have zero mean and a common covariance matrix $\Omega = B^{-1}\Sigma^{-1}B'^{-1}$. When we need to, we assume that the *a priori* (identifying) restrictions on A are of the exclusion (zero order) type that may be represented as follows:

$$s - S\alpha = \text{Vec } A. \tag{5}$$

Here Vec denotes stacking by *rows*, α is the vector of the non-zero elements of A, $S = \text{diag}(S_1, \ldots, S_n)$ such that $XS_i = X_i$ represents the selection from X of only those columns that appear on the right hand side of the ith equation. The selection vector s represents the "normalization" restriction since its ith subvector, s_i, is such that $Xs_i = y_i$, the left hand side endogenous variable of the ith equation.

We define the following estimators of P:

$$\hat{P} = (Y'Z)(Z'Z)^{-1} \tag{6}$$

is the Unrestricted Least Squares (ULS) estimator of P in (4). Under the classical assumptions it is unbiased but less efficient than some Restricted Reduced Form (RRF) estimators,

$$P^+ = -B^{+^{-1}}\Gamma^+, \tag{7}$$

where B^+ and Γ^+ may be such full information estimators as 3SLS or FIML. In this paper I assume that the estimator A^+ satisfies the following property:

$$A4: \ \Delta A = A^+ - A = O_p(T^{-1/2}).$$

Define ΔB and $\Delta\Gamma$ as ΔA was defined. Then:

$$\begin{aligned} P^+ &= -(B+\Delta B)^{-1}(\Gamma + \Delta\Gamma) \\ &= -(B^{-1}\Delta B + I)^{-1}B^{-1}(\Gamma + \Delta\Gamma). \end{aligned} \tag{8}$$

Expanding $(B^{-1}\Delta B + I)^{-1}$,

$$\begin{aligned} (B^{-1}\Delta B + I)^{-1} = I - B^{-1}\Delta B + B^{-1}\Delta B \cdot B^{-1}\Delta B \\ + O_p(T^{-3/2}) \end{aligned} \tag{9}$$

and using this in (8), we have:

$$\begin{aligned} \Delta P &= P^+ - P \\ &= -B^{-1}\Delta\Gamma + B^{-1}\Delta B\ B^{-1}\Delta\Gamma + B^{-1}\Delta B \cdot B^{-1}\Gamma \\ &\quad - B^{-1}\Delta B \cdot B^{-1}\Delta B \cdot B^{-1}\Delta\Gamma \\ &\quad - B^{-1}\Delta B \cdot B^{-1}\Delta B \cdot B^{-1}\Gamma + O_p(T^{-2}). \end{aligned} \tag{10}$$

If one defines $Q = \left(\frac{P}{Im}\right)$ and rearrange, one finds that:

$$\Delta P = -B^{-1}\Delta A \cdot Q + B^{-1}\Delta B \cdot B^{-1}\Delta A \cdot Q + O_p(T^{-3/2}). \tag{11}$$

Let $b^+ = E$ (Vec ΔP) and $V(p^+)$ denote, respectively, the bias and the variance-covariance matrix of $p^+ =$ Vec P^+. Then:

$$V(p^+) = E\left[(\text{Vec}\Delta P)(\text{Vec}\Delta P)'\right] - b^+b^{+\prime}, \tag{12}$$

where the first term on the r.h.s. of (12) is the MSE of p^+. Approximations to b^+ are obtained by taking expectations of (11), term by term. For the first term, we note that

$$\text{Vec}\left(-B^{-1} \cdot \Delta A \cdot Q\right) = -\left(B^{-1} \otimes Q'\right) \text{Vec}\Delta A, \tag{13}$$

where $\otimes$ denotes the Kronecker product. The r.h.s. of (13) is the basic relationship used by (e.g.) Dhrymes (1973) and Goldberger *et al.* (1961) to obtain the asymptotic properties of P^+ from those of A^+. We will employ higher order expansions of E(Vec ΔA) in (13), and will combine these with the expectation of the second term of (11). To obtain the latter expectation, it is easier to work first with simpler linear functions of the required terms. We will thus first obtain:

$$E\left[\text{tr}\left(\Phi B^{-1}\Delta B\cdot B^{-1}\Delta A\cdot Q\right)\right] \tag{14}$$

for a known arbitrary $n\times n$ matrix Φ, and then recover the desired expectation from (14). We first note that:

$$\begin{aligned}(14) &= E\left\{\text{tr}\left[B'^{-1}\Delta B'\left(Q\Phi B^{-1}\right)'\Delta A'\right]\right\}\\ &= E\left\{(\text{Vec }\Delta A)'\left(B'^{-1}\otimes Q\Phi B^{-1}\right)(\text{Vec }\Delta B')\right\}.\end{aligned} \tag{15}$$

Let Π be a permutation matrix such that, for a matrix D,

$$\text{Vec } D' = \Pi\cdot\text{Vec } D. \tag{16}$$

We also define the following "slash" product, $\oslash$, for any two matrices F_1 and F_2:

$$(F_1\oslash F_2) = (F_1\otimes F_2)\Pi. \tag{17}$$

Subsequently, using (15)–(17), we conclude:

$$\begin{aligned}(14) &= E\left[(\text{Vec }\Delta A)'\left(B'^{-1}\oslash Q\Phi B^{-1}\right)(\text{Vec }\Delta B)\right]\\ &= \text{tr}\left\{\left(B'^{-1}\oslash Q\Phi B^{-1}\right)\cdot E\left[(\text{Vec }\Delta B)(\text{Vec }\Delta A)'\right]\right\}.\end{aligned} \tag{18}$$

The expression inside { } and, in particular, the covariance between ΔB and ΔA is to be evaluated. Let the asymptotic variance of ΔA be denoted by G. Then (an approximation to) the variance of ΔA is given by

$$E\left[(\text{Vec }\Delta A)(\text{Vec }\Delta A)'\right] \doteq \frac{1}{T}G, \tag{19}$$

where G is $n(n+m)\times n(n+m)$ and, (e.g.) for 3SLS, we have

$$G = S\left[S'\left(\Sigma^{-1}\otimes\overline{R}\right)S\right]^{-1}S', \tag{20}$$

where $\overline{R} = QMQ'$; see Sargan (1978) or Maasoumi (1978). [It is not necessary at this stage to consider approximations to the variance of ΔA. But this approximation is inevitable at the final stage.] Define $\overline{\psi} = (I_n 0_m)$ and $\psi^* = (I\otimes\overline{\psi})$. It follows that:

$$\text{Vec }\Delta B = \psi^*\cdot\text{Vec }\Delta A$$

and

$$E\left[(\text{Vec }\Delta B)(\text{Vec }\Delta A)'\right] \doteq \frac{1}{T}\psi^{*}G = \overline{G}, \quad \text{say}. \tag{21}$$

Using (21) in (18), we may assert:

$$\begin{aligned}(14) &\doteq \operatorname{tr}\left\{\left[B'^{-1} \oslash Q\Phi B^{-1}\right]\overline{G}\right\} \\ &= \operatorname{tr}\left\{\left[B'^{-1} \otimes Q\Phi B^{-1}\right]\Pi_n^n\overline{G}\right\},\end{aligned} \tag{22}$$

where

$$\Pi_n^n\overline{G} = \begin{pmatrix} \overline{G}_{11} & \cdots & \overline{G}_{1n} \\ \vdots & & \vdots \\ \overline{G}_{n1} & \cdots & \overline{G}_{nn} \end{pmatrix} \tag{23}$$

with each submatrix $\overline{G}_{ij}$ having $n \times (n+m)$ dimensions. From (22) it may be verified that:

$$(14) \doteq \sum_{j=1}^{n} b^{jj} \cdot \operatorname{tr}\left(Q\Phi B^{-1}\overline{G}_{jj}\right), \tag{24}$$

where b^{ii} is the ith diagonal element of B'^{-1}, and noting that $E[\operatorname{tr}(\cdot)] = \operatorname{tr}[E(\cdot)]$, we have:

$$E\left(B^{-1}\Delta B \cdot B^{-1}\Delta A \cdot Q\right) \doteq \sum_{j=1}^{n} b^{jj} B^{-1}\overline{G}_{jj}Q. \tag{25}$$

Gathering terms from (25) and (13), we have:

$$b^{+} = (B^{-1} \otimes Q')\left[\sum_{j=1}^{n} b^{jj}(\text{Vec }\overline{G}_{jj}) - E(\text{Vec }\Delta A)\right] + \ldots. \tag{26}$$

Approximations to Vec $\overline{G}_{jj}$ and $E(\text{Vec }\Delta A)$ should retain all terms of order T^{-1} or larger in order to correspond to (11). Lower order approximations for b^{+} will necessarily omit the first term of (26). It remains to replace $E(\text{Vec }\Delta A)$ with the approximate bias of A^{+}. For a variety of structural estimators, such as k-class, FIML and 3SLS, these approximations have been given in the literature. For 3SLS, in particular, these are given by Sargan (1976a). To order T^{-2}, the bias is as follows:

$$\begin{aligned}E(\text{Vec }\Delta A) \doteq\ & mG\left(\Sigma^{-1} \otimes I\right)q + \Psi'\left(I \otimes \overline{Q}_{s}\right)q \\ & + \overline{Q}q - 2G\left(\Sigma^{-1} \otimes I\right)\Psi^{T}q - G\left[\left(\Sigma^{-1}H\Sigma^{-1}\right) \otimes I\right]q \\ & - G\left(\Sigma^{-1} \otimes \Psi_{s}\right)q,\end{aligned} \tag{27}$$

where Ψ^T is the block transpose of $\Psi = (\Sigma^{-1} \otimes \overline{X}'\overline{X})G$, $\Psi_s = \sum_j \Psi_{jj}$, $\overline{X} = (\overline{Y}\ Z)$, $\overline{Y} = Y - V$, $q = (q_1', q_2', \ldots, q_n')'$, $q_i = 1/T\ E(V^{*\prime}u_i)$, $V^* = (UB'^{-1}0)$ is $T \times (n+m)$, u_i is the T-element vector of disturbances in the ith equation, $H = \text{tr}[G(I \otimes \overline{X}'\overline{X})]$, and $\overline{Q} = \text{diag.}(\overline{Q}_1, \overline{Q}_2, \ldots, \overline{Q}_n)$, where $\overline{Q}_i$ is obtained from $(\overline{X}_i'\overline{X}_i)^{-1}$ by adding rows and columns of zeros corresponding to the excluded variables in the ith equation, and $\overline{Q}_S = \sum_{i=1}^n \overline{Q}_i$.

The above techniques may be used to obtain higher order approximations to the second and higher order moments of P^+. These expressions can be used to obtain almost unbiased estimators, and are also useful in comparing higher order efficiency of the estimators. We will proceed to demonstrate their use by considering an "optimal" mixture of the ULS estimator ($\hat{P}$) and the 3SLS (P^+) reduced form estimators under quadratic loss. The latter (or its expectation) is approximated using the moment expansions given above.

2. APPROXIMATE MINIMUM EXPECTED LOSS (MELO) COMBINATIONS OF OLS-3SLS

This and the next section are based on Maasoumi (1985).

Let $\hat{P} = Y'Z(Z'Z)^{-1}$ denote the ULS estimator of P and $P^+ = -B^{+-1}\Gamma^+$ another estimator derived from such restricted estimators of B and Γ as 2SLS, 3SLS, FIML, etc. We propose the following mixed estimator of P:

$$P^* = \lambda\hat{P} + (1-\lambda)P^+ \tag{28}$$
$$= P^+ + \lambda(\hat{P} - P^+). \tag{29}$$

In the remainder of this paper lower case letters p, $\hat{p}$, p^+ denote vec P, vec $\hat{P}$ and vec P^+, respectively. Let the bias in p^* be denoted by b^*. It is readily seen that, since $E(\hat{p}) - p = 0$,

$$|b^*| = |E(p^*) - p| = |(1-\lambda)b^+| \leq |b^+|, \qquad \text{if } 0 \leq \lambda \leq 1. \tag{30}$$

Using the following identity, the variance matrix of p^*, $V(p^*)$, is obtained in terms of the variances of $\hat{p}$, p^+ and their covariance:

$$p^* - E(p^*) = \lambda(\hat{p} - p) + (1-\lambda)(p^+ - E(p^+)); \tag{31}$$

$$V(p^*) = \lambda^2 V(\hat{p}) + (1-\lambda)^2 V(p^+) + 2\lambda(1-\lambda)\text{cov}(\hat{p}, p^+). \tag{32}$$

Under the standard assumptions of our model, and since the mixing parameter λ is a constant, the bias and variance of p^* would be finite if p^+ has finite moments. For instance, when p^+ represents the FIML estimator, b^*

and $V(p^*)$ are finite so long as $T - n - m \geq 2$; see Sargan (1976b). If p^+ represents either 2SLS or 3SLS, then b^* and $V(p^*)$ will not be finite unless $\lambda = 1(p^* \equiv \hat{p})$. For exactly identified models $P^* = \hat{P}$ since $P^+ \equiv \hat{P}$ in that case.

Since $\hat{p}$ is a consistent estimator under our assumptions, it is seen that p^* will be consistent if p^+ is consistent. If not, the inconsistency in p^* will be smaller than that in p^+ as long as $\lambda \in [0, 1]$. If both $\hat{p}$ and p^+ are inconsistent but have the same limit in probability, then p^* will also be inconsistent with the same plim as $\hat{p}$ (or p^+).

As for asymptotic efficiency, the derivations given in the next section may be used to verify that:

$$\begin{aligned} AV(p^*) &= \lambda^2 AV(\hat{p}) + (1 - \lambda^2) AV(p^+) \\ &= AV(p^+) + \lambda^2 \left[AV(\hat{p}) - AV(p^+)\right], \end{aligned} \tag{33}$$

where $AV(\cdot)$ denotes the asymptotic variance. From (33) it is clear that p^* is more efficient than the ULS so long as $\hat{p}$ is less efficient than the restricted estimator p^+. While this is the case for the full information estimators such as the 3SLS and FIML, it is not always so for the limited information estimators such as 2SLS and LIML; see Dhrymes (1973). The latter statement holds even if p^+ is replaced by the Partially Restricted Reduced Form (PRRF) estimator of Amemiya (1966) and Kakwani and Court (1972). For while PRRF has finite moments (see Knight, 1977), it is not necessarily more asymptotically efficient than 2SLS. On the other hand, p^* is less efficient than 3SLS and FIML, but can be more efficient than (e.g.) 2SLS whenever $\hat{p}$ is.

3. MIXED PREDICTION UNDER QUADRATIC LOSS

Let $Y_f^* = P^* Z_f$ be the predictor of Y_f conditional on Z_f under the assumption that $Y_f = PZ_f + v_f$, where v_f denotes the forecast period random disturbance with the same properties as v_t, $t = 1, \ldots, T$. The forecast error and a general quadratic loss are given as follows:

$$Y_f^* - Y_f = (P^* - P)Z_f - v_f, \tag{34}$$

and

$$\begin{aligned} L(Y_f^*, Y_f) &= (Y_f^* - Y_f)' W^* (Y_f^* - Y_f) \\ &= \operatorname{tr}\left[W^* (Y_f^* - Y_f)(Y_f^* - Y_f)'\right], \end{aligned} \tag{35}$$

where W^* is a symmetric, positive definite matrix of known weights. From (34)–(35) the expected loss (risk) is derived as follows:

$$\begin{aligned} R(Y_f^*) &= E[L(\cdot)] \\ &= \operatorname{tr}[W\mathrm{MSE}(P^*)] + \operatorname{tr}[W\Omega], \end{aligned} \tag{36}$$

where $W = (W^* \otimes Z_f Z_f')$ and $\mathrm{MSE}(P^*)$ is the MSE matrix of p^*. Since the second term of (36) is common to all conditional forecasts, we focus on the first term which is a well known *estimation* risk function. Consequently, minimization of $R(Y_f^*)$ is equivalent to minimization of $R(p^*) = \operatorname{tr}[W\mathrm{MSE}(p^*)]$ with respect to λ. We note that, from (30) to (32):

$$\begin{aligned} R(p^*) = {} & \lambda^2 \operatorname{tr}[WV_0] + (1-\lambda)^2 \operatorname{tr}[WV^+] \\ & + 2\lambda(1-\lambda)\operatorname{tr}[W\operatorname{cov}(\hat{p}, p^+)] + (1-\lambda)^2 b^{+\prime} W b^+, \end{aligned} \tag{37}$$

where $V_0 = V(\hat{p})$ and $V^+ = V(p^+)$. To minimize $R(p^*)$ with respect to λ consider

$$\begin{aligned} \frac{\partial R(\cdot)}{\partial \lambda} = {} & 2\lambda \operatorname{tr}[WV_0] - 2(1-\lambda)\operatorname{tr}[WV^+] \\ & + 2(1-2\lambda)\operatorname{tr}[W \operatorname{cov}(\cdot)] - 2(1-\lambda) b^{+\prime} W b^+ \end{aligned} \tag{38}$$

and

$$\begin{aligned} \frac{\partial^2 R(\cdot)}{\partial \lambda^2} &= 2\operatorname{tr}[WV_0] + 2\operatorname{tr}[WV^+] - 4\operatorname{tr}[W \operatorname{cov}(\cdot)] + 2b^{+\prime} W b^+ \\ &= 2\operatorname{tr}[WV(\hat{p} - p^+)] + 2b^{+\prime} W b^+ \geq 0, \end{aligned} \tag{39}$$

where $V(\hat{p} - p^+)$ denotes the variance of $(\hat{p} - p^+)$. From (38), the optimal value λ_1^* of λ is obtained by solving $\partial R(\cdot)/\partial \lambda = 0$,

$$\lambda_1^* = \frac{b^{+\prime} W b^+ + \operatorname{tr}[W(V^+ - \operatorname{cov}(\cdot))]}{b^{+\prime} W b^+ + \operatorname{tr}[WV(\hat{p} - p^+)]}. \tag{40}$$

It may be observed that the denominator of λ_1^* is non-negative, and $\lambda_1^* \leq 1$ if $[V_0 - \operatorname{cov}(\hat{p}, p^+)]$ is positive semidefinite. In what follows we demonstrate that this condition, as well as the range of possible values for λ_1^*, depends on the level of approximation considered for the otherwise unknown moments entering in (40). Equivalently, these issues depend on the order of finite sample approximations for the $L(\cdot)$ and $R(\cdot)$ functions.

Strictly speaking, since 2SLS and 3SLS reduced form estimators have no finite moments, $\lambda_1^* = 1 \Rightarrow p^* = \hat{p}$ is the only member of the corresponding mixtures that has finite quadratic risk. For FIML, on the other hand, all the corresponding mixtures have finite risk if $T-n-m \geq 2$. In either case, when we consider Nagar-type approximations to these moments we are in effect evaluating the risk functions with respect to finite sample *approximations* to the exact sampling distributions of the p^+ estimator.

We note the following well known results:

$$\operatorname*{plim}_{T\to\infty} T(\hat{p}-p)(\hat{p}-p)' = (\Omega \otimes M^{-1}) = \lim\ TV_0 = O(1). \tag{41}$$

In other words, under our standard assumptions, $V_0 = O(T^{-1})$. When p^+ is the 3SLS estimator, it follows that:

$$\begin{aligned}\operatorname*{plim}_{T\to\infty} T(p^+-p)(p^+-p)' &= (B^{-1}\otimes Q')G(B'^{-1}\otimes Q)\\ &= O(1)\\ &= \lim\ TV_a^+, \quad \text{say.}\end{aligned} \tag{42}$$

Consequently $V_a^+ = O(T^{-1})$. Maasoumi (1978) has shown that, if p^+ denotes the 3SLS estimator,

$$\operatorname*{plim}_{T\to\infty} T(p^+-p)(\hat{p}-p^+)' = 0 \tag{43}$$

and

$$\operatorname*{plim}_{T\to\infty} T(p^+-p)(\hat{p}-p)' = \lim\ TV_a^+ = O(1). \tag{44}$$

The *asymptotic* properties given in (43)–(44) hold for both the 3SLS and FIML estimators and may also be deduced from a Rao-Blackwell lemma; see, for example, Hausman (1978). They do not hold for the asymptotically less efficient 2SLS or LIML reduced form estimators.

In λ_1^*, if we replace all terms with their $O(T^{-1})$ approximations and utilize the results in (41)–(44), we find

$$\lambda_2^* = b_a^{+\prime}Wb_a^+ / \left\{\operatorname{tr}[W(V_0 - V_a^+)] + b_a^{\prime +}Wb_a^+\right\}, \tag{45}$$

where b_a^+ is the approximate bias of p^+ obtained by retaining terms of $O_p(T^{-1/2})$ in the expansion of $p^+ - p$. We note that, if $b_a^+ = O(T^{-1/2})$, $\lambda_2^* = O(1)$ since $V_0 = (\Omega \otimes (Z'Z)^{-1}) = O(T^{-1})$ and

$$\begin{aligned}V_a^+ &= (B^{-1}\otimes Q')\, S\left[S'\left(\Sigma^{-1}\otimes \hat{R}\right)S\right]^{-1} S'\,(B'^{-1}\otimes Q)\\ &= O(T^{-1}),\end{aligned} \tag{46}$$

where $\hat{R} = (X'Z)(Z'Z)^{-1}(Z'X)$, and under these conditions we have:

$$0 \leq \lambda_2^* \leq 1 \tag{47}$$

whenever $V_0 - V_a^+$ is non-negative definite. This last condition is clearly satisfied for the full information estimators which permitted the simple formula in (45).

The mixing parameter λ_2^* has several desirable properties:

(i) As the efficiency gain of the restricted estimator over the ULS decreases, $\lambda_2^* \rightarrow 1$ and the corresponding mixed estimator (p^*) moves closer to the simple ULS estimator.

(ii) As the bias of the efficient estimator increases, $\lambda_2^* \rightarrow 1$ and $p^* \rightarrow \hat{p}$. This is evidently desirable since this bias would be large either due to structural misspecification or due to poor finite sample properties of the efficient estimator (even as judged by its approximate distribution), or both. On the other hand, $p^* \rightarrow p^+$ as $\lambda_2^* \rightarrow 0$ which occurs as $b^+ \rightarrow 0$.

(iii) The formula for λ_2^* is seen to provide a mechanism for pooling of estimators (predictors) which accounts for the efficiency-bias trade-offs.

(iv) Under correct specification p^+ is a consistent estimator. Therefore, when the sample size is "sufficiently" large it is reasonable to expect b_a^+ to be close to zero. This will also pool the mixed estimator toward the asymptotically desirable estimator (p^+). This pattern of large sample behavior for b^+ has been confirmed by numerous Monte Carlo studies—for example, see Maasoumi (1977), and Rhodes and Westbrook (1980).

The formula given for λ_1^* in (40) may of course be approximated at a higher level. It can be verified that the next possible degree of approximation will retain terms of $O(T^{-1})$. The resulting value for λ will behave more like λ_1^* while exhibiting only some of the properties enumerated for λ_2^*. While these higher order expressions may be computed from Section 2 of this paper and the moments given by Sargan (1976a), *improved* approximation is by no means guaranteed by the additional terms. Some have argued that if $O(T^{-1})$ terms are of significance then the sample size is too small to allow reliable inferences in reasonably sized simultaneous systems. Nevertheless, there is a higher level of approximation for b^+ that results in an interesting variant of λ_2^*. This is obtained from λ_2^* by replacing b_a^+ with an $O(T^{-1})$ approximation of b^+ given in Section 2, and maintaining the $O(T^{-1})$ approximations for variances and covariances. This approximation produces a mixing parameter, λ^*, which is $O(T^{-1})$, and therefore a mixed estimator (predicator) which is asymptotically equivalent to the asymptotically desirable method based on p^+. A further justification for this choice of λ is that, if p^+ is consistent, the odd order terms ($O_p(T^{-1/2})$, $O_p(T^{-3/2})$ etc.) in the expansion of $p^+ - p$ have zero expectations under the normality assumption and may be dropped in obtaining an $O(T^{-1})$ approximation for b^+.

ACKNOWLEDGMENTS

I wish to thank the Centre of Decision Sciences and Econometrics, Department of Economics, The University of Western Ontario, for their invitation and financial support to attend the Symposia in Statistics and the Festschrift in honor of Professor V. M. Joshi. In particular, I should like to thank Professors R. A. L. Carter and Aman Ullah for their hospitality. I am also grateful to Mrs. Alice Gilmore for her skillful typing.

REFERENCES

Amemiya, T. (1966), "On the use of principal components of independent variables in two-stage least-squares estimation". *International Economic Review* **7**, 283–303.

Dhrymes, P. (1973), "Restricted and unrestricted reduced forms: asymptotic distribution and relative efficiency". *Econometrica* **41**, 119–134.

Goldberger, A., A. L. Nagar, and H. S. Odeh (1961), "The covariance matrices of reduced form coefficients and of forecasts in a structural model". *Econometrica* **29**, 556–573.

Hausman, J. (1978), "Specification tests in econometrics". *Econometrica* **46**, 1251–1271.

Kakwani, N. C., and R. H. Court (1972), "Reduced form coefficient estimation and forecasting from a simultaneous equation model". *Australian Journal of Statistics* **14**, 143–160.

Knight, J. L. (1977), "On the existence of moments of the partially restricted reduced-form estimators from a simultaneous-equation model". *Journal of Econometrics* **5**, 315–321.

Maasoumi, E. (1977), "A study of improved methods of estimating the reduced form coefficients based upon 3SLS". Ph.D. Thesis, London School of Economics and Political Science.

Maasoumi, E. (1978), "A modified Stein-like estimator for the reduced form parameters of simultaneous equations". *Econometrica* **46**, 695–703.

Maasoumi, E. (1985), "Mixing forecasts in linear simultaneous equations under quadratic loss". Mimeograph, Indiana University, Bloomington.

Maasoumi, E. (1986), "Reduced form estimation and prediction from uncertain structural models: a generic approach". *Journal of Econometrics,* **31**, 3–29.

McCarthy, M. (1972), "A note on the forecasting properties of 2SLS restricted reduced forms". *International Economic Review* **13**, 757–761.

Nagar, A. L. (1959), "The bias and moment matrix of the general *k*-class estimator of the parameters in simultaneous equations". *Econometrica* **27**, 575–595.

Nagar, A. L., and R. Carter (1974), "Minimum second moment estimation in simultaneous equation systems". *International Economic Review* **15**, 31–38.

Nagar, A. L., and S. N. Sahay (1978), "The bias and mean squared error of forecasts from partially restricted reduced form". *Journal of Econometrics* **7**, 227–243.

Rhodes, G. F., and D. Westbrook (1980), "An experimental investigation of the properties of simultaneous equations estimators". (July), Colorado State University, Fort Collins, technical report (mimeo).

Rothenberg, T. (1984), "Approximating the distributions of econometric estimators and test statistics". In *Handbook of Econometrics*, Volume 2, ed. Z. Griliches and M. Intriligator, Ch. 15. Amsterdam: North-Holland.

Sargan, J. D. (1970), "The finite sample distribution of the FIML estimator". Presented at the World Congress of Econometric Society, Cambridge, England.

Sargan, J. D. (1976a), "Econometric estimators and Edgeworth approximations". *Econometrica* **44**, 421–448.

Sargan, J. D. (1976b), "The existence of the moments of estimated reduced form coefficients". Discussion Paper No. A6, LSE Econometrics Programme.

Sawa, T. (1973a), "Almost unbiased estimator in simultaneous equations systems". *International Economic Review* **14**, 97–106.

Sawa, T. (1973b), "The mean square error of a combined estimator and numerical comparison with the TSLS estimator". *Journal of Econometrics* **1**, 115–132.

Zellner, A., and W. Vandaele (1975), "Bayes-Stein estimators for *k*-means, regression and simultaneous equation models". In *Studies in Bayesian Econometrics and Statistics in Honor of Leonard J. Savage*, ed. S. E. Feinberg and A. Zellner, pp. 627–653. Amsterdam: North-Holland.

Michael R. Veall [1]

BOOTSTRAPPING AND FORECAST UNCERTAINTY: A MONTE CARLO ANALYSIS

ABSTRACT

For certain kinds of problems such as capacity planning, the estimate of forecast uncertainty may be as important as the prediction itself. Earlier research (Veall, 1985) has applied Efron's bootstrapping technique to a linear regression forecast of peak demand for Ontario Hydro. This paper presents a limited Monte Carlo analysis to assess the potential accuracy of bootstrapping for this example.

1. INTRODUCTION

Since the invention of the bootstrap by Efron (1979), there have been a number of applications of the technique to problems in applied econometrics (for example, Finke and Theil, 1984; Freedman and Peters, 1984c; Korajczyk, 1985; Vinod and Raj, 1984). One particular example of great interest is the work with energy forecasting of Freedman and Peters (1984a,b), which is closely related to the simulation methods of Fair (1979, 1980) for evaluating macroeconomic predictive accuracy. Standard methods of estimating forecast uncertainty are inadequate because they rely on normality assumptions and/or on the assumption that values of the independent variables for the forecast period are known with certainty. Largely because of their nonparametric features, the bootstrap and other methods of computationally intensive statistics represent potential solutions to these difficulties.

This paper will describe a simple Monte Carlo evaluation of the success of bootstrapping in assessing forecast uncertainty. Because of the large computer costs involved in using a computer-intensive technique to analyze another computer-intensive technique, the Monte Carlo experiment is neces-

[1] Department of Economics, The University of Western Ontario, London, Ontario N6A 5C2

I. B. MacNeill and G. J. Umphrey (eds.), Time Series and Econometric Modelling, 373–384.

sarily very limited. However, the case chosen is of special interest as it relates to the forecasting of peak electricity demand in Ontario, where the uncertainty estimate is critical with respect to capacity planning (Ellis, 1980). If demand is very uncertain, the risk associated with generation types which have high lead times, high fixed costs but low variable costs may be too great and capacities with lower lead times and fixed costs may be desirable even though these have higher running costs. If demand estimates have low uncertainty, all other things equal, capital-intensive, high-lead-time generating capacity will be more worthwhile.

The example chosen is based on Veall (1985). To the author's knowledge, the research here is the first sampling experiment on bootstrap estimation of the probability distribution of the forecast error.

2. BOOTSTRAPPING FORECAST UNCERTAINTY: AN EXAMPLE

Explanations of bootstrapping are given by Efron (1979) and Efron and Gong (1983), with some description of the forecasting problem given by Freedman and Peters (1984a,b). A detailed discussion of the example here is presented by Veall (1985) so that only the briefest possible explanation is provided in the following.

The model here is very simple, so that attention can be focussed on the estimation of the forecast uncertainty rather than the point estimates. Estimated by Ordinary Least Squares (OLS), the result for the first equation is:

$$\log(PEAK_t) = 0.8953 + 0.9482 \log(AMW_t) + \hat{\varepsilon}_t \tag{1}$$

with standard errors 0.1257 and 0.0141 respectively, an R^2 of .9960 and a Durbin-Watson statistic of 1.7709. For the second equation:

$$\begin{aligned}\log(AMW_t) = 1.2331 - 0.3887\ \log(P_t) + 0.4336\ \log(Y_t)\\ + 0.0293\ TIME_t + \hat{\eta}_t\end{aligned} \tag{2}$$

with standard errors 1.7291, 0.0495, 0.1083 and 0.0046 respectively, an R^2 of .9985 and a Durbin-Watson statistic of 1.7526. All data are for the period 1963–1982 for the East System of Ontario Hydro (which comprises about 90 percent of total provincial demand), $PEAK_t$ is peak demand in megawatts (mW), AMW_t is average demand over the year, also in mW, P_t is the real average price of electricity, Y_t is total Ontario real income and $TIME_t$ is a linear time trend. The data are described by Veall (1985) and are available upon request.

A number of diagnostic tests were performed on the model given by (1)–(2). While the details are given by Veall (1985), summarizing the main results, it is found at the 5 percent level:

(1) The null hypothesis of no serial correlation could not be rejected for either disturbance (Durbin-Watson test, Godfrey (1978) test).

(2) The null hypothesis of homoscedasticity could not be rejected for either disturbance (White (1980) test, Breusch and Pagan (1979) test, Engle (1982) ARCH test).

(3) The coefficients of additional variables added to (2), specifically weather variables and the log of the real price of natural gas, were not significantly different from zero.

(4) The null hypothesis that ε and η were uncorrelated could not be rejected (Hausman (1978) test).

(5) The null hypothesis of normality of ε and η could each not be rejected (Shapiro and Wilk (1965) test, Shapiro and Francia (1972) test, Kiefer and Salmon (1983) test, all as applied to the OLS residuals as suggested by White and MacDonald (1980)).

While the power of all these tests in such a small sample is questionable, the fact that there is no evidence of heteroscedasticity, serial correlation and simultaneous equations bias is encouraging for the following bootstrap application. As the bootstrap is nonparametric, the normality tests are not as important.

For forecasting the peak demand for the year 1990, it is assumed that the real price of electricity will be at its 1982 level and real income will grow at 2 percent per year. To follow what is perhaps the most complex example given by Veall (1985), it is in addition assumed that there is a subjective uncertainty attached to each of the log real price and log real income forecasts corresponding to a random normal variable with mean 0 and standard deviation 0.1. One of the justifications for the bootstrap is that as $PEAK_{1990}$ is a function of AMW_{1990}, which is in turn a function of P_{1990} and Y_{1990}, then even if ε, η, Y_{1990} and P_{1990} are all normal, $PEAK_{1990}$ will not be. This suggests that the nonparametric aspect of the bootstrap may be very important.

The actual bootstrapping is performed as follows:

(i) An ordinary regression forecast for 1990, $\widehat{PEAK}_{1990}$, is calculated using the estimates from (1) and (2) and the point predictions of P_{1990}, Y_{1990} and $TIME_{1990}$, and then taking the exponent.

(ii) Artificial residual sets $\hat{\eta}^i$ and $\hat{\varepsilon}^i$ for 1963–1982 are created by drawing randomly and with replacement 20 times from $\hat{\eta}$ and $\hat{\varepsilon}$. This is repeated $i = 1, \ldots, B$ times. Random numbers throughout the procedure are generated by routines due to Wickman and Hill (1982) and (for normal random numbers), Beasley and Springer (1977).

(iii) B artificial samples are created by replacing $\hat{\eta}_t$ with $\hat{\eta}_t^i$ in (2) and taking the resulting log (AMW_t^i) and using it along with $\hat{\varepsilon}_t^i$ in (1).

(iv) A forecast $\widehat{PEAK}^i_{1990}$ is calculated by estimating the model (1^i) and (2^i) on each bootstrap sample $i = 1, \ldots, B$ and following the process described in (i) above, using the estimates from the bootstrap sample.

(v) A simulated actual $PEAK^i_{1990}$ is calculated by first constructing $\log(P^i_{1990})$ and $\log(Y^i_{1990})$ which are B independent artificially-generated $N(0, 0.01)$ variables added to $\log(P_{1990})$ and $\log(Y_{1990})$ respectively. Then $\log(AMW^i_{1990})$ is calculated by putting $\log(P^i_{1990})$ and $\log(Y^i_{1990})$ into (2) and replacing $\hat{\eta}$ with $\hat{\eta}^i_{1990}$, another residual drawn randomly and with replacement from $\hat{\eta}$. Continuing, $PEAK^i_{1990}$ is calculated by putting $\log(AMW^i_{1990})$ in (1) and replacing $\hat{\varepsilon}$ with $\hat{\varepsilon}^i_{1990}$ (another residual drawn randomly and with replacement from $\hat{\varepsilon}$) and taking the exponent.

(vi) The simulated forecast error SFE^i is calculated as

$$PEAK^i_{1990} - \widehat{PEAK}^i_{1990}.$$

(vii) The distribution of $PEAK_{1990}$, conditional on the model and current information is modeled as:

$$\widehat{PEAK}_{1990} + SFE^i. \tag{3}$$

In words, forecasts from the bootstrap samples are compared to "simulated actuals" computed using random draws from the residuals $\hat{\varepsilon}$ and $\hat{\eta}$ as well as assumed subjective probability distributions for P_{1990} and Y_{1990}. The differences are estimated forecast errors and are used to estimate the uncertainty in the single regression forecast $\widehat{PEAK}_{1990}$.

3. MONTE CARLO ANALYSIS

As mentioned, the Monte Carlo analysis of the bootstrap will be necessarily simple and limited, because of computational cost. The first step was to calculate a "true" distribution for 1990 peak demand, conditional on the model and current information. This is done by maintaining the assumption that $\log(P_{1990})$ and $\log(Y_{1990})$ are $N(0, 0.01)$ and in addition assuming ε and η are also normal with standard deviations equal to .0206420 and .0142578 respectively. These standard deviations are the standard errors of regressions (1) and (2). It can be seen that the 1990 peak will then be distributed lognormally with location and scale parameters that are easy to calculate.

The next step is to determine how well the bootstrap estimate of the distribution could estimate the true distribution just calculated. To do this, 200 Monte Carlo data sets are calculated for 1963–1982 using the estimates and the variables P_t, Y_t and $TIME_t$ (with no randomness) from (1) and (2) with $\hat{\varepsilon}_t$ and $\hat{\eta}_t$ replaced with random normal disturbances with standard deviations equal again to the standard errors of their respective regressions. Each Monte Carlo data set is then used as a basis for a bootstrap procedure and the resulting set of bootstrap estimates of the probability distribution is compared to the "true" distribution above. A variety of methods are attempted, such as using 100, 500 or 1000 bootstraps, changing the number of Monte Carlo data sets from 200 to 500 and employing Efron's smoothed bootstrap (1982, p. 30) with various smoothing constants. The results are presented in Table 1. The analysis will be largely from the perspective of the upper tail, as presumably this is most important to the capacity planner.

With 100 bootstraps, the average bootstrap estimate of the .90, .95 and .99 probability points is quite accurate, in all cases within about .5 percent of the truth. The variation across Monte Carlo runs can be analyzed somewhat crudely by looking at the standard deviations of these bootstrap estimates: these are typically of moderate size, about 600–800 mW or about 3 percent of the true value. For this example the lower tail does not fit as well but both the median of the bootstrap and the bootstrap forecast (the average of the bootstrap forecasts) are quite accurate estimates of the actual median 22096 (which is also the actual forecast with the original data). Finally, the bootstrap estimates of the bias (equal to the forecast from each Monte Carlo 1963–1982 data set subtracted from the average of the bootstrap forecasts) are all tiny.

Increasing the number of bootstraps to 500 or 1000 doesn't change these results very much, improving the fit in the lower tail and, surprisingly for 1000, worsening the fit slightly in the upper tail. Returning to 100 bootstraps but increasing the number of Monte Carlo runs to 500 has no important effect.

In different kinds of bootstrap problems than the one studied here, Efron (1982, p. 32) reported some success with smoothing, which is accomplished by adding independent draws from pseudo-random variates $N(0, k^2\hat{\sigma}_n^2)$ and $N(0, k^2\hat{\sigma}_\varepsilon^2)$ to the bootstrap residuals $\hat{\eta}_t^i$ and $\hat{\varepsilon}_t^i$ respectively, where k is an arbitrary smoothing constant. These disturbances are then scaled back to their estimated variances by dividing each result by $(1 + k^2)^{1/2}$, and the rest of the bootstrap proceeds as before. Compared to the unsmoothed case with 100 bootstraps, a smoothing constant k of .1 helps very slightly in the tails but there seems to be little gain from further smoothing. As $k \to \infty$, the smoothed bootstrap becomes the so-called parametric bootstrap which replaces the residual distribution entirely with a parametric (in this case

Table 1. *Monte Carlo Results on Bootstrap Estimate of Distribution of 1990 Peak Demand (Random 1990 Price and Income)*

	Percentage Probability Points									Average	
	.01	.05	.10	.25	.50	.75	.90	.95	.99	Bootstrap Forecast	Bias Estimate
Actual Distribution	19196	20004	20448	21212	22096	23017	23877	24407	25434	- - -	- - -
Bootstrap Estimate (100 Bootstraps)	18388 (971)	19626 (779)	20146 (749)	21021 (721)	22004 (672)	23034 (665)	23954 (665)	24472 (683)	25320 (753)	22030 (648)	–15 (72)
Bootstrap Estimate (500 Bootstraps)	18673 (735)	19617 (692)	20117 (688)	20987 (675)	21977 (664)	22986 (646)	23904 (640)	24467 (636)	25520 (666)	21983 (655)	–7 (27)
Bootstrap Estimate (1000 Bootstraps)	18766 (748)	19707 (729)	20222 (715)	21093 (695)	22076 (679)	23092 (662)	24027 (650)	24605 (656)	25692 (658)	22079 (678)	–8 (22)
Bootstrap Estimate (100 Bootstraps/500 Monte Carlo runs)	18534 (991)	19658 (819)	20179 (792)	21067 (762)	22053 (757)	23040 (741)	23960 (763)	24496 (786)	25422 (874)	22055 (729)	–7 (65)

Standard errors are in parentheses. All results based on 200 Monte Carlo replications unless stated otherwise.

Table 1. *(Continued)*

	Percentage Probability Points									Average Bootstrap	Bias
	.01	.05	.10	.25	.50	.75	.90	.95	.99	Forecast	Estimate
Smoothed Bootstrap (100 Bootstraps)											
$k = .10$	18401	19551	20110	21004	22000	23008	23919	24430	25329	22009	–5
	(917)	(826)	(817)	(769)	(747)	(728)	(725)	(735)	(811)	(711)	(67)
$k = .25$	18536	19588	20123	21004	21990	23006	23903	24444	25301	22006	–9
	(932)	(791)	(761)	(723)	(705)	(701)	(707)	(747)	(872)	(686)	(75)
$k = .50$	18486	19555	20119	21007	22009	23013	23919	24493	25353	22020	–10
	(931)	(850)	(800)	(772)	(745)	(746)	(727)	(736)	(754)	(726)	(67)
Parametric Bootstraps	18389	19597	20147	21031	22061	23098	24038	24587	25510	22071	–11
	(980)	(823)	(820)	(778)	(746)	(733)	(717)	(728)	(808)	(729)	(70)

Standard errors in parentheses. All results based on 200 Monte Carlo replications unless stated otherwise.

normal) distribution. Again somewhat surprisingly the Monte Carlo results indicate that there is little gain from this procedure for this example.

To isolate the residual/bootstrap part of the exercise, the calculations in Table 2 are based on another case (Veall, 1985) with the assumption that 1990 price and income are known and hence nonrandom. Again for this example the bootstrap is quite accurate in the upper tail, although perhaps not as accurate as before. The bootstrap continues to overestimate the dispersion in the lower tail and the pattern is not changed much by smoothing.

Finally, Table 3 is based on the assumption that the peak distribution is uniform, so that the nonparametric nature of the bootstrap may be explored. Maintaining the assumption of nonrandom 1990 price and income, the actual 1990 peak probability distribution can be calculated by exponentiating the appropriate convolution of the normal and uniform. The table shows that the upper tail is again fit fairly well by the bootstrap. Smoothing again is not very helpful but is not especially deleterious, despite the fact that it is employing a normal kernel and the actual distribution is nonnormal.

4. CONCLUSIONS

For this series of Monte Carlo experiments on this example, the accuracy of the bootstrap in estimating the upper tail of the probability distribution of future peak demand is remarkable. This is particularly so given that most of the evidence is based on methods which employ 100 bootstrap samples. While the upper tail is most important for the capacity planning example at hand, the lower tail is not fit as well, suggesting caution in the application of these findings to other situations. In this example there is no evidence that increasing the number of bootstrap samples or using the smoothed or parametric bootstrap will offer much improvement.

ACKNOWLEDGMENTS

Thanks are due to D. Fretz, G. Green and W. Cheng for research assistance and to B. Efron, R. Ellis, I. McLeod, R. Tibshirani, F. Trimnell and A. Ullah for valuable advice. The assistance of the Centre for the Study of International Economic Relations at The University of Western Ontario, Ontario Hydro and the Social Science and Humanities Research Council of Canada is acknowledged.

Table 2. *Monte Carlo Results in Bootstrap Estimate of Distribution of 1990 Peak Demand (Nonrandom 1990 Price and Income)*

	Percentage Probability Points									Average Bootstrap Forecast	Bootstrap Bias Estimate
	.01	.05	.10	.25	.50	.75	.90	.95	.99		
Actual Distribution	20864	21217	21408	21731	22096	22467	22806	23011	23401	- - -	- - -
Bootstrap Estimate	20038	20667	20976	21473	22006	22543	23004	23265	23673	22007	–4
	(812)	(733)	(730)	(725)	(734)	(760)	(789)	(814)	(854)	(729)	(69)
Smoothed Bootstrap											
$k = .10$	20063	20695	21019	21535	22087	22629	23091	23354	23762	22085	–9
	(738)	(696)	(692)	(684)	(696)	(713)	(737)	(748)	(775)	(682)	(61)
$k = .25$	19966	20610	20915	21431	21991	22540	23015	23282	23737	21998	–6
	(679)	(640)	(638)	(639)	(654)	(682)	(723)	(751)	(810)	(659)	(61)
$k = .50$	20103	20735	21059	21559	22115	22663	23153	23414	23858	22123	–5
	(725)	(670)	(661)	(661)	(678)	(718)	(754)	(774)	(851)	(689)	(64)

Standard errors are in parentheses. All results based on 200 Monte Carlo replications and 100 bootstraps.

Table 3. *Monte Carlo Results on Bootstrap Estimate of Distribution of 1990 Peak Demand (Nonrandom 1990 Price and Income; ε Uniform)*

	Percentage Probability Points									Average Bootstrap Forecast	Bootstrap Bias Estimate
	.01	.05	.10	.25	.50	.75	.90	.95	.99		
Actual Distribution	20942	21215	21377	21687	22096	22513	22840	23014	23314	- - -	- - -
Bootstrap Estimate	20048	20666	20967	21461	22001	22554	23017	23281	23670	22007	–5
	(802)	(721)	(728)	(726)	(742)	(766)	(797)	(814)	(854)	(734)	(70)
Smoothed Bootstrap											
$k = .10$	20005	20636	20946	21461	22031	22586	23061	23306	23706	22025	–2
	(770)	(695)	(687)	(687)	(692)	(728)	(753)	(761)	(817)	(689)	(64)
$k = .25$	20070	20707	21010	21538	22096	22655	23146	23429	23841	22105	–10
	(715)	(673)	(674)	(676)	(708)	(745)	(783)	(796)	(823)	(697)	(67)
$k = .50$	20088	20725	21052	21555	22110	22671	23158	23422	23824	22131	–12
	(735)	(646)	(659)	(684)	(701)	(730)	(765)	(797)	(833)	(690)	(63)

Standard errors are in parentheses. All results based on 200 Monte Carlo replications and 100 bootstraps.

REFERENCES

Beasley, J. D., and S. G. Springer (1977), "The percentage points of the normal distribution". *Applied Statistics* **26**, 118–121.

Breusch, T. S., and A. R. Pagan (1979), "A simple test for heteroscedasticity and random coefficient variation". *Econometrica* **47**, 1287–1294.

Efron, B. (1979), "Bootstrapping methods: another look at the jackknife". *Annals of Statistics* **7**, 1–26.

Efron, B. (1982). *The Jackknife, The Bootstrap and Other Resampling Plans.* Philadelphia: SIAM.

Efron, B., and G. Gong (1983), "A leisurely look at the bootstrap, the jackknife and cross-validation". *The American Statistician* **37**, 36–48.

Ellis, R. (1980), "Optimal pricing and investment decisions by electric utilities under demand uncertainty". M.I.T. Energy Lab Working Paper No. MIT-EL 80-048 WP, November.

Engle, R. F. (1982), "Autoregressive conditional heteroskedasticity with estimates of the variance of United Kingdom inflations". *Econometrica* **50**, 987–1008.

Fair, R. (1979), "An analysis of the accuracy of four macroeconomic models". *Journal of Political Economy* **87**, 701–718.

Fair, R. (1980), "Estimating the predictive accuracy of econometric models". *International Economic Review* **21**, 355–378.

Finke, R., and H. Theil (1984), "Bootstrapping for standard errors of instrumental variables estimates". *Economics Letters* **14**, 297–301.

Freedman, D. A., and S. C. Peters (1984a), "Using the bootstrap to evaluate forecasting equations". Technical Report No. 20, Department of Statistics, University of California, Berkeley, January.

Freedman, D. A., and S. C. Peters (1984b), "Bootstrapping an econometric model: some empirical results". *Journal of Business and Economic Statistics* **2**, 150–158.

Freedman, D. A., and S. C. Peters (1984c), "Bootstrapping a regression equation: some empirical results". *Journal of the American Statistical Association* **79**, 97–106.

Godfrey, L. G. (1978), "Testing against general autoregressive and moving average error models when the regressors include lagged dependent variables". *Econometrica* **46**, 1293–1302.

Hausman, J. A. (1978), "Specification tests in econometrics". *Econometrica* **46**, 1251–1271.

Kiefer, N. M., and M. Salmon (1983), "Testing normality in econometric models". *Economics Letters* **11**, 123–128.

Korajczyk, R. A. (1985), "The pricing of forward contracts for foreign exchange". *Journal of Political Economy* **93**, 346–368.

Shapiro, S. S., and R. S. Francia (1972), "An approximate analysis of variance test for normality". *Journal of the American Statistical Association* **67**, 215-216.

Shapiro, S. S., and M. B. Wilk (1965), "An analysis of variance test for normality (complete samples)". *Biometrika* **52**, 591–611.

Veall, M. R. (1985), "Bootstrapping the probability distribution of peak electricity demand". Mimeograph, The University of Western Ontario.

Vinod, H. D., and B. Raj (1984), "Bootstrapping confidence intervals for arbitrary functions of regression parameters". Mimeograph, Wilfrid Laurier University.

White, H. (1980), "A heteroskedasticity-consistent covariance matrix estimator and a direct test for heteroskedasticity". *Econometrica* **48**, 817–838.

White, H., and G. M. MacDonald (1980), "Some large sample tests for nonnormality in the linear regression model". *Journal of the American Statistical Association* **75**, 16–28.

Wickman, B. A., and I. D. Hill (1982), "An efficient and portable pseudo-random number generator". *Applied Statistics* **31**, 188–190.

Hiroki Tsurumi [1]

USE OF THE MEAN SQUARED ERRORS OF FORECASTS IN TESTING FOR STRUCTURAL SHIFT: A COMPARISON WITH THE CHOW TEST FOR AN UNDERSIZED CASE

1. INTRODUCTION

The mean squared-errors of forecast (MSEF) is a statistic used to evaluate post-sample prediction performance. The MSEF has been used as a descriptive measure, but its exact distribution can be derived either from a sample theoretical or from a Bayesian perspective if the MSEF is computed from a linear regression model. In this paper, sampling and Bayesian distributions of the MSEF are derived, and it is suggested that the MSEF may be used as a statistic to test for structural shift. The powers of the MSEF are compared with those of the Chow test for a case where the sample size of the second regime is less than the number of the regression coefficients (i.e., the undersized case).

2. SAMPLING AND BAYESIAN DISTRIBUTIONS OF THE MSEF

Let the linear model be given by

$$y = X\beta + u, \tag{1}$$

where y is an $(n \times 1)$ vector of observations on the dependent variable, X is an $(n \times k)$ matrix of observations on the explanatory variables with rank k, u is an $(n \times 1)$ vector of error terms, and β is a $(k \times 1)$ vector of unknown regression coefficients. Assume that $u \sim N(0, \sigma^2 I_n)$ and that β is estimated by $\hat{\beta} = (X'X)^{-1}X'y$.

The mean-squared-error for the post-sample period, $n+1, \ldots, n+m$ is computed using the post-sample actual observations on y and X. Let y_* and

[1] Department of Economics, Rutgers University, New Brunswick, New Jersey 08903

I. B. MacNeill and G. J. Umphrey (eds.), Time Series and Econometric Modelling, 385–394.

X_* be, respectively, an $(m \times 1)$ vector and an $(m \times k)$ matrix of post-sample observations and assume that the rank of X_* is $\min(m, k)$. Then the MSEF is

$$MSEF = \frac{1}{m}(\hat{y}_* - y_*)'(\hat{y}_* - y_*), \tag{2}$$

where $\hat{y}_* = X_*\hat{\beta}$. Given equation (1) and $\hat{\beta} = \beta + (X'X)^{-1}X'u$, equation (2) can be written as

$$MSEF = \frac{1}{m}\epsilon_*' B' B \epsilon_* = \frac{1}{m}\sum_{i=1}^{m} \mu_i \epsilon_i^2, \tag{3}$$

where: $\epsilon_* = (u', u_*')'$, $B = (A, -I_m)$, $A = X_*(X'X)^{-1}X'$, and the μ_i's are the nonzero characteristic roots of $B'B$. The ϵ_i are elements of $\epsilon = c'\epsilon_*$, where c is the matrix of characteristic vectors of $B'B$. In passing, let us note that the μ_i's are given by $\mu_i = 1 + \lambda_i$, $i = 1, \ldots, m$ for $m \leq k$, and $\mu_i = 1 + \lambda_i$, $i = 1, \ldots, k$; $\mu_i = 1$, $i = k+1, \ldots, m$, for $m > k$, where λ_i is the ith nonzero characteristic root of AA'.

Since $\epsilon_i \sim NID(0, \sigma^2)$, $m \cdot MSEF$ is a quadratic form in normal variables. The distribution of quadratic forms or ratios of quadratic forms has been investigated by many; some of the earlier works are by McCarthy (1939), von Neumann (1941), and Bhattacharyya (1943). Bhattacharyya (1945) and Hotelling (1948) employed Laguerre expansions, and Gurland (1953) and Johnson and Kotz (1970) refined further the convergent Laguerre expansions. Tsurumi (1985) used the degenerate hyperbolic function, which is convenient for computational purposes. Since we will use the distribution of a quadratic form in normal variables, let us give Tsurumi's result as a lemma.

Lemma 1. Let $x = m \cdot MSEF/\sigma^2$. Then the distribution of x is given by

$$f(x) = \frac{e^{-x/(2\mu_m)} x^{m/2-1}}{2^{m/2}\pi^{1/2}\prod_{i=1}^{m}\mu_i^{1/2}} \sum_{p=0}^{\infty} c(m,p) x^p, \tag{4}$$

where $c(m, p)$ is the recursive coefficient given by

$$c(m,p) = \frac{\Gamma(p + \frac{m-1}{2})}{\Gamma(p + \frac{m}{2})} \sum_{j=0}^{p} \frac{c(m-1, j) a_m^{p-j}}{(p-j)!}, \quad \text{for } m \geq 2, \tag{5}$$

and $c(1,0) = 1$, $c(1,j) = 0$ for $j \geq 1$; $a_m = \frac{1}{2}\left(\frac{1}{\mu_m} - \frac{1}{\mu_{m-1}}\right)$, $\mu_1 \geq \mu_2 \geq \cdots \geq \mu_m$, for $m \geq 2$, $a_1 = (1/2_{\mu_1})$, and $a_i^0 = 1$ for all $i = 1, \ldots, m$.

Proof. See Tsurumi (1985). Using this lemma, we will establish the sampling distribution of the MSEF.

Theorem 1. Let z be the MSEF. Then $u = z/(\mu_m s^2)$ has the probability density function given by

$$f(u \mid \mu_1, \ldots, \mu_m, m, \nu) = \text{const} \cdot \frac{u^{m/2-1}}{(1+\frac{m}{\nu}u)^{(m+\nu)/2}} \cdot \sum_{p=0}^{\infty} \Gamma\left(\frac{m+\nu}{2}\right) 2^p c(m,p) \mu_m^p \left(\frac{m}{\nu}\right)^p \left(\frac{u}{1+mu/\nu}\right)^p, \tag{6}$$

where

$$\text{const} = \frac{(m/\nu)^{m/2} \mu_m^{m/2}}{\pi^{1/2} \prod_{i=1}^{k} \mu_i^{1/2} \Gamma\left(\frac{\nu}{2}\right)}$$

and

$$\nu s^2 = y'[I - X(X'X)^{-1}X']y, \quad \nu = n - k.$$

Proof. Equation (4) in Lemma 1 is the probability density function of $x = m \cdot MSEF/\sigma^2$. The probability density function (pdf) of $z = MSEF$ may be obtained by transforming $z = (\sigma^2/m)\, x$, and this becomes

$$f(z \mid m, \sigma^2) = c_1 z^{m/2-1} \sigma^{-m} \exp\left(-\frac{m}{2\mu_m} z\sigma^{-2}\right) \sum_{p=0}^{\infty} c(m,p) m^p z^p \sigma^{-2p}, \tag{7}$$

where

$$c_1 = \frac{m^{m/2}}{2^{m/2} \pi^{1/2} \prod_{i=1}^{k} \mu_i^{1/2}}.$$

On the other hand the pdf of s^2 is given by

$$f(s^2 \mid \sigma^2, \nu) = \frac{\nu \sigma^{-2}}{2^{1/2} \Gamma\left(\frac{\nu}{2}\right)} \left(\frac{\nu s^2}{\sigma^2}\right)^{v/2-1} \exp\{-\nu s^2/(2\sigma^2)\}$$

and it is easy to show that z and s^2 are independent. Hence, the joint pdf of z and s^2 is

$$f(z, s^2 \mid m, \sigma^2) = c_3 \nu^{\nu/2} z^{m/2-1} (s^2)^{\nu/2} \sigma^{-(m+\nu)} \cdot \exp\left\{-\frac{\nu s^2}{2\sigma^2}[1 + mz/(\nu \mu_m s^2)]\right\} \sum_{p=0}^{\infty} c(m,p) m^p z^p \sigma^{-2p},$$

where $c_3 = c_1[2^{\nu/2}\Gamma(\nu/2)]^{-1}$. Changing the variables z and s^2 to $u = z/(\mu_m s^2)$ and $y = s^2$, and integrating out y, we obtain the desired result.

The Bayesian predictive density of the MSEF may be derived in two ways. We may first reduce the MSEF as a quadratic form in normal variables by subtracting out β and obtain equation (3). Realizing that this quantity contains a nuisance parameter, σ^2, we integrate it out by utilizing the posterior pdf for σ^2. This approach was used by Tsurumi (1985). Alternatively, we may start with the joint density for y_*, β, and σ^2, and integrate β first, and transform y_* into the MSEF given σ^2. Finally we integrate σ^2 out. This second approach is used in the theorem below.

Theorem 2. The Bayesian predictive density of the MSEF is given by

$$\begin{aligned} p(z \mid s^2, \nu, m) \propto & \frac{z^{m/2-1}}{(\nu s^2 + mz/\mu_m)^{(m+\nu)/2}} \\ & \cdot \sum_{p=0}^{\infty} \Gamma\left(\frac{m+\nu}{2} + p\right) 2^p c(m,p) \left(\frac{mz}{\nu s^2 + mz/\mu_m}\right)^p . \end{aligned} \tag{8}$$

Proof. The predictive pdf of y_* is given by

$$p(y_* \mid y, X, X_*) = \int\int p(y_* \mid \beta, \sigma^2, X_*) p(\beta, \sigma \mid y, X) d\sigma d\beta, \tag{9}$$

where

$$p(y_* \mid \beta, \sigma^2, X_*) \propto \sigma^{-m} \exp\left[-\frac{1}{2\sigma^2}(y_* - X_*\beta)'(y_* - X_*\beta)\right],$$

and we shall use the posterior pdf of β and σ that is given by

$$p(\beta, \sigma \mid y, X) \propto \sigma^{-(n+1)} \exp\left\{-\frac{1}{2\sigma^2}[\nu s^2 + (\beta - \hat{\beta})'X'X(\beta - \hat{\beta})]\right\}.$$

Thus $p(y_* \mid \beta, \sigma^2, X_*)p(\beta, \sigma \mid y, X)$ becomes

$$\begin{aligned} p(y_* \mid \beta, \sigma^2, X_*)p(\beta, \sigma \mid y, X) \propto & \, \sigma^{-(n+m+1)} \\ & \cdot \exp\left\{-\frac{1}{2\sigma^2}\left[y_*'(I - X_*X_*^+)y_* + (\beta - \hat{\beta}_*)'X_*'X_*(\beta - \hat{\beta}_*)\right]\right\} \\ & \cdot \exp\left\{-\frac{1}{2\sigma^2}\left[\nu s^2 + (\beta - \hat{\beta})'X'X(\beta - \hat{\beta})\right]\right\}, \end{aligned} \tag{10}$$

where X_*^+ is the Moore-Penrose generalized inverse of X_*, and $\hat{\beta}_* = X_*^+ y_* + (I - X_* X_*^+) r$, where r is an arbitrary $m \times 1$ vector. Arranging the sum of two quadratic forms in β into a quadratic form in β, and integrating β out, we derive from equation (9)

$$p(y_* \mid y, X, X_*) \propto \int \sigma^{-(\nu+m+1)} \exp\left(-\frac{\nu s^2}{2\sigma^2}\right) \cdot \exp\left[-\frac{1}{2\sigma^2}(y_* - X_*\hat{\beta})' H (y_* - X_*\hat{\beta})\right] d\sigma, \tag{11}$$

where $H = [I + X_*(X'X)^{-1}X_*']^{-1}$. The right-hand side of equation (11) shows that given σ, $y_* - X_*\hat{\beta}$ is distributed as $N(0, \sigma^2 H^{-1})$. Let $w = y_* - X_*\hat{\beta}$ and $H = R'R$, where R is a nonsingular matrix. Then $m \cdot z = w'w - \eta'(RR')^{-1}\eta$, with $\eta = Rw \sim N(0, \sigma^2 I_m)$. Since the nonzero characteristic roots of $H^{-1} = (R'R)^{-1}$ are the same as those of $(RR')^{-1}$, we see that $m \cdot z = \sum_{i=1}^{m} \mu_i \epsilon_i$, with $\epsilon_i \sim N(0, \sigma^2)$, and μ_i is the ith characteristic root of H^{-1}. Hence, given σ, $m \cdot z$ has the distribution of a quadratic form in normal variables that is given in equation (4). Thus

$$p(z \mid s^2, m) \propto \int \sigma^{-(\nu+m+1)} p(z \mid \sigma^2, m) \exp\left(-\frac{\nu s^2}{2\sigma^2}\right) d\sigma, \tag{12}$$

and integrating σ out, we obtain the desired result.

3. USE OF THE MSEF FOR TESTING FOR A PARAMETER SHIFT

The MSEF may be used for testing for a structural shift or for comparing non-nested linear models. Tsurumi (1985) presented numerical examples of the use of the MSEF for testing for a structural shift and for comparing non-nested linear models using actual data. In this section, let us focus our attention on the use of the MSEF for testing for a structural shift when the number of post-sample observations, m, is smaller than the number of regression coefficients, k. We assume that the join point is known and the switch from the first to second regime is abrupt. Using sampling experiments, we compare the performances of the MSEF with those of the Chow test. The performances are judged by empirical powers.

The hypotheses of a parameter shift may be given as follows:

$$H_0 : \beta = \beta_* \quad \text{versus} \quad H_1 : \beta \neq \beta_*, \tag{13}$$

where β_* is the vector of regression coefficients in regime 2. The Chow test for an undersized case ($m \leq k$) is given by the following F criterion:

$$F(m, \nu) = \frac{(\tilde{e}'\tilde{e} - \nu s^2)/m}{s^2}, \quad \nu = n - k, \tag{14}$$

where $\tilde{e}'\tilde{e} = (\tilde{y} - \tilde{X}\tilde{\beta})'(\tilde{y} - \tilde{X}\tilde{\beta})$, $\tilde{y} = (y', y_*')'$, $\tilde{X} = (X', X_*')'$, and $\tilde{\beta} = (X'X)^{-1}X'y$. Thus, $\tilde{e}'\tilde{e}$ is the sum of squared residuals from the pooled sample of size $n + m$. For $m = 1$, the sampling MSEF test criterion, $u = z/(\mu_m s^2)$, is identical to the Chow test, and this is given in the following lemma.

Lemma 2. For $m = 1$, the sampling MSEF test criterion, $u = z/(\mu_m s^2)$, is identical to the Chow test.

Proof. For any $m \geq 1$, the sum of squared residuals from the pooled sample becomes

$$\begin{aligned}\tilde{e}'\tilde{e} &= \tilde{y}'[I - \tilde{X}(\tilde{X}'\tilde{X})^{-1}\tilde{X}']\tilde{y} \\ &= \tilde{y}'\tilde{y} - (y'XB^{-1}X'y + 2y_*'X_*B^{-1}X'y + y_*'X_*B^{-1}X_*'y_*),\end{aligned}$$

where $B^{-1} = (X'X + X_*'X_*)^{-1}$. Using the identity

$$(X'X + X_*'X_*)^{-1} = (X'X)^{-1} - (X'X)^{-1}X_*HX_*'(X'X)^{-1},$$

where $H = [I + X_*(X'X)^{-1}X_*']^{-1}$, we see that

$$\begin{aligned}y'XB^{-1}X'y &= \hat{\beta}'X'X\hat{\beta} - \hat{\beta}'X_*HX_*'\hat{\beta}, \\ y_*'X_*B^{-1}X'y &= y_*'HX_*\hat{\beta},\end{aligned}$$

and

$$y_*'X_*B^{-1}X_*'y_* = y_*'(I - H)y_*.$$

Hence, $\tilde{e}'\tilde{e}$ becomes

$$\tilde{e}'\tilde{e} = \nu s^2 + (y_* - X_*\hat{\beta})'H(y_* - X_*\hat{\beta}),$$

the Chow test statistic becomes

$$F(m, \nu) = (y_* - X_*\hat{\beta})'H(y_* - X_*\hat{\beta})/(ms^2)$$

and for $m = 1$, $H^{-1} = \mu_1 = 1 + X_*(X'X)^{-1}X_*'$, we have $F(1, \nu) = z/(\mu_1 s^2)$.

The MSEF test criterion, $u = z/(\mu_m s^2)$, can be interpreted either from the sampling or Bayesian perspective. In the sampling theory view, both z and s^2 are regarded as random variables, whereas in the Bayesian view z is the random variable and s^2 is a given realized random variable. The predictive density of z is obtained given s^2, and inference about z is made conditionally on s^2. The predictive density of z, then, will vary if s^2 varies. If we wish to use the Bayesian MSEF criterion to test for a parameter shift such as is given in H_0 versus H_1 in (13), the power of the test will have to be obtained given a particular value of s^2. If the realized value of s^2 is much smaller than the value of σ^2 under which the data, y and y^* are generated, the power associated with this small s^2 may tend to be higher than the powers of the sampling MSEF criterion or of the Chow test. In sampling experiments, we shall derive the power of tests for various values of s^2 and identify that value of s^2 which yields roughly the same power as the sampling MSEF. If this value of s^2 comes from, say, the 70th percentile point of the sampling distribution of s^2, then one may argue that at least 70% of the time, the Bayesian MSEF criterion produces a better power than the sampling MSEF criterion.

Sampling experiments are made under the following model and design:

$$Y_i = \beta_1 + \beta_2 X_{i2} + \beta_3 X_{i3} + \beta_4 X_{i4} + \beta_5 X_{i5} + \epsilon_i, \quad i = 1, \ldots, n. \tag{15}$$

X_{i2} and X_{i3} are drawn from uniform distributions, $U(0,3)$, and are independent of each other. Since economic data often suffer from multicollinearity, we design X_{i4} and X_{i5} as follows:

$$\begin{aligned} X_{i4} &= X_{i2} + v_{i1}, & v_{i1} &\sim U(0,.5), \\ X_{i5} &= X_{i2} + X_{i3} + v_{i2}, & v_{i2} &\sim U(0,1), \end{aligned}$$

and v_{i1} and v_{i2} are independent of each other. The values of β_i $(i = 1, \ldots, 5)$ under the null hypothesis, H_0, are

$$H_0 : (\beta_1 = 10,\ \beta_2 = .4,\ \beta_3 = .6,\ \beta_4 = 1,\ \beta_5 = 2)$$

and under the alternative hypothesis, H_1, β_2 and β_3 are scalar multiples of $\beta_2 = .4$ and $\beta_3 = .6$, while β_1, β_4, and β_5 stay the same as in H_0. The error term of the regression equation, ϵ_i, is drawn from $\epsilon_i \sim NID(0,1)$, and we set the sample size, n, to be 20. The values of X_{ij}'s are drawn once and for all, and 500 replications are made for each set of $(\beta_1, \ldots, \beta_5)$. The determinant of the matrix of simple correlation coefficients among $X_{i2}, \ldots, X_{i5}$ is .0006195, showing a high degree of multicollinearity. The average coefficient of determination, R^2, for each set of 500 replications is .72 .

As discussed earlier, the Bayesian interpretation of the MSEF is that the predictive density of z is derived given a realized sample of size n, $y = (Y_1, \ldots, Y_n)'$. To derive empirical powers of the Bayesian MSEF criterion, $z/(\mu_m s^2)$, we generated y_* for the second regime in the following way. Since the predictive density of y_* given X and X_* is the multivariate Student t,

$$p(y_* \mid y, X, X_*) \propto [\nu s^2 + (y_* - X_*\hat{\beta})' H (y_* - X_*\hat{\beta})]^{-(\nu+m)/2}.$$

We generated the values of y_* from the multivariate Student t pdf by using the estimate $\hat{\hat{\beta}}$ that is obtained by $\hat{\hat{\beta}} = \beta + (X'X)^{-1}X'\epsilon_y$, where ϵ_y are those values of the error terms that give rise to a particular value of s^2, through $y = (Y_1, \ldots, Y_n)' : y = X\beta + \epsilon_y$. For the Chow test and sampling MSEF criterion, on the other hand, y_* is drawn from $y_* = X_*\beta_* + \epsilon_*$, $\epsilon_* \sim N(0,1)$. The matrix X_* was generated in the same way as the matrix X.

Empirical powers are presented in Table 1 for $m = 1, \ldots, 5$. The MSEF statistics are given by $u = z/(\mu_m s^2)$, both for the sampling and Bayesian interpretations. When $m = 1$, the Chow and sampling MSEF criteria are identical, and the Bayesian MSEF that is evaluated at the 75th percentile value of s^2 produces power that is comparable to the Chow and sampling MSEF statistics. Any value of s^2 that is less than the 75th percentile value produces higher powers than the Chow test does. For $m = 2$, the Bayesian MSEF produces equal or better power than that of the Chow or the sampling MSEF criterian for about 65% of the time. As m increases to 3, 4, and 5, the percentage of the times in which the Bayesian MSEF does equally well or better than the Chow or the sampling MSEF statistics declines slowly to the 55 percentile for $m = 5$. Comparing the Chow test statistic and the sampling MSEF criterion, we notice that for $m = 2$, the latter has a slightly better power than the former, but for $m = 3, 4, 5$, the Chow test performs better than the sampling MSEF.

4. CONCLUSIONS

In this paper, sampling and Bayesian distributions of the MSEF were derived, and using sampling experiments it was demonstrated that the MSEF criterion may be used as a statistic to test for a parameter shift. When the number of post-sample observations is less than the number of regression coefficients ($m \leq k$), the MSEF criterion, especially the Bayesian MSEF, tends to perform better than or as well as the Chow test. For the case of $m > k$, there is a Chow test statistic that is more powerful than the one given in equation (14) and thus the Chow test outperforms the sampling as well as the Bayesian MSEF criteria.

Table 1. *Empirical Powers of the Chow, Sampling MSEF and Bayesian MSEF. Under $H_0 : c = 1$, $\alpha = .05$.*

	Values of c for $c\beta_2, c\beta_3$ under H_1					
	1.5	2.0	2.5	3.0	3.5	4.0
$m = 1$						
Chow=Sampling MSEF	.13	.31	.65	.86	.97	1.00
Bayesian MSEF(75%)*	.21	.35	.66	.88	.98	1.00
$m = 2$						
Chow	.12	.26	.51	.70	.86	.95
Sampling MSEF	.17	.33	.58	.77	.90	.98
Bayesian MSEF(65%)	.12	.27	.48	.71	.91	.96
$m = 3$						
Chow	.06	.14	.27	.52	.78	.89
Sampling MSEF	.07	.10	.22	.48	.61	.82
Bayesian MSEF(58%)	.05	.17	.34	.53	.77	.90
$m = 4$						
Chow	.09	.18	.40	.65	.87	.97
Sampling MSEF	.07	.09	.17	.32	.60	.78
Bayesian MSEF(55%)	.10	.23	.38	.71	.91	.99
$m = 5$						
Chow	.12	.23	.48	.80	.93	.99
Sampling MSEF	.09	.18	.41	.75	.90	.98
Bayesian MSEF(55%)	.08	.28	.60	.86	.97	.99

(1) Sample Size, $n = 20$, 500 replications.

(2) Sampling and Bayesian MSEF is $u = z/(u_m s^2)$.

(3) *Bayesian MSEF(75%) means that those values of $\varepsilon_y = (\varepsilon_1, \ldots, \varepsilon_n)'$ that gave rise to s^2 at the 75th percentile of the distribution of s^2 are used to generate y_*.

REFERENCES

Bhattacharyya, A. (1945), "A note on the distribution of the sum of chi-squares". *Sankhya* **7**, 27–28.

Bhattacharyya, B. C. (1943), "On an aspect of the Pearson system of curves, and a few analogies". *Sankhya* **6**, 415–448.

Gurland, J. (1953), "Distribution of quadratic forms and ratios of quadratic forms". *Annals of Mathematical Statistics* **24**, 416–427.

Hotelling, H. (1948), "Some new methods for distributions of quadratic forms" (Abstract). *Annals of Mathematical Statistics* **19**, 119.

Johnson, N. L., and S. Kotz (1970), *Continuous Univariate Distributions—2.* New York: Wiley and Sons.

McCarthy, M. D. (1939), "On the application of the z-test to randomized blocks". *Annals of Mathematical Statistics* **10**, 337–359.

Tsurumi, H. (1985), "A Bayesian predictive density for the mean squared errors of forecast from a linear model". Mimeograph.

von Neumann, J. (1941), "Distribution of the ratio of the mean square successive difference to the variance". *Annals of Mathematical Statistics* **12**, 367–395.

THE UNIVERSITY OF WESTERN ONTARIO SERIES IN PHILOSOPHY OF SCIENCE

A Series of Books in Philosophy of Science, Methodology, Epistemology, Logic, History of Science, and Related Fields

1. J. Leach, R. Butts, and G. Pearche (eds.), *Science, Decision and Value.* 1973, vii + 219 pp.
2. C. A. Hooker (ed.), *Contemporary Research in the Foundations and Philosophy of Quantum Theory.* 1973, xx + 385 pp.
3. J. Bub, *The Interpretation of Quantum Mechanics.* 1974, ix + 155 pp.
4. D. Hockney, W. Harper, and B. Freed (eds.), *Contemporary Research in Philosophical Logic and Linguistic Semantics.* 1975, vii + 332 pp.
5. C. A. Hooker (ed.), *The Logico-Algebraic Approach to Quantum Mechanics.* 1975, xv + 607 pp.
6. W. L. Harper and C. A. Hooker (eds.), *Foundations of Probability Theory, Statistical Inference, and Statistical Theories of Science.* 3 Volumes. Vol. I: *Foundations and Philosophy of Epistemic Applications of Probability Theory.* 1976, xi + 308 pp. Vol. II: *Foundations and Philosophy of Statistical Inference.* 1976, xi + 455 pp. Vol. III: *Foundations and Philosophy of Statistical Theories in the Physical Sciences.* 1976, xii + 241 pp.
7. C. A. Hooker (ed.), *Physical Theory as Logico-Operational Structure.* 1979, xvii + 334 pp.
8. J. M. Nicholas (ed.), *Images, Perception, and Knowledge.* 1977, ix + 309 pp.
9. R. E. Butts and J. Hintikka (eds.), *Logic, Foundations of Mathematics, and Computability Theory.* 1977, x + 406 pp.
10. R. E. Butts and J. Hintikka (eds.), *Foundational Problems in the Special Sciences.* 1977, x + 427 pp.
11. R. E. Butts and J. Hintikka (eds.), *Basic Problems in Methodology and Linguistics.* 1977, x + 321 pp.
12. R. E. Butts and J. Hintikka (eds.), *Historical and Philosophical Dimensions of Logic, Methodology and Philosophy of Science.* 1977, x + 336 pp.
13. C. A. Hooker (ed.), *Foundations and Applications of Decision Theory.* 2 volumes. Vol. I: *Theoretical Foundations.* 1978, xxiii + 442 pp. Vol. II: *Epistemic and Social Applications.* 1978, xxiii + 206 pp.
14. R. E. Butts and J. C. Pitt (eds.), *New Perspectives on Galileo.* 1978, xvi + 262 pp.
15. W. L. Harper, R. Stalnaker, and G. Pearce (eds.), *Ifs. Conditionals, Belief, Decision, Chance, and Time.* 1980, ix + 345 pp.
16. J. C. Pitt (ed.), *Philosophy in Economics.* 1981, vii + 210 pp.
17. Michael Ruse, *Is Science Sexist?* 1981, xix + 299 pp.

18. Nicholas Rescher, *Leibniz's Metaphysics of Nature*. 1981, xiv + 126 pp.
19. Larry Laudan, *Science and Hypothesis*. 1981, x + 258 pp.
20. William R. Shea, *Nature Mathematized*. Vol. I, 1983, xiii + 325 pp.
21. Michael Ruse, *Nature Animated*. Vol. II, 1983, xiii + 274 pp.
22. William R. Shea (ed.), *Otto Hahn and the Rise of Nuclear Physics*. 1983, x + 252 pp.
23. H. F. Cohen, *Quantifying Music*. 1984, xvii + 308 pp.
24. Robert E. Butts, *Kant and the Double Government Methodology*. 1984, xvi + 339 pp.
25. James Robert Brown (ed.), *Scientific Rationality: The Sociological Turn*. 1984, viii + 330 pp.
26. Fred Wilson, *Explanation, Causation and Deduction*. 1985, xviii + 385 pp.
27. Joseph C. Pitt (ed.), *Change and Progress in Modern Science*. 1985, viii + 398 pp.
28. Henry B. Hollinger and Michael John Zenzen, *The Nature of Irreversibility*. 1985, xi + 340 pp.
29. Kathleen Okruhlik and James Robert Brown (eds.), *The Natural Philosophy of Leibniz*. 1985, viii + 342 pp.
30. Graham Oddie, *Likeness to Truth*. 1986, xv + 218 pp.
31. Fred Wilson, *Laws and Other Worlds*. 1986, xv + 328 pp.
32. John Earman, *A Primer on Determinism*. 1986, xiv + 273 pp.
33. Robert E. Butts (ed.), *Kant's Philosophy of Physical Science*. 1986, xii + 363 pp.
34. Ian B. MacNeill and Gary J. Umphrey (eds.), Vol. I, *Applied Probability, Stochastic Processes, and Sampling Theory*. 1987, xxv + 329 pp.
35. Ian B. MacNeill and Gary J. Umphrey (eds.), Vol. II, *Foundations of Statistical Inference*. 1987, xvii + 287 pp.
36. Ian B. MacNeill and Gary J. Umphrey (eds.), Vol. III, *Time Series and Econometric Modelling*. 1987, xix + 394 pp.
37. Ian B. MacNeill and Gary J. Umphrey (eds.), Vol. IV, *Stochastic Hydrology*. 1987, xv + 225 pp.
38. Ian B. MacNeill and Gary J. Umphrey (eds.), Vol. V, *Biostatistics*. 1987, xvi + 283 pp.
39. Ian B. MacNeill and Gary J. Umphrey (eds.), Vol. VI, *Actuarial Science*. 1987, xvi + 250 pp.

GPSR Compliance

The European Union's (EU) General Product Safety Regulation (GPSR) is a set of rules that requires consumer products to be safe and our obligations to ensure this.

If you have any concerns about our products, you can contact us on ProductSafety@springernature.com

In case Publisher is established outside the EU, the EU authorized representative is:

Springer Nature Customer Service Center GmbH
Europaplatz 3
69115 Heidelberg, Germany

Batch number: 10761599

Printed by Printforce, the Netherlands